OKAVANGO

AND THE SOURCE OF LIFE

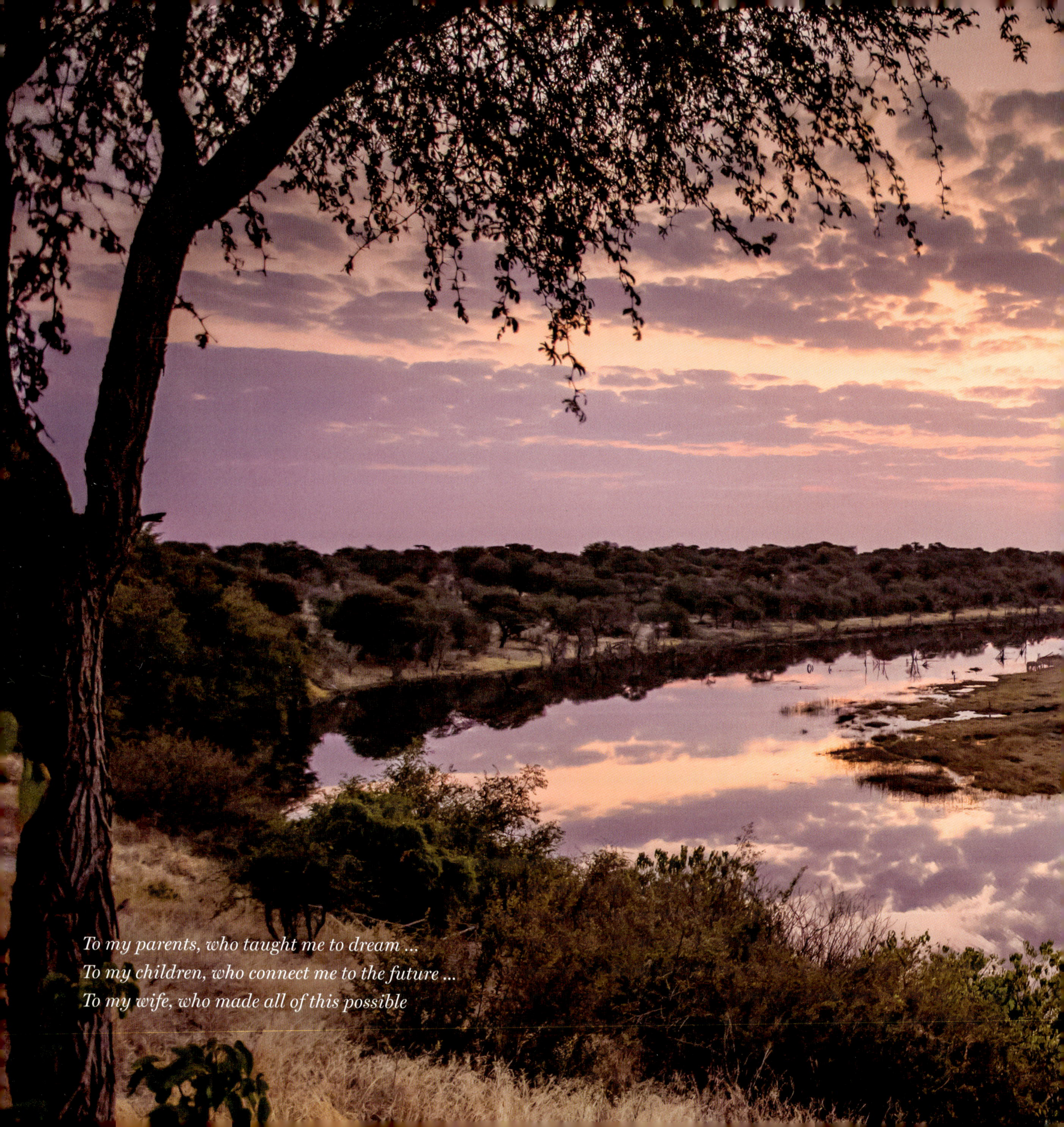

To my parents, who taught me to dream ...
To my children, who connect me to the future ...
To my wife, who made all of this possible

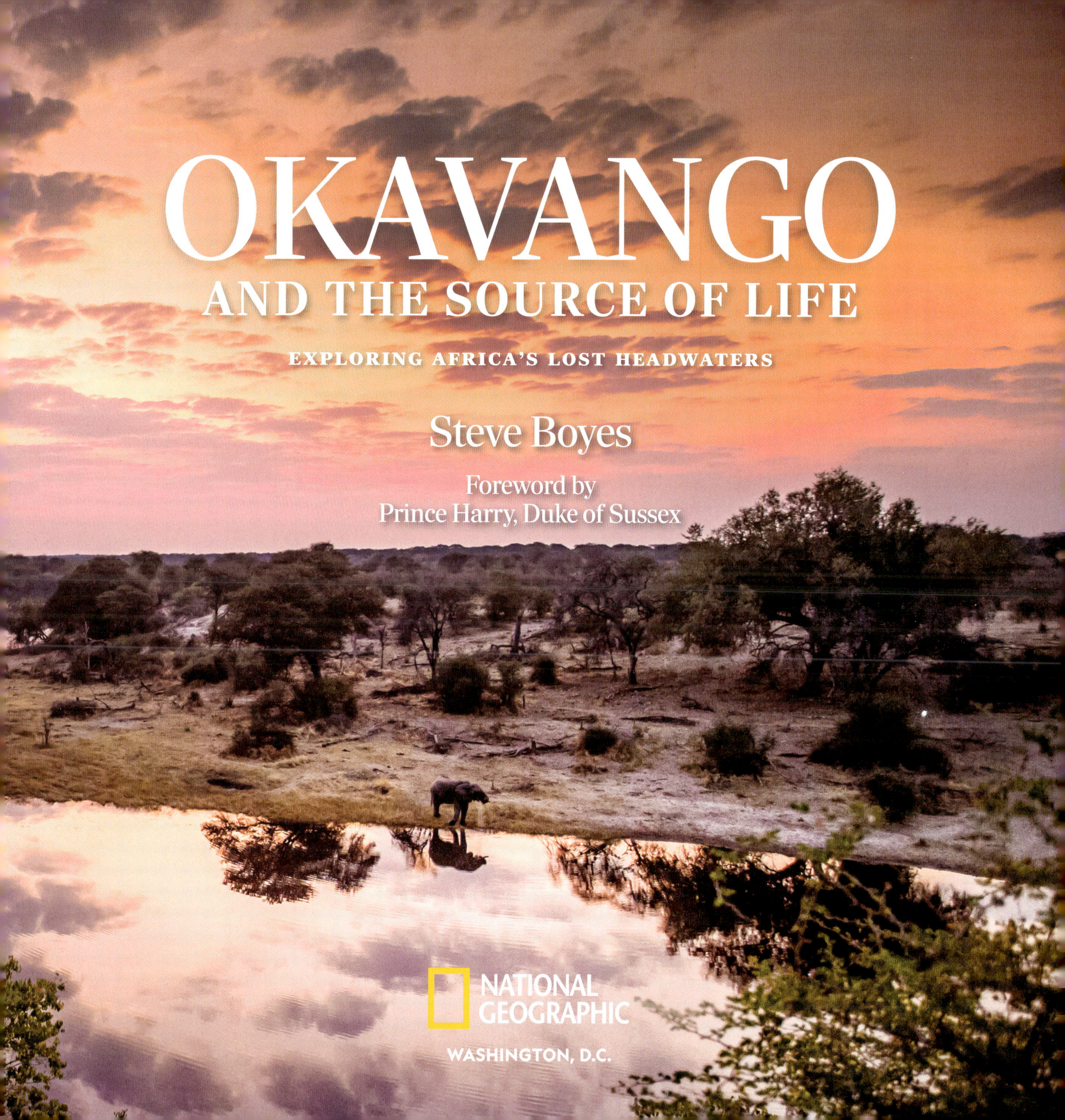
OKAVANGO
AND THE SOURCE OF LIFE
EXPLORING AFRICA'S LOST HEADWATERS
Steve Boyes
Foreword by
Prince Harry, Duke of Sussex
NATIONAL GEOGRAPHIC
WASHINGTON, D.C.

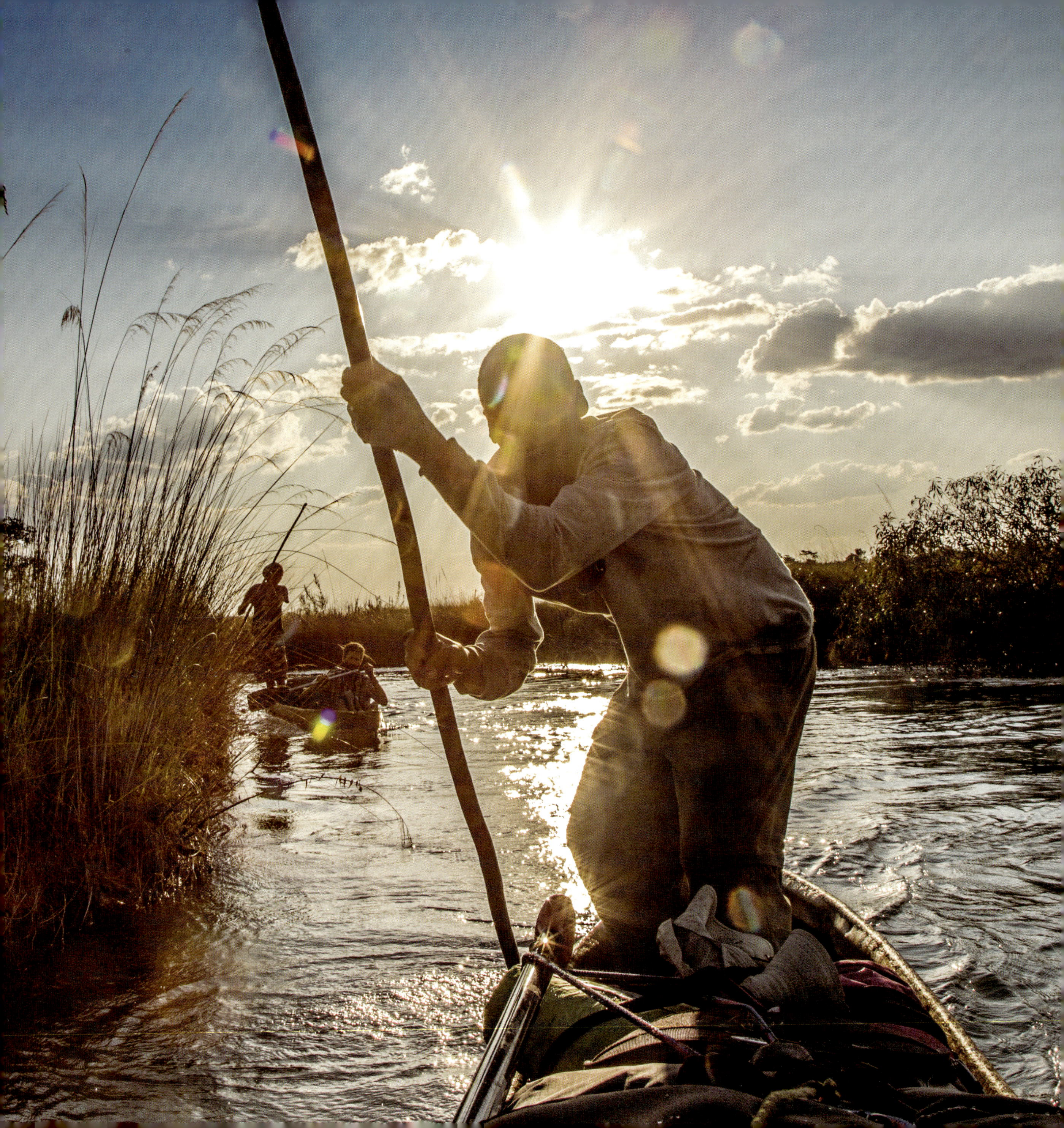

Contents

River by river, we had no idea what to expect as we poled our mokoros*—dugout canoes—down undocumented rivers in the remote Angolan Highlands. Mr. Water, an elder who has lived his whole life in the delta, told us to trust that the floodwater would take us home.* **PAGES 2–3:** *These rivers, the originating waters of the Okavango Delta, once represented lifelines for Africa's largest wildlife migrations.*

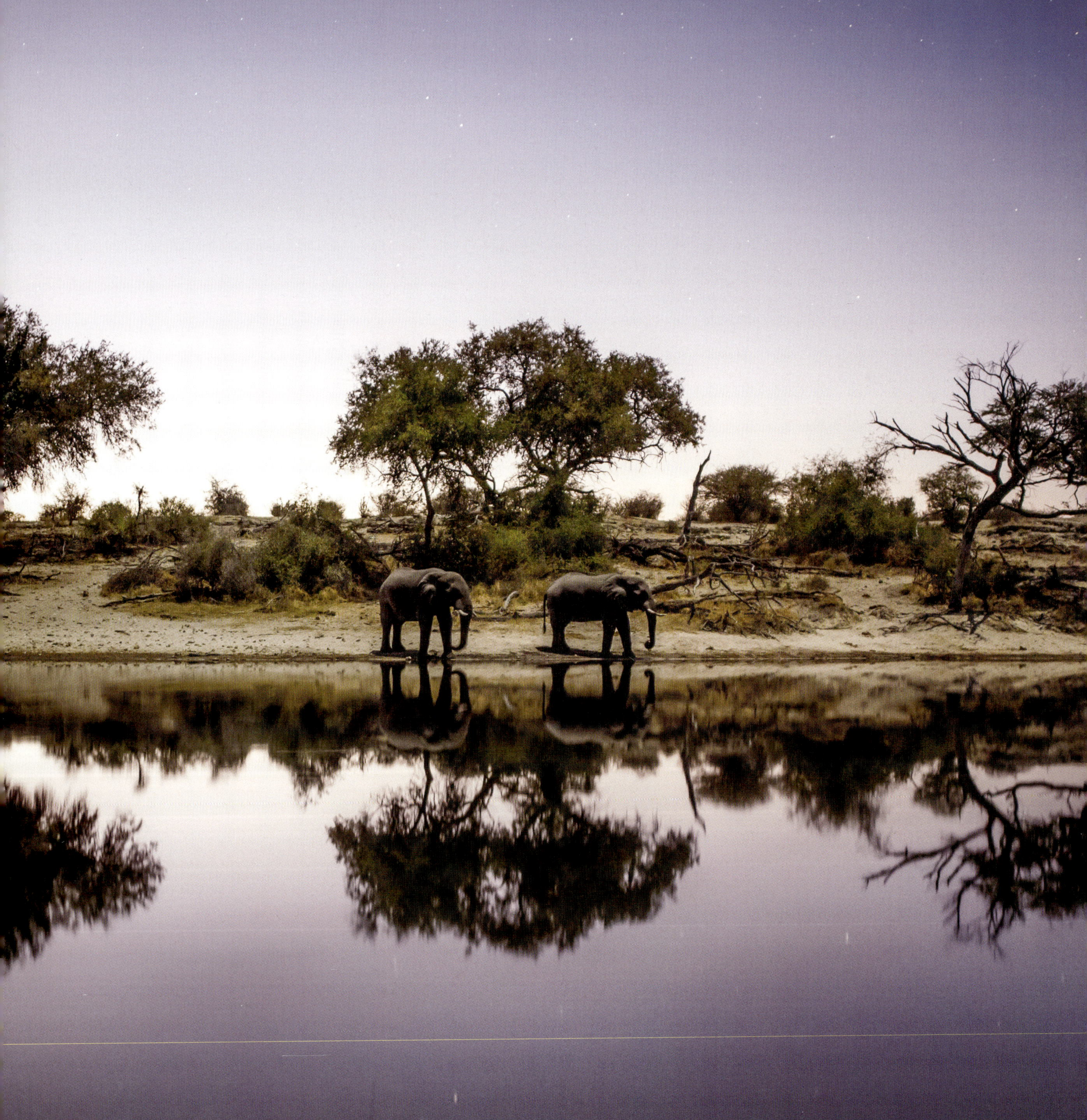

Foreword

PRINCE HARRY, DUKE OF SUSSEX

THERE ARE SOME places on Earth that are so vast, beautiful, and alive, they truly open your eyes and mind.

The Okavango Delta is one of those places, an enigmatic wetland that disappears into the Kalahari Desert and is home to the world's largest remaining elephant population—a wilderness beyond comparison. This paradise has been my second home for more than 25 years, a place to escape and be enveloped by nature's sheer magnificence.

Back in 1997, in Huambo, Angola, just a few miles from one of the sources of the Okavango, my mother walked through a live minefield being cleared by the HALO Trust, a humanitarian land mine clearance charity. That famous walk was a turning point in the fight against these lethal devices. By 1999, the use, production, and transport of antipersonnel land mines had been banned globally. Today, there are more than 80,000 amputees in Angola, a tragic legacy of 27 years of the Angolan Civil War, which ended in 2002.

When I first visited the Okavango Delta, shortly after my mother's death, the war was still raging around the Okavango's headwaters. Africa's largest tank battle since World War II was fought over a bridge at the confluence of the Cuito and Cuanavale Rivers, the Okavango's two major tributaries in the eastern Angolan Highlands. This battle left behind one of the world's largest minefields, effectively preventing all safe access. Nearly 20 years later, in 2015, I heard about a National Geographic expedition supported by the HALO Trust that had found a way through the land mines to the undocumented source of the Cuito River. I was intrigued.

There the expedition team found a crystal clear, acidic source lake sustained by vast, previously undocumented peatlands. The intact miombo woodlands surrounding the lake, seemingly endless, were bigger than the whole of England.

This book is the story of the expeditions and discoveries that followed. Over the coming years, a team of 57 scientists and local guides explored all the major rivers and tributaries of the Okavango Delta, traversing this vast landscape in dugout canoes, on motorbikes, on foot, and in armored vehicles. They discovered hundreds of new species and documented 29 source lakes in what the local Luchazi people call Lisima lya Mwono—the Source of Life, where the floodwaters of the Okavango Delta come from.

As wildfires rage and hurricanes tear apart entire peninsulas, we're faced with a harsh reality: a climate crisis and a mass extinction that can no longer be ignored. Appreciating, preserving, and protecting these last wild ecosystems is essential to our collective survival.

The best way for us to understand where we came from, and where we need to get to, is by visiting places like the Okavango Delta—my source of life.

Elephants visit the edge of the Boteti River in Botswana's Makgadikgadi salt pans, part of the larger Okavango Basin.

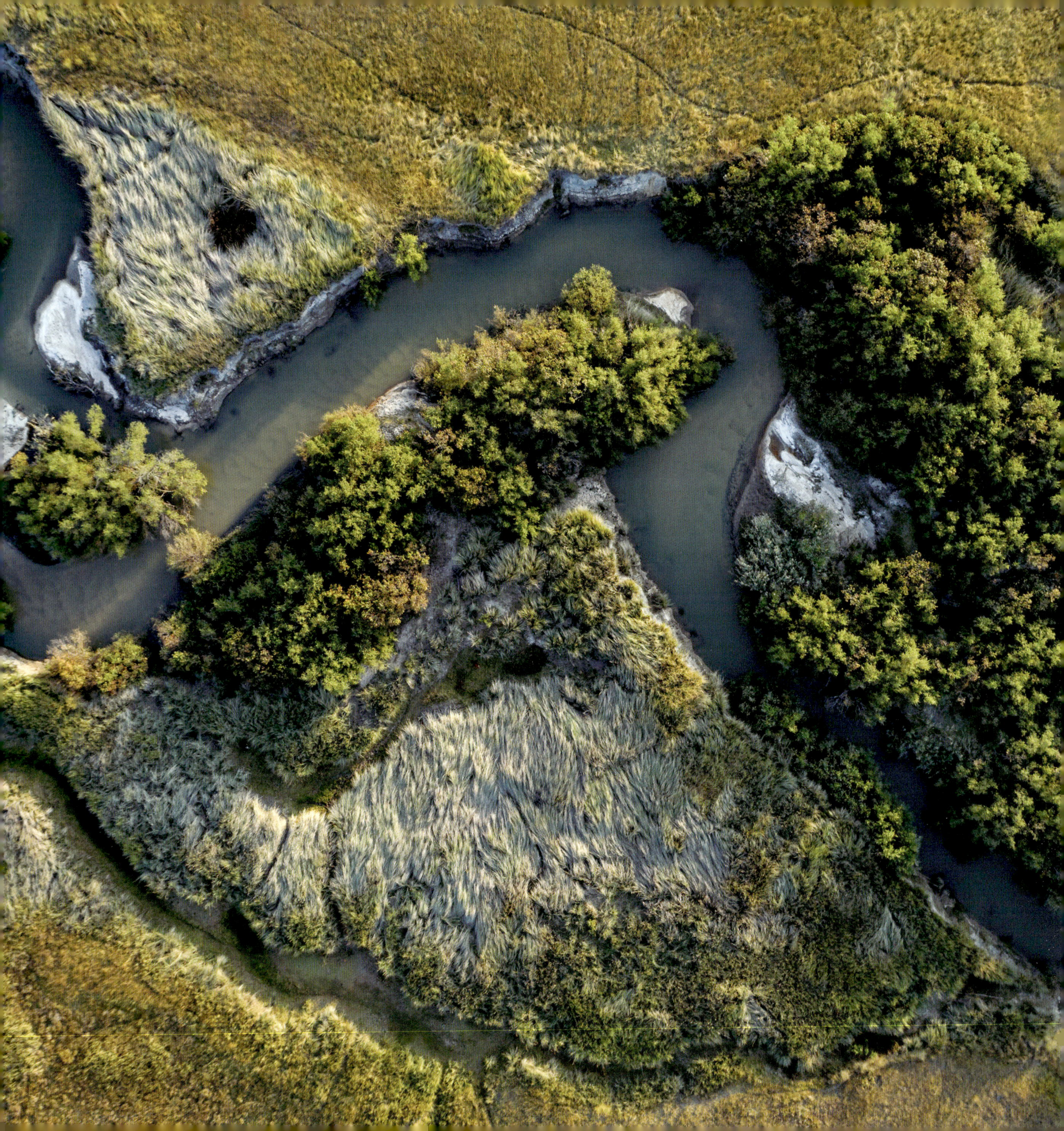

Preface

JILL TIEFENTHALER, CHIEF EXECUTIVE OFFICER
NATIONAL GEOGRAPHIC SOCIETY

A PALETTE OF ORANGES and pinks blanketed the sky as I stood knee-deep in the cool waters of the Okavango River. Though I knew this river was among the most important on Earth, sustaining countless species, communities, and livelihoods, it was the first time I'd seen it with my own eyes. The water flowing past me began its journey thousands of kilometers away in the highlands of Angola, weaving through channels, tributaries, and peatlands before reaching northern Botswana, where I stood that evening in 2022.

This was my first trip in the field with National Geographic Explorers, and I'll never forget the awe I felt: the rustling reeds, the calls of fish eagles, and the overwhelming sense of being immersed in a living, breathing wonder. The experience deepened my understanding of our interconnectedness with nature and underscored the urgency of our mission at the National Geographic Society to illuminate and protect the wonder of our world.

Few efforts exemplify our mission more than the National Geographic Okavango Wilderness Project, led by National Geographic Explorer Steve Boyes. His life's work is rooted in a profound connection to Africa, forged through years spent living and working in this extraordinary oasis where his passion for conservation was born.

In *Okavango and the Source of Life: Exploring Africa's Lost Headwaters*, Steve shares a deeply personal account of his journey across the Okavango River Basin and the powerful movement working to protect it. Today, Steve's story is inseparable from those who stand alongside him: fellow Explorers, government leaders, local partners, and communities who have sustained this landscape for generations. Grounded in collaboration and community-led conservation, this impressive network uses science, traditional knowledge, education, and storytelling to secure lasting protection for the Okavango Delta and its source waters. Steve recounts undertaking grueling expeditions spanning thousands of kilometers, establishing critical scientific baselines, and documenting dozens of species new to science.

Okavango and the Source of Life is a story of resilience and hope as well as a call to action—inviting us to learn from those who have devoted their lives to preserving this vital ecosystem.

As I stepped out of the water during that 2022 trip, one thought stayed with me: The Okavango River Basin is irreplaceable. I hope these pages, filled with breathtaking photography and vivid storytelling, not only captivate you with the Okavango's beauty but also inspire you to reflect on the role we play in protecting the world's last wild places.

During the 2015 expedition, Boyes and his team members camped most nights on burned ground—the gray swaths of land in this aerial photo—along the upper reaches of the Cuito River, the eastern source of the Okavango Delta. Local Luchazi people told the expedition team that hunters light fires to chase animals into snares and traps and to open up access to the river.

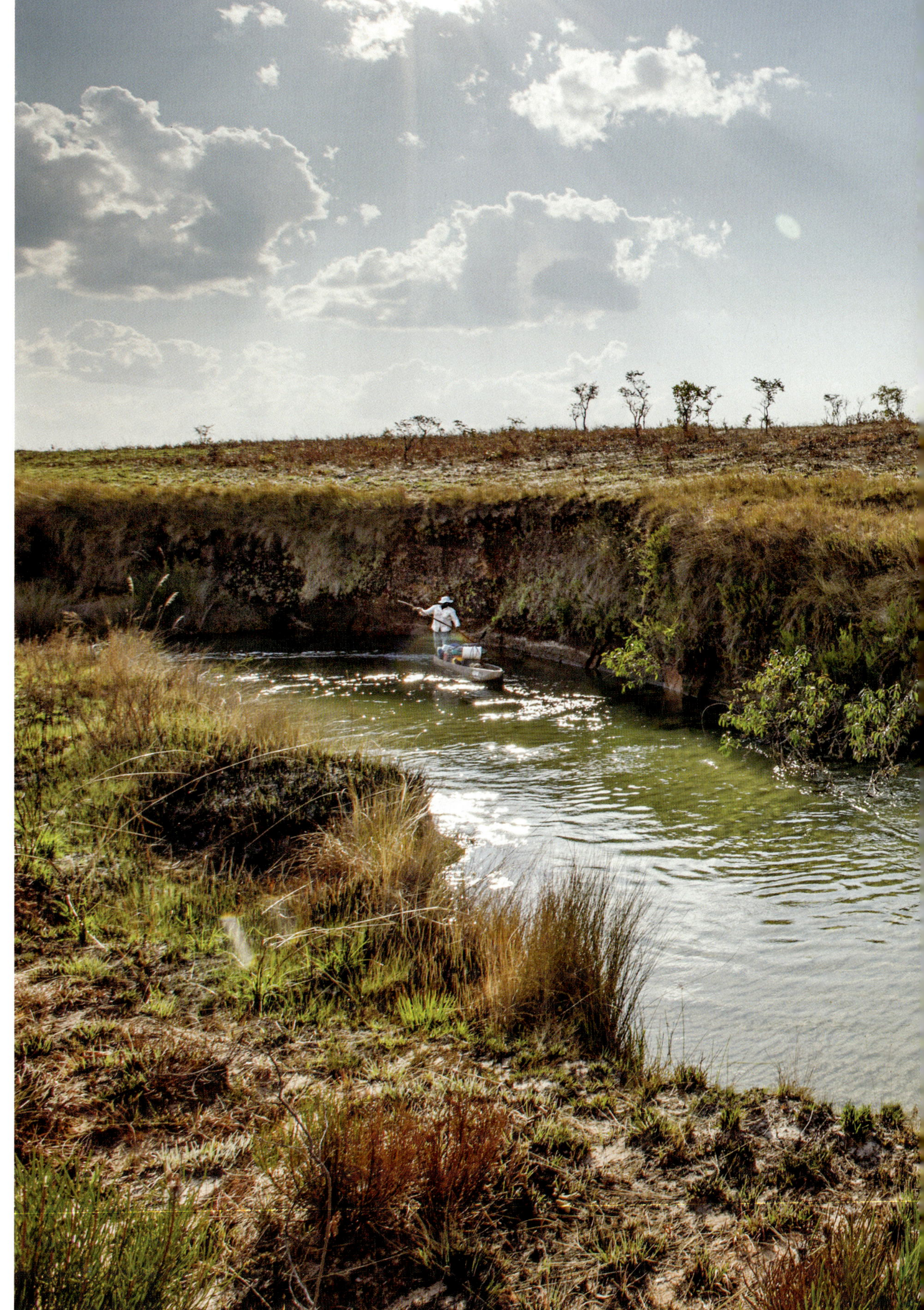

The first few weeks poling our mokoros down the crystal clear Cuito River, near its source lake hidden deep in unending miombo woodlands, were among the best weeks of my life. The sense of this abandoned, burned place was other-worldly. I fell in love with the Lisima Landscape—the Source of Life.

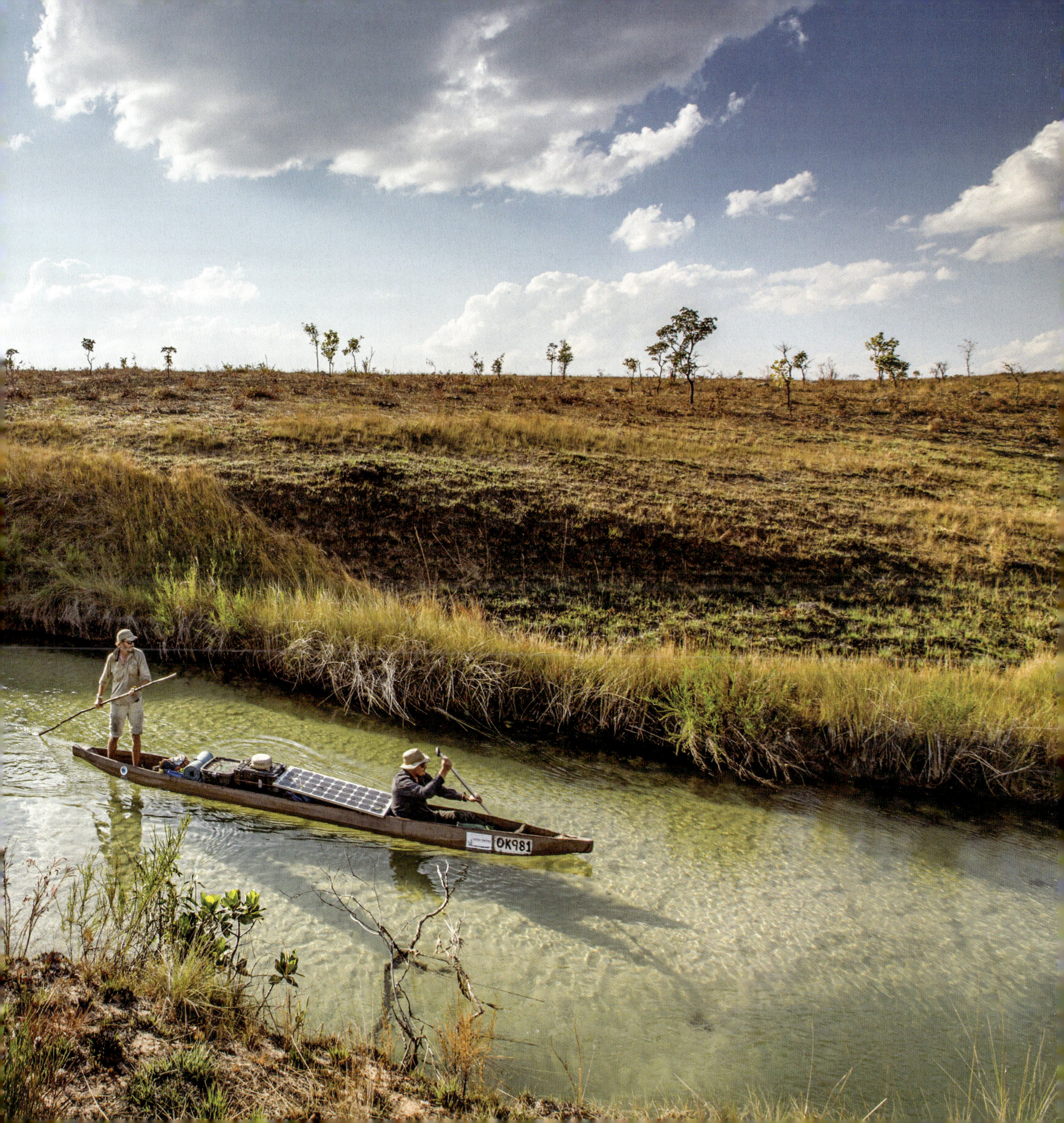
OK981

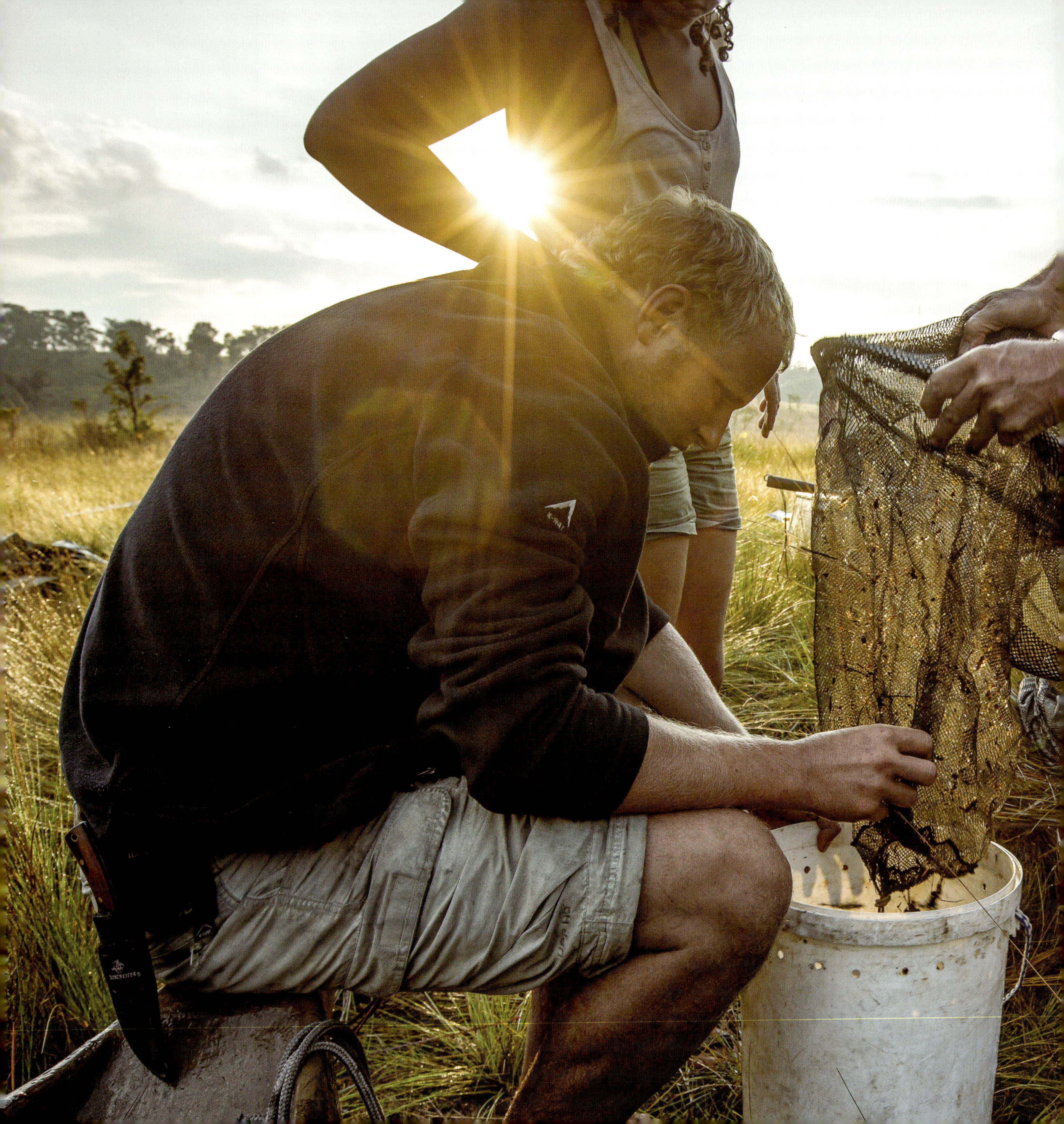

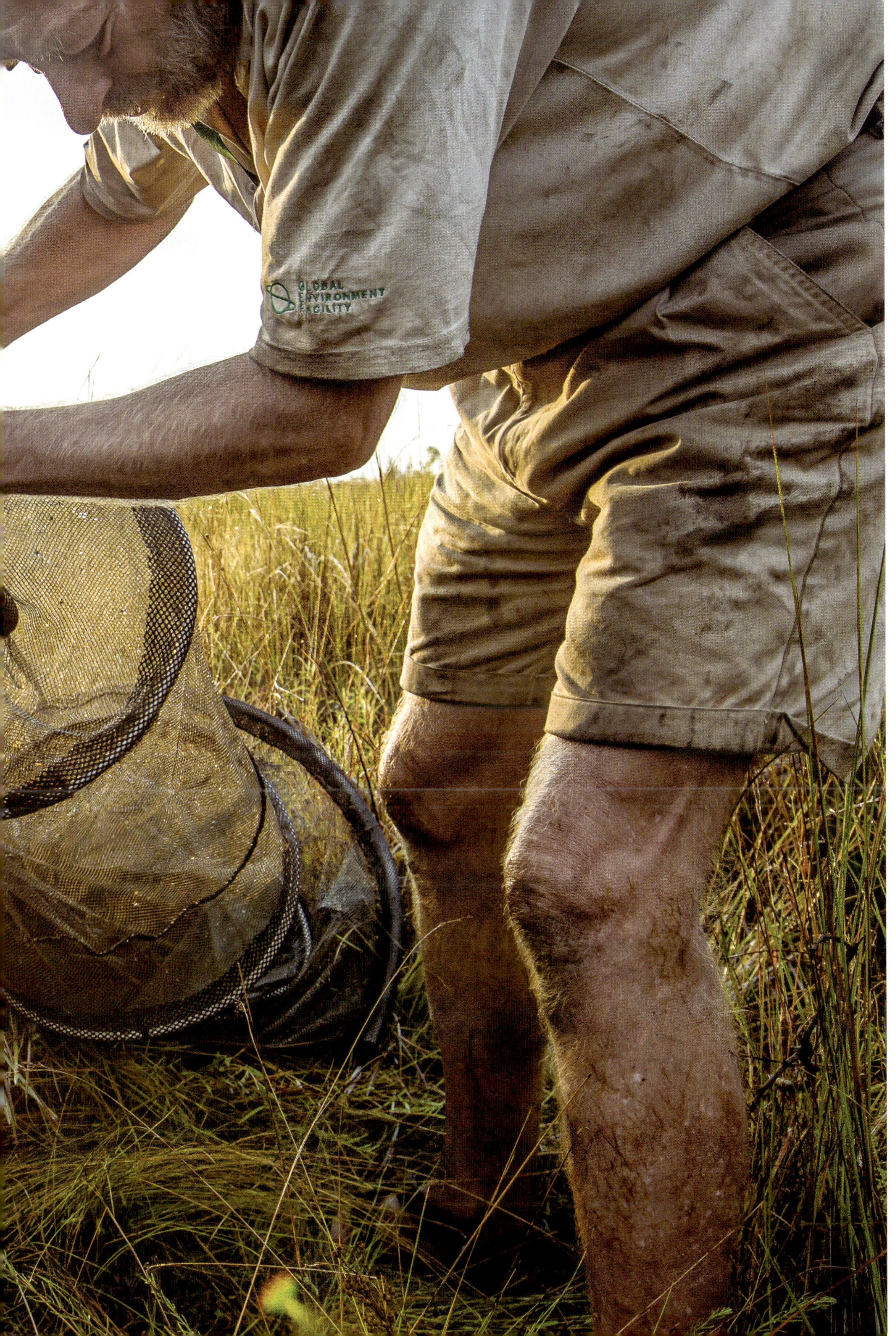

Within the first few mornings on the Cuito River, our team had already found several species new to science. Here Götz Neef (left), Adjany Costa (standing at left), and Paul Skelton (standing at right) sift through fish caught in a net set overnight.

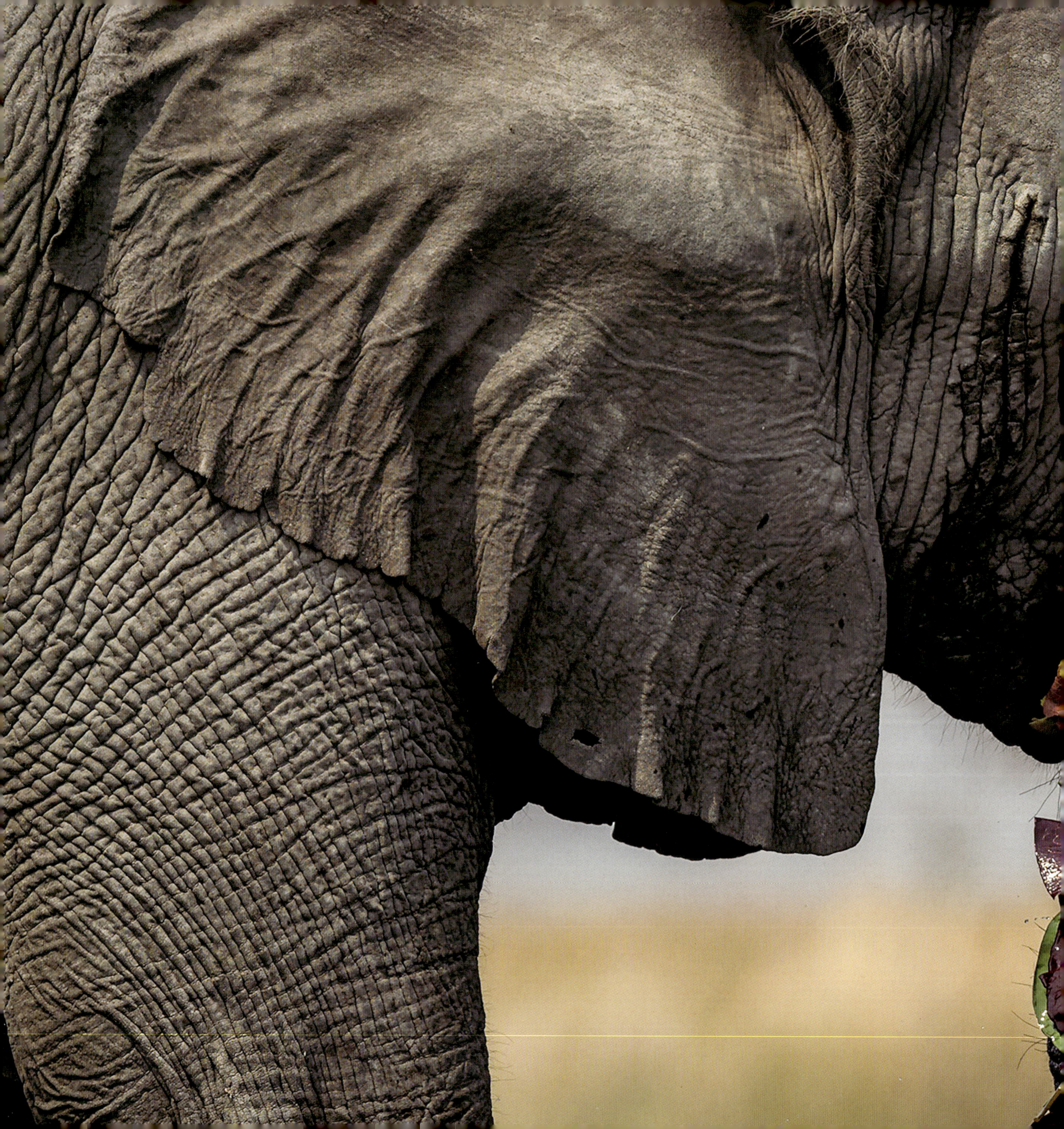

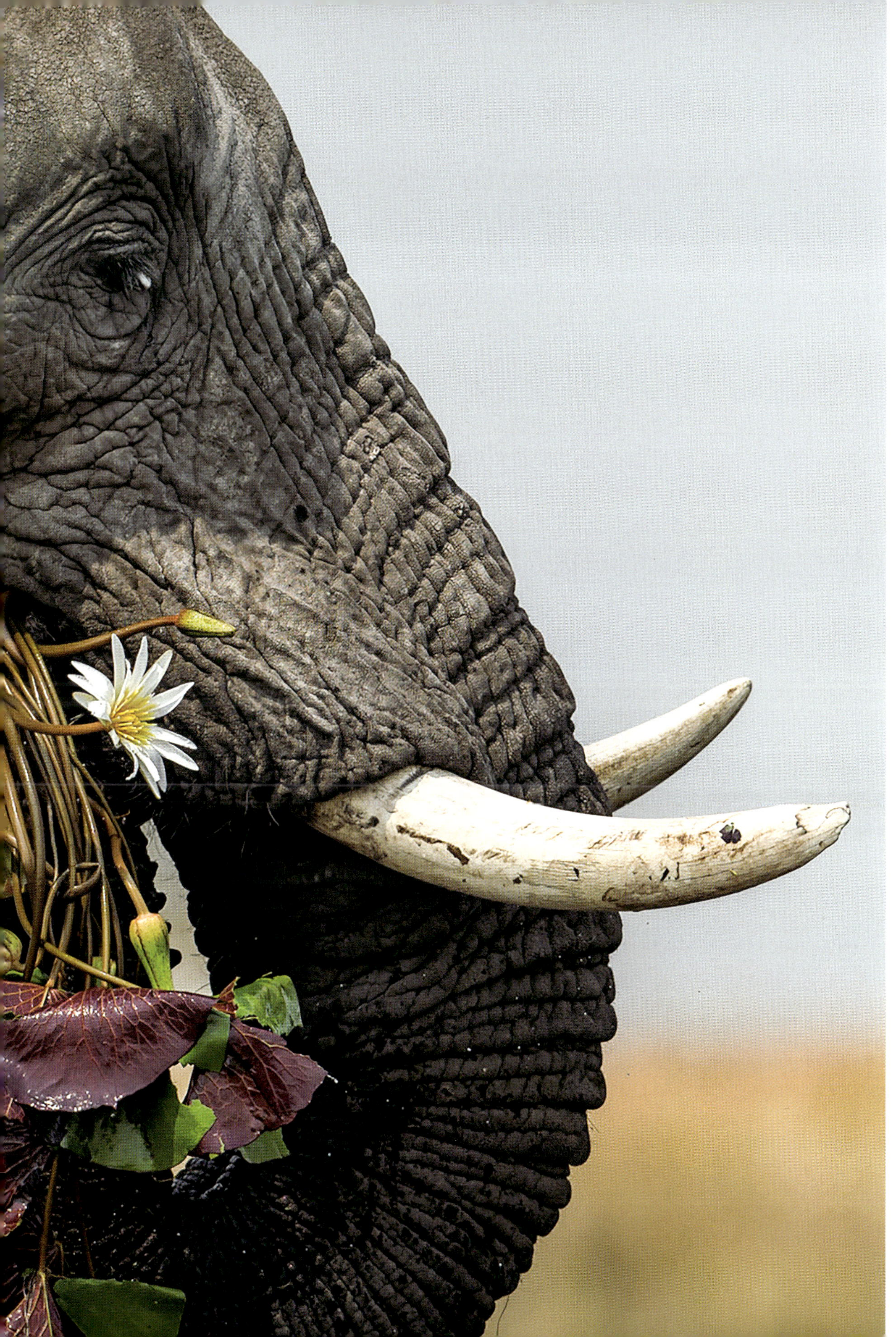

Called the "Jewel of the Kalahari," the Okavango Delta is an elephant paradise where they can grow old by feeding on soft foods, like water lilies, which help preserve their last sets of teeth. When we moved slowly on the water, as the local Wayeyi people have for centuries, the elephants began to accept us.

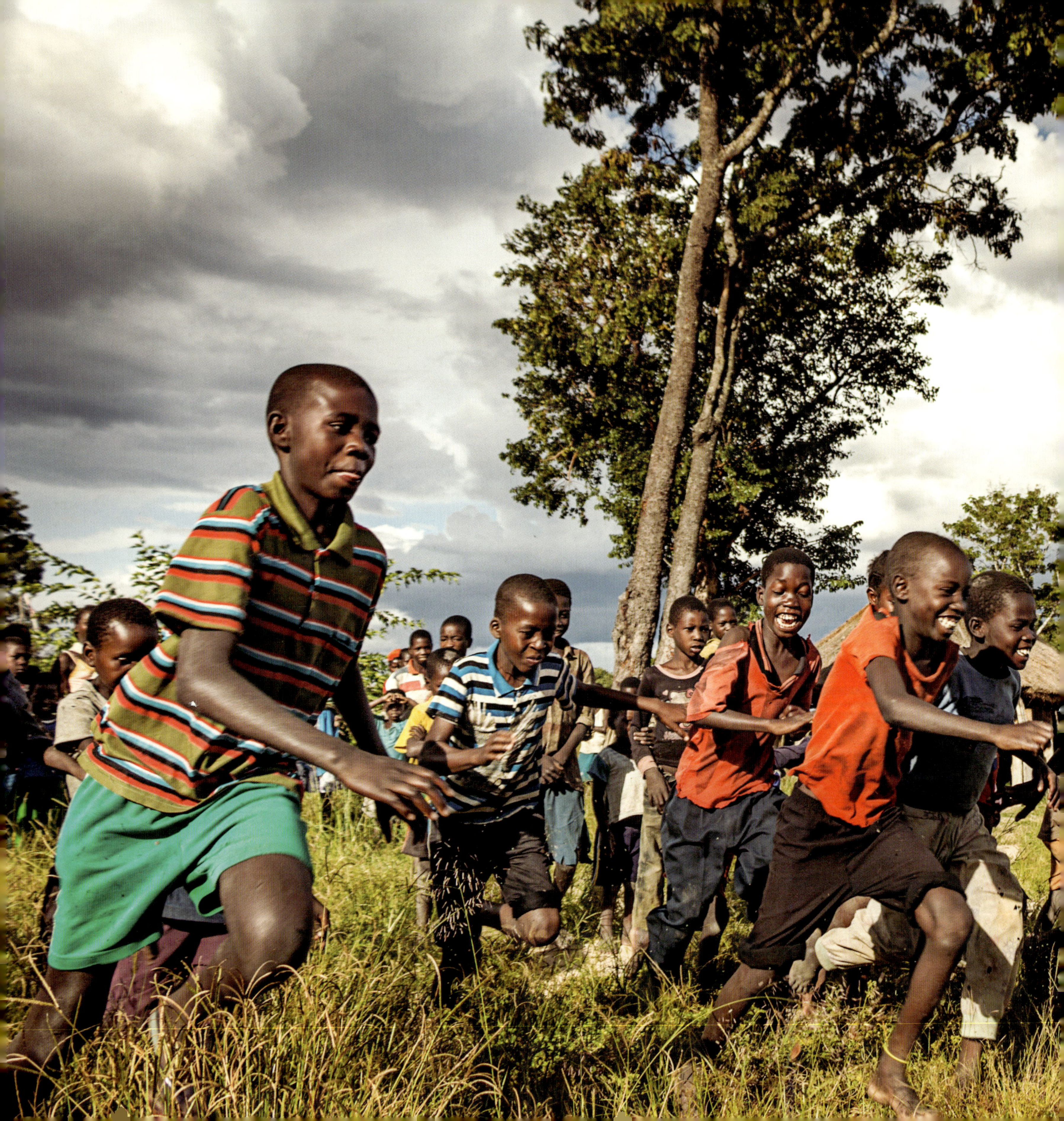

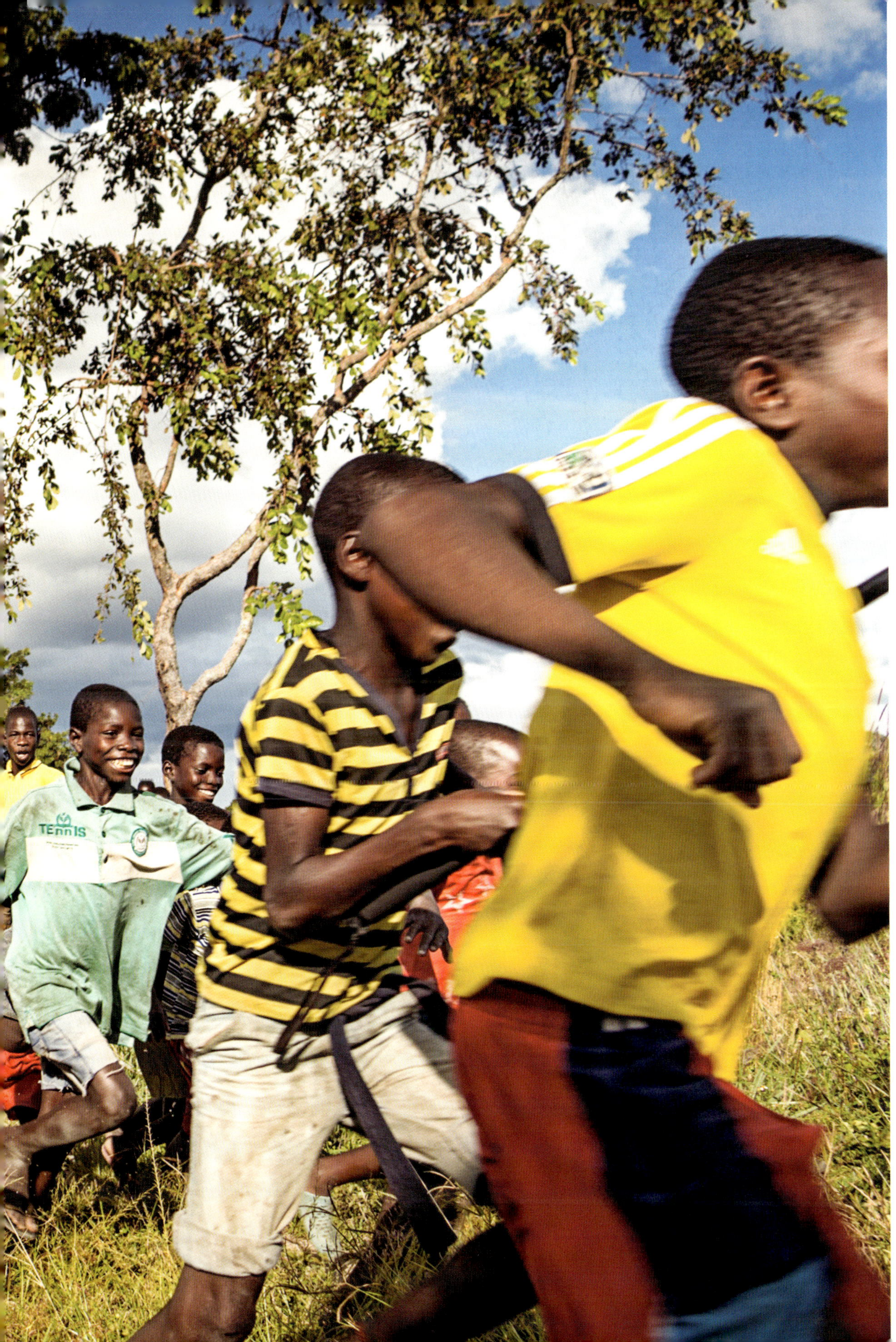

These children in the Angolan Highlands watched in amazement as a photographic drone flew nearby. Then they took off, thrilled to be chased by it. We can only imagine the opportunities that will come to these young boys as new technologies, like artificial intelligence and satellite connectivity, come to the Okavango Basin.

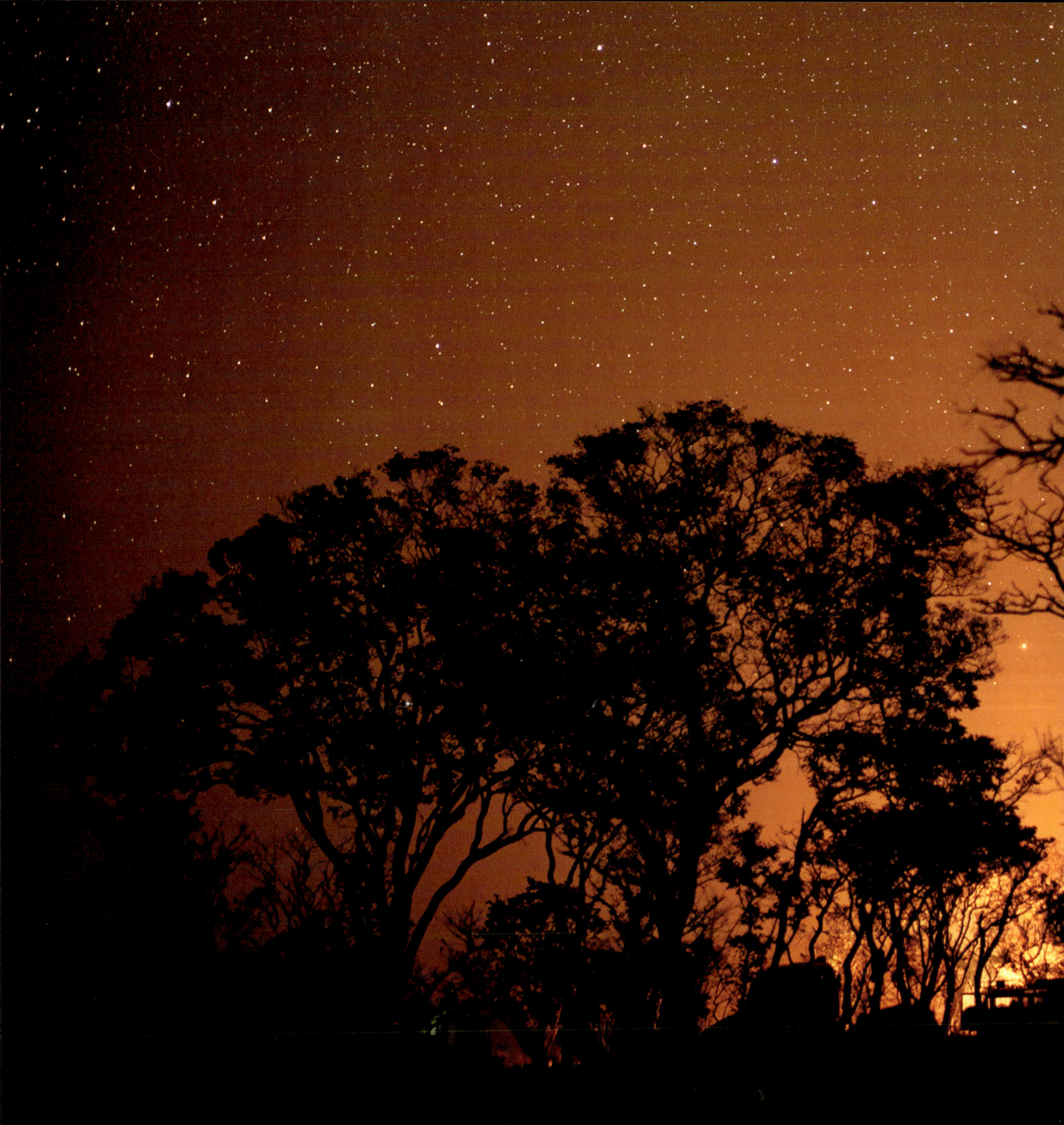

INTRODUCTION

Beginnings

THE OKAVANGO DELTA, 2001 TO PRESENT

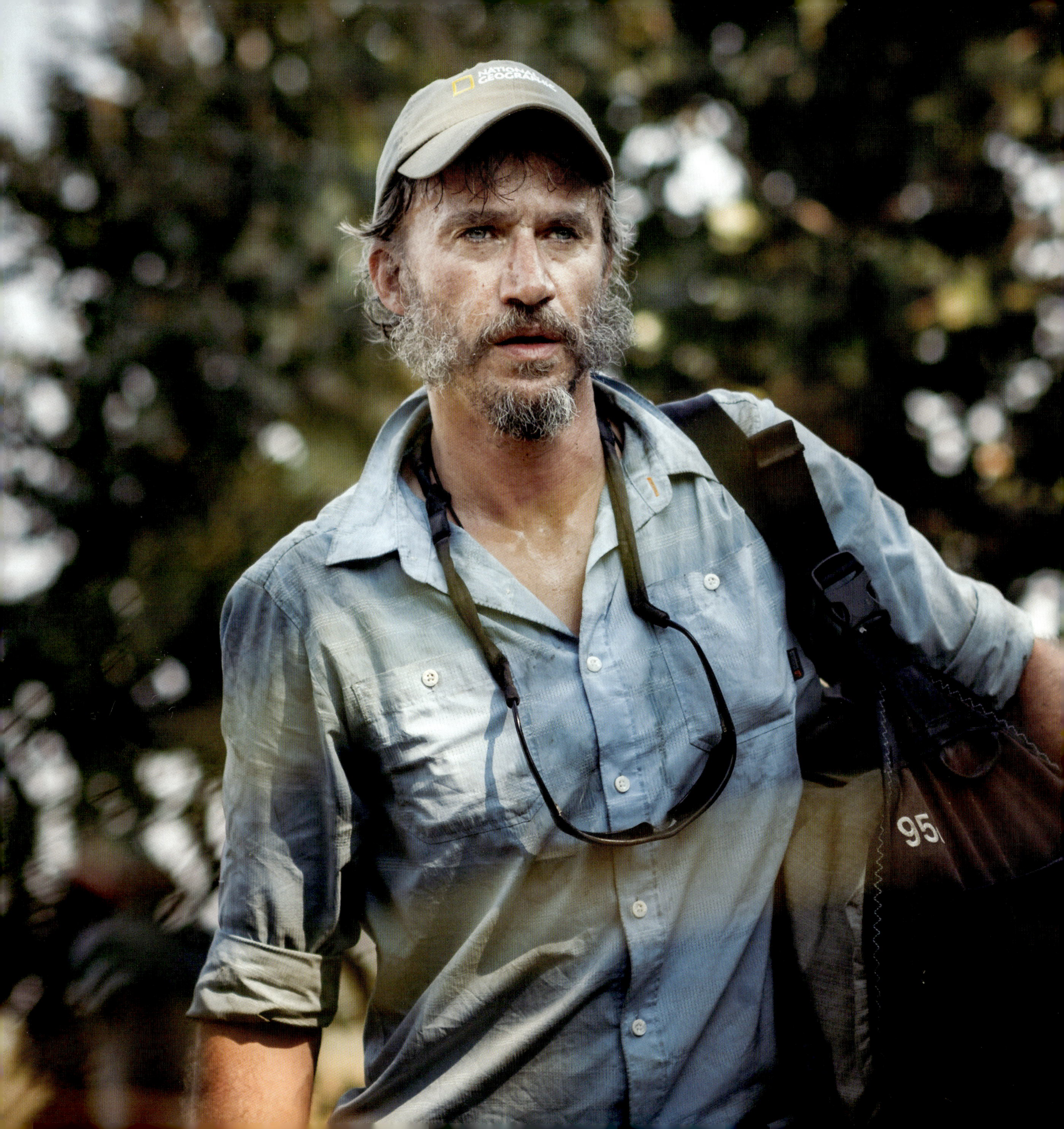

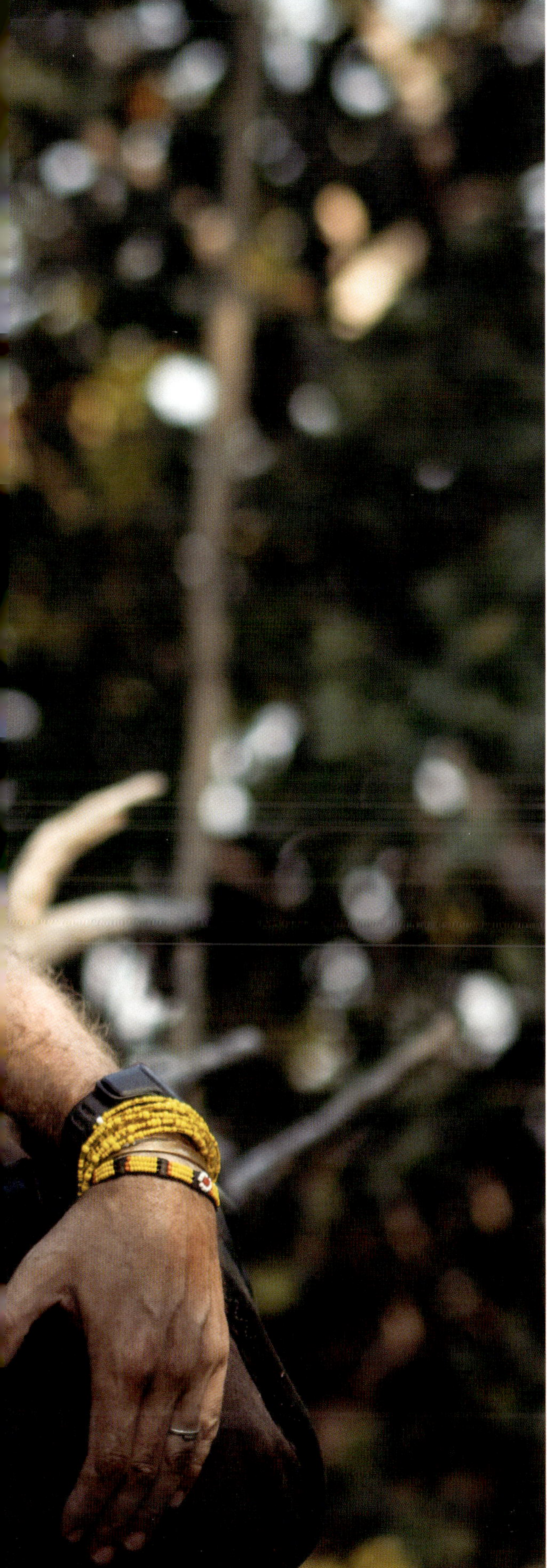

Every quest into the wild is a warning. Wild landscapes are sacred places, where people live connected to millennia of death, reverence, and struggle. Explorers know that our last wild places and the people living in them hold the humility, resilience, and interconnectedness of all living things. They call us to stop, learn, protect a vision of being that is abundant, tragic, wonderful, and true. By entering to document them, do we destroy these last wildernesses, or can we remember how to be part of them?

It was a Sunday in August, my day off, and I had been walking alone in the Okavango Delta since dawn. I was barefoot, carrying only a small spear, tea supplies, milk powder, and biscuits for the day. The weather was cool, and there were ripe jackalberries on the ground for breakfast. I was looking for a safe place to have lunch when I ended up under a sausage tree, startling away a group of impalas munching on the trees' beautiful maroon flowers. I made a little fire and watched red lechwe, an aquatic antelope, far off in the floodplain with my binoculars.

I had been there for 20 minutes or so when some Meyer's parrots—small gray-brown birds with blue-green bellies and yellow flashes—arrived and began screeching loudly in the uppermost branches of the tree, feeding on the flower part from which the sausage fruit would eventually grow.

As the parrots progressed down into the tree canopy in search of flowers, they became increasingly agitated. I started to feel they were trying to tell me something. Only after scanning the

It's now 25 years since I first went on expedition in the Okavango Delta. No matter what happens, I will continue poling across the delta until I die. The floodwater is a source of life for me too. **PAGES 18–19:** *The Okavango Delta is flatter than a billiard table, making sunsets spectacular. They linger, giving color to a starry sky. Tourists come here just to see the sunsets.*

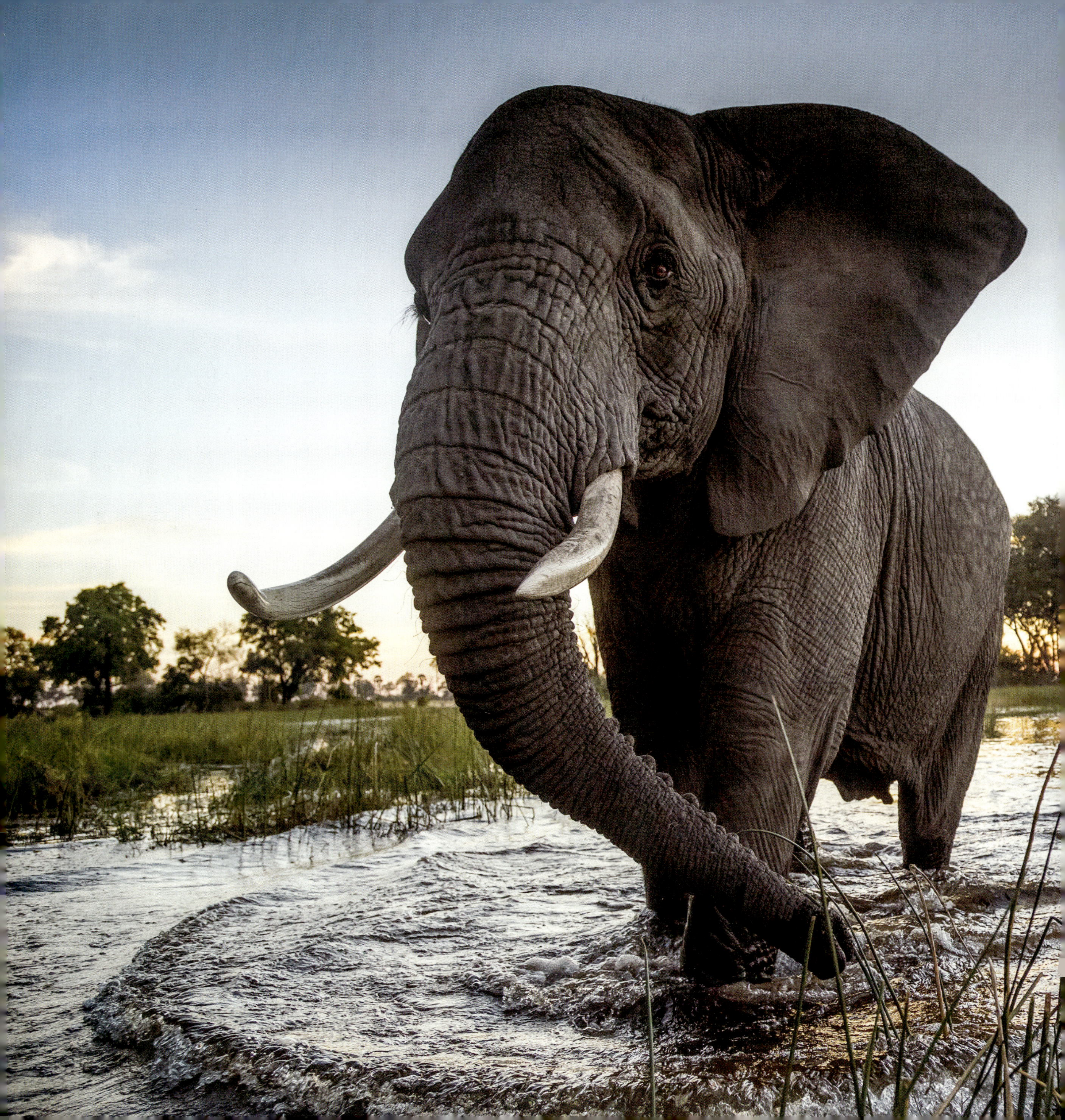

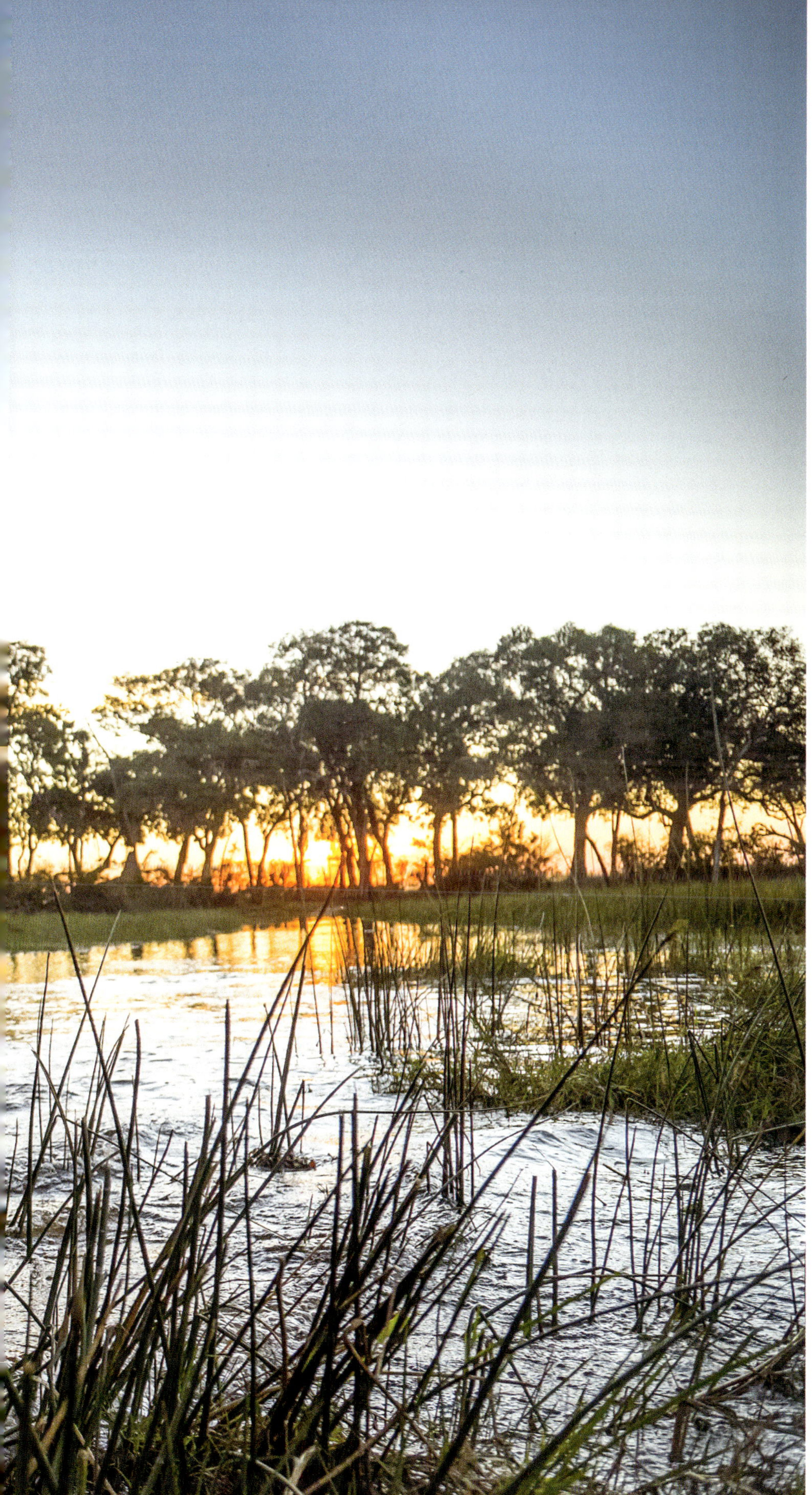

canopy above did I realize that I had someone else's attention. Hidden in the mottled light of the tree's dense foliage was a young male leopard I knew: "Little Gideon," who often hunted baboon here on Vundumtiki Island. Completely still, he had been looking at me all this time, from just over my head.

Meyer's parrots would teach me many things in the years to come. This was my first lesson: Leopards wait for impalas to feed on sausage tree flower petals that the parrots drop.

My heart was racing as I put out the fire and packed up. Gideon was still relaxed when I looked at him out of the corner of my eye. But when I picked up my spear, he jumped out of the tree, moving in complete silence. It felt like all life was holding its breath. He landed in front of me and looked into my eyes, beyond my soul. I had never felt so connected to the present moment. I had been in the delta for just two months and had now met two of the animals that would become part of my story in the Okavango Basin.

A YEAR BEFORE, in 2001, a school friend had asked me to join him and his fiancée on a trip to the Okavango Delta and Chobe River. The three of us drove into the delta in an old Toyota Land Cruiser and spent the next week stuck in flooded roads, digging and winching ourselves out of the mud, jamming branches and tree trunks under the tires, and camping under the stars. We bathed with crocodiles, got sunstroke, were raided nightly by hyenas, and fell silent between the exhilarating laughter of wildness and rum. It was heaven.

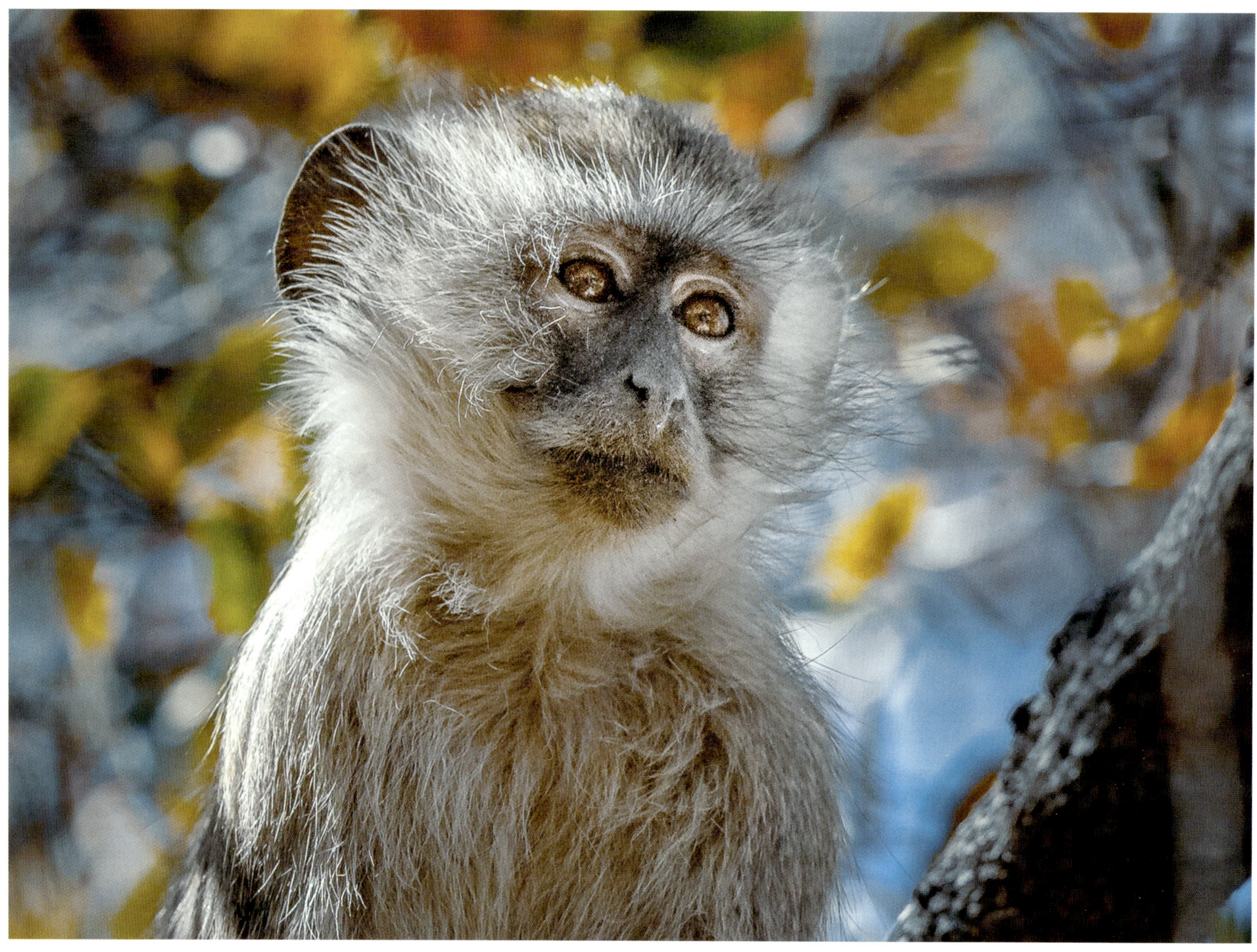

The wildlife was extraordinary: thousands of lechwe, zebras, buffalo, giraffes, and birds. The air was filled with flying ants, mosquitoes, and biting flies, with birds and bats darting in between. I had never before seen such abundance of life. The ever present call of woodland kingfishers and African fish eagles followed us everywhere. Every night, the lions called, hyenas cackled, and scops owls screeched.

I had found the place I had been looking for all my life. The decision was made: I was coming back to live in northern Botswana. This is where I would get completely and entirely lost—and where I would come to find myself.

I was born in Cape Town, South Africa, and my family moved northeast to Johannesburg in the early 1980s to be closer to the rest of Africa. Because we were white South Africans, apartheid had provided undeniable privilege in my youth. Yet the people of the continent we loved, bereft of opportunity and support, were fighting for freedom, and we often felt the deep-seated torture of powerlessness to circumstance.

We lived every day to be African, and so we loved every minute we could spend in the bush. We could be ourselves in the wild, away from the madness of rioting, protests, and unfairness of urban life. We were always on long-distance safari drives, traveling to see the Serengeti's wildebeest migrations; to cruise the Chobe River in Botswana; to walk the dry, rocky Richtersveld or Kaokoveld deserts; or to explore remote valleys filled with baobab trees in southern Tanzania. We were Africans stuck between old and new.

My brother, Chris, and I spent long afternoons integrating ourselves into baboon troops, investigating the inner workings of termite mounds, mimicking birdcalls, and playing with banded mongoose in the baking sun. We learned a deep respect for the dangers of the African wilderness. The beauty of my parents was that they were as passionate and naive as we were. We all just wanted to be in the wild, immersed in its beauty.

I am proud to have lived my life on this grand primordial continent. My children are seventh-generation South Africans. I have no other passport, heritage, or identity. I will never forget my privilege, or past inequalities and injustices brought upon my fellow South Africans and Africans in general. I work tirelessly to correct them. My heart beats and my lungs open for Africa.

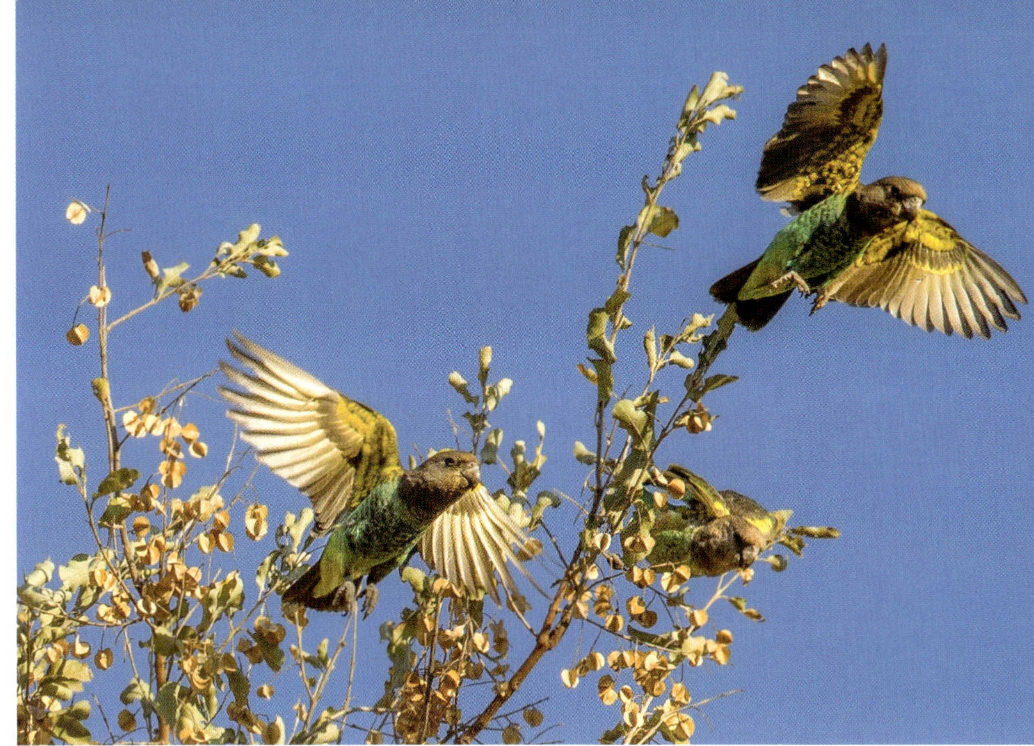

After I'd spent the first two decades of my life searching for my place in the African wild, my friend's offer to take me to the Okavango Delta would change my life forever. I was 22 years old, completing a master's degree in environmental management at the University of KwaZulu-Natal in Durban, specializing in protected areas management. Yet when I returned from the delta, all I wanted to do was go back. I resigned from my job at a forestry company, where I was working to pay my way while collecting data for my thesis, and took a job with a well-known company called Wilderness Safaris. I told my professors I'd try to complete my master's some other way.

In early 2002, the safari company flew me in a small plane to Vundumtiki Camp, one of the remotest camps in the Okavango Delta, on a tiny island in the northeast part of the delta. I became "head of housekeeping." My life has had meaning ever since.

Meyer's parrots feeding on russet pods of bush willow are common winter sightings. **OPPOSITE:** *Vervet monkeys are just as interested in us as we are in them. They bark, screech, and chatter in alarm whenever they spot a leopard.* **PAGES 22–23:** *If you wait at an elephant crossing, sitting quietly, elephants will eventually come. They will smell you, then see you, and then, completely in control of what's going on, walk past you, just meters away.*

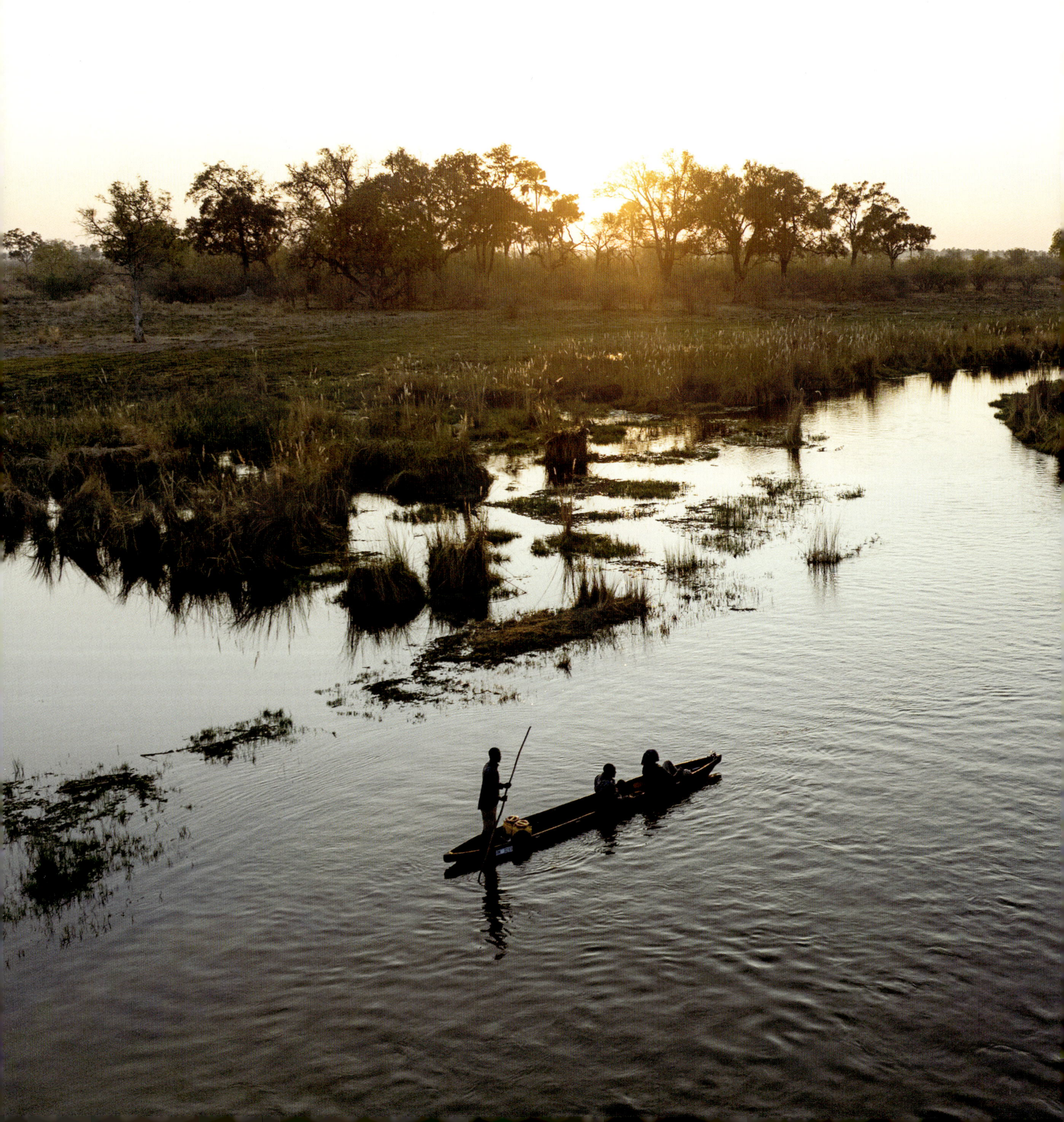

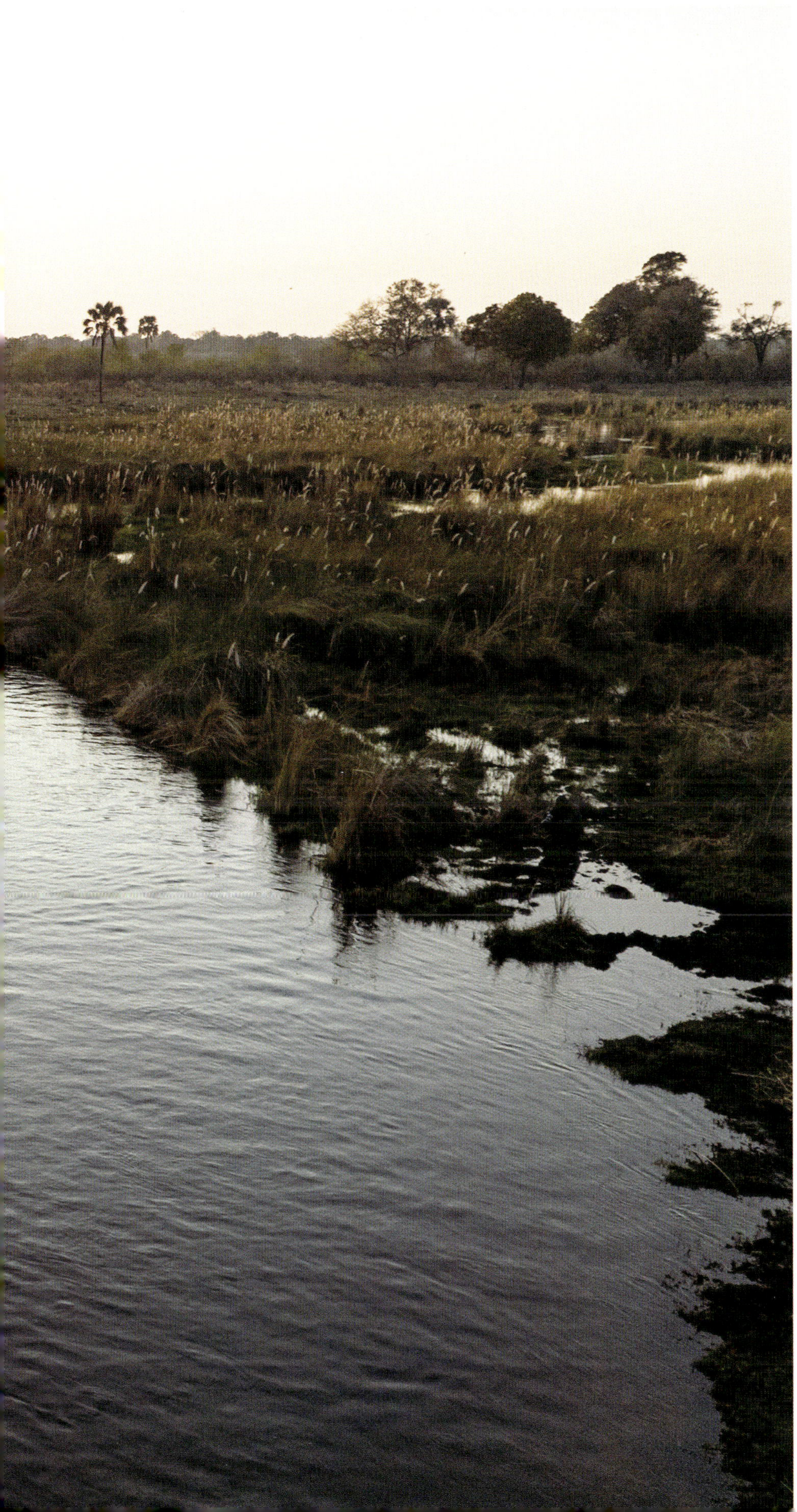

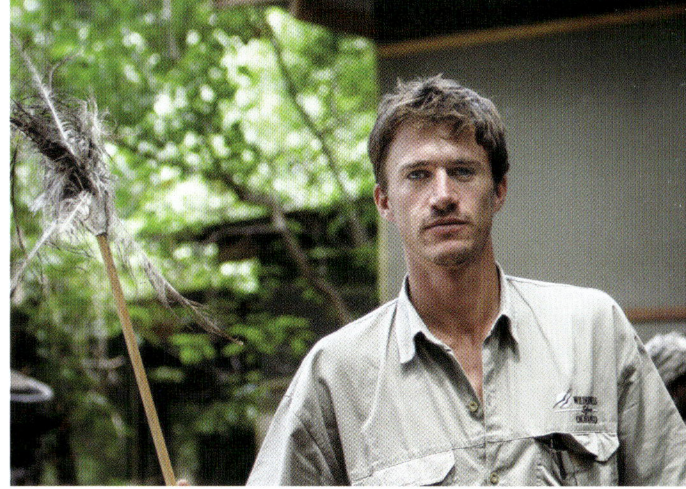

Within six months, I had finished my master's dissertation on census techniques for the southern reedbuck, a common African antelope. I submitted it both via bushmail, an internet connection using radio frequencies, and as a printed copy in a mailbag with some departing guests. The method I developed was published in the *South African Journal of Wildlife Research,* and the university offered me the opportunity to register for a Ph.D. in zoology. I decided to do my thesis on the ecology of Meyer's parrots, Africa's least known and most abundant parrots. A healthy population could be studied in the Okavango. Wise beyond their diminutive size, these little parrots became my guides. Following them tree to tree, my knowledge of the delta grew exponentially.

I started doing daily 26-kilometer (16 mi) vehicle-based transects, describing and mapping the birds' habitat types along sandy tracks around Vundumtiki. I stopped every time I heard the parrots calling, then grabbed a clipboard and my binoculars to go find them. They

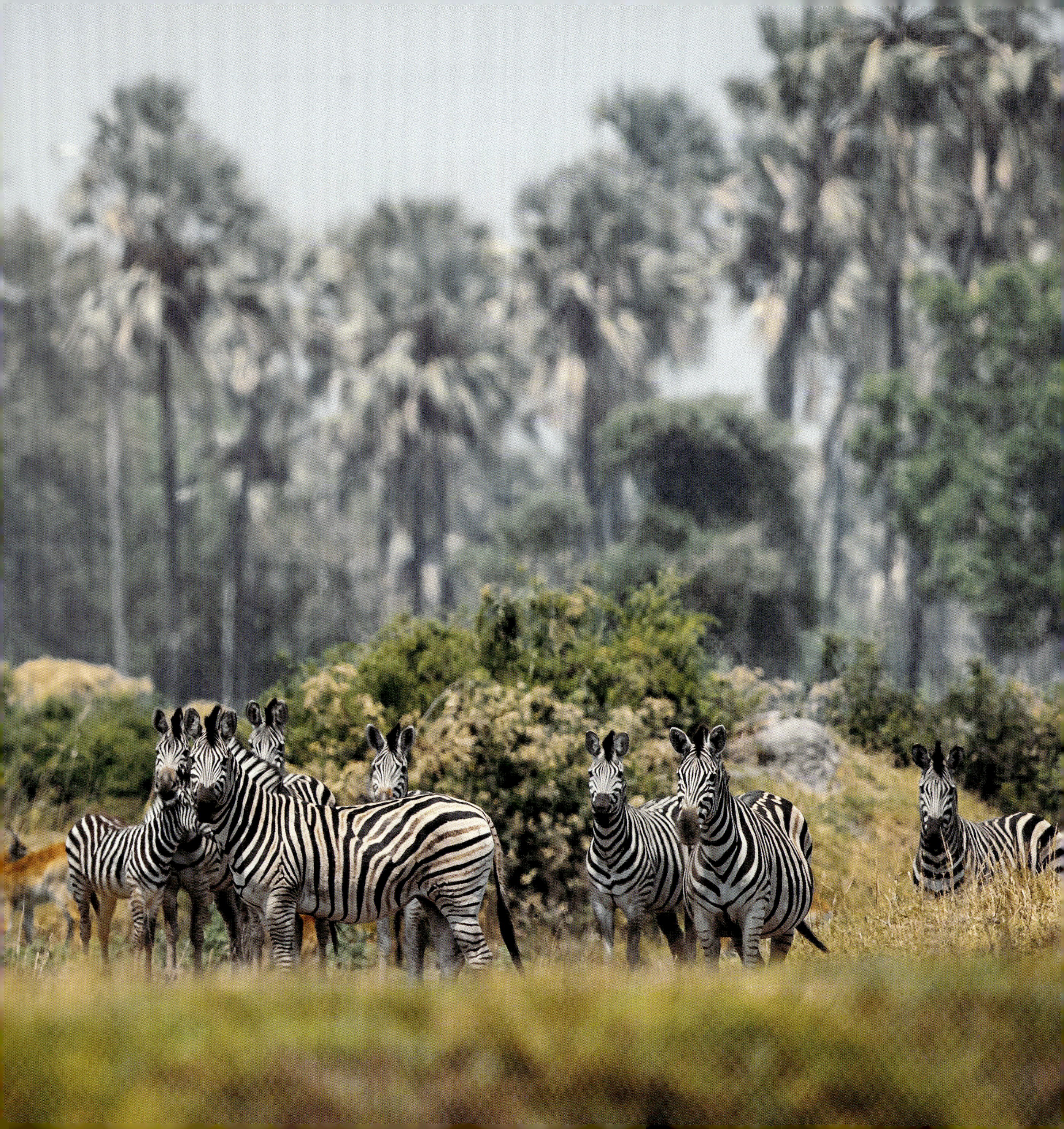

had language and tried to tell me so much in their different pitches. I still react emotionally to their calls today.

Eventually, the parrots led me onto the channels and floodplains of the delta. In March each year, as the floodwaters arrive, the birds go out to small palm islands. In 2002, I was exploring a remote channel system, alone on an old motorboat, when I found a sunken *mokoro:* a 5.5-meter (18 ft) traditional dugout canoe specifically designed for sub-Saharan Africa's diverse river basins—with a flat bottom and the width of a hippo's rump. It must have been one of the first mokoros made of fiberglass; by 2002, the Botswanan government had outlawed carving dugouts from large hardwood trees. I spent an entire afternoon getting the canoe off the bottom, wondering where it came from. It was old and battered, but it was sound. To this day, I still use that same mokoro on all my expeditions. I had followed my parrots into the delta and found a mokoro that would play a starring role in the next few decades of my life.

ALL I WANTED was to learn more about this grand wilderness, and the best way was with the Wayeyi, the people of the delta. They taught how to make my own *nkashi,* a forked 3.7-meter (12 ft) pole, and how to use it to propel my mokoro across floodplains and down small channels. They showed me how the termite mounds had built Vundumtiki, how salts accumulated on the old palm islands, and how certain tree species hold islands together. Historically, the Wayeyi were hippo hunters, so these wonderful beasts played a central role in their culture. The only way to navigate the delta's labyrinth is by following hippo paths, which is why mokoros are hippo-wide.

Poling my mokoro around the Okavango Delta, I discovered something obvious: Wilderness is our natural habitat. Alone in the wilderness, you feel alive and present, as if for the first time. I learned to listen: The game birds' cackling, the distant baboons' plaintive barks, the squirrels' purposeful jabbering, and the sharp snorts of impalas and kudu were my chorus of alertness, of concern for our collective safety. The most connected to this living, blue-green planet that I have ever felt is when I was learning how to survive in this wild.

Vundumtiki means "one small fish," from a story about an old fisherman who spent a day and a night fishing and caught nothing. The next morning, a pied kingfisher hovering above dropped a small fish next to him. When I arrived, I felt exactly like the "one small fish" that fell from the sky. I had landed in a place with a people for whom I would risk my life to help.

I began to build my career around making sure the Okavango Delta will survive for future generations. This would be my life's work. This was the reason I later founded the National Geographic Okavango Wilderness Project. It was the reason I went upstream to explore the major rivers and tributaries that sustain this wetland. Every year, I go back to Vundumtiki Island, where it all began—a place that has become my spiritual home. My children connect me to the future, and I can't imagine a future without the Okavango Delta.

Zebras and lechwe are abundant in the central wilderness of the Okavango Delta, where the skyline is punctuated by palm trees. **PAGES 26–27:** *Arriving in the delta in the 1770s, the Wayeyi tribe introduced the* mokoro, *a dugout canoe, and the* nkashi, *a specialized pole used to propel it. Both were quickly adopted by the Batawana, Basarwa, and Hambukushu of the region.* **PAGE 27:** *My first job in the Okavango Delta was as "head of housekeeping" at Vundumtiki Camp.*

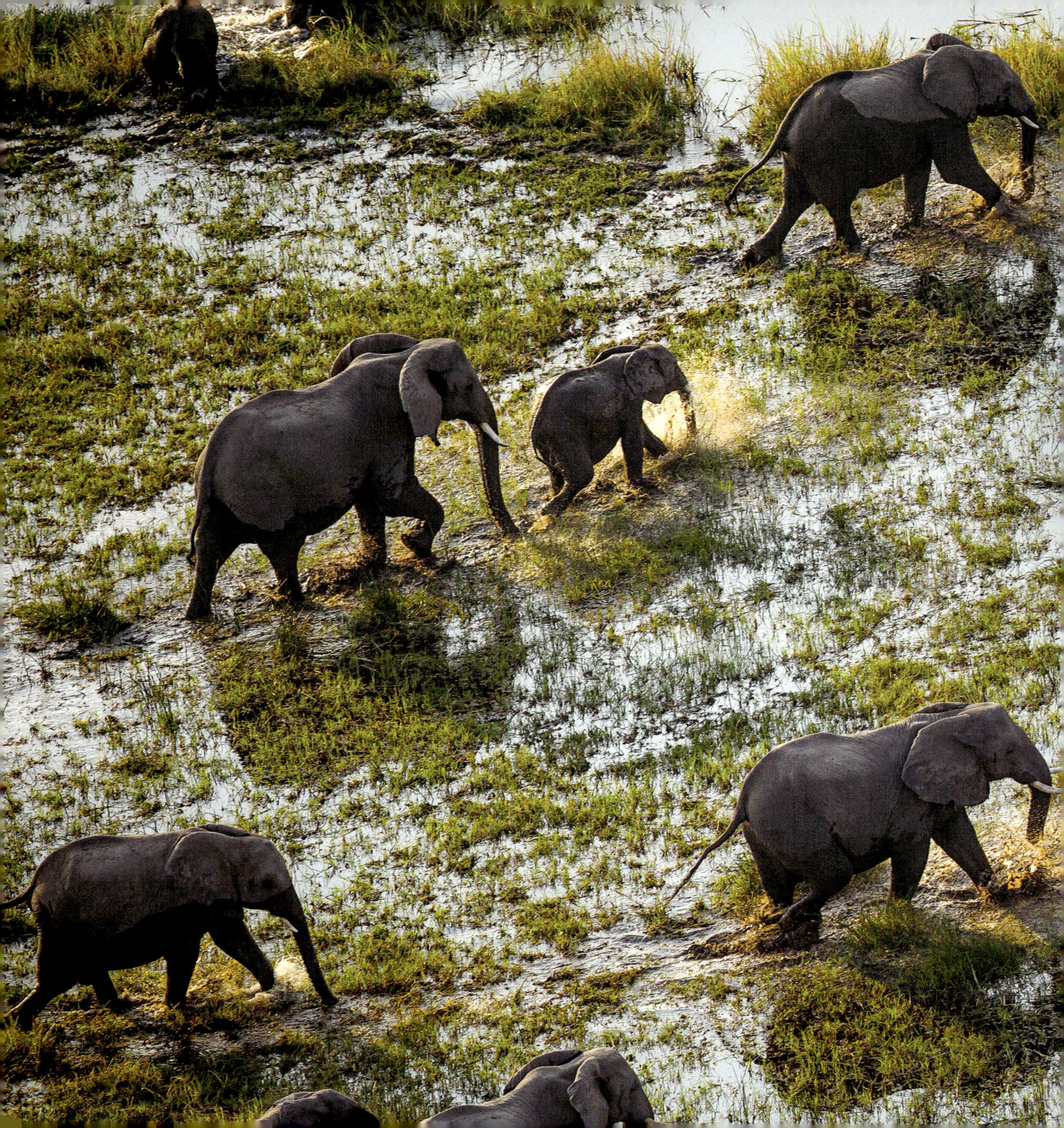

CHAPTER 1

The Vision

THE PLACE, THE PEOPLE, THE MISSION

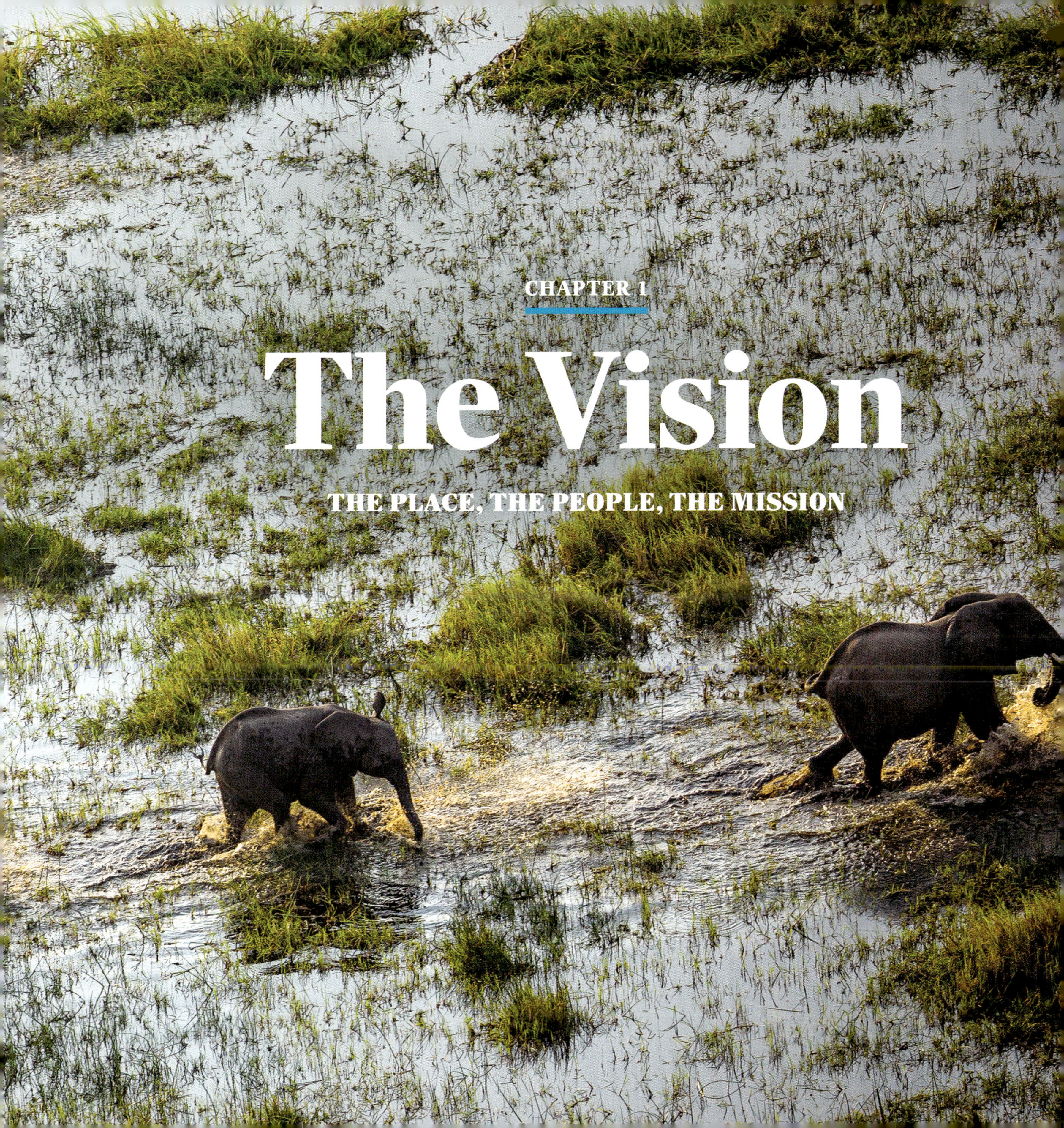

THE OKAVANGO DELTA is our planet's last remaining intact wetland wilderness. Located in Botswana, it is southern Africa's largest wetland, and Africa's second largest—so large that it's visible from space. It's a vast alluvial fan formed at the end of the Okavango River, covering about 21,000 square kilometers (8,100 sq mi) of the semiarid Kalahari Desert; another of its names is the "Jewel of the Kalahari."

Every year, rain that falls upstream in Angola between November and April makes its way slowly down the Kalahari's dry riverbeds to flood this delta. As the water arrives in May, it transforms the landscape into a mosaic dominated by papyrus, reedbeds, and sedges, with channels, floodplains, lagoons, and thousands upon thousands of islands.

In August, as the floodwaters slow and start to recede, the Okavango horizon becomes hazy due to a strong and consistent northeasterly wind blowing from the Namib and Kalahari Deserts, depositing another 250,000 metric tons (275,000 tn) of sand and silt on the islands, floodplains, and waterways. Under natural circumstances, this ongoing sedimentation gives channels an expiration date and sets up channel switching: the process by which old river channels collapse and water spills out into new ones. Random sequences of strong flood pulses and small earthquakes can kick-start the "sandbox" into these rapid, somewhat random shifts; seasonal fires also play a role. The fanning effect of channel switching eventually spreads sand evenly across the delta. As a result, the alluvial fan is mathematically flatter than a billiard

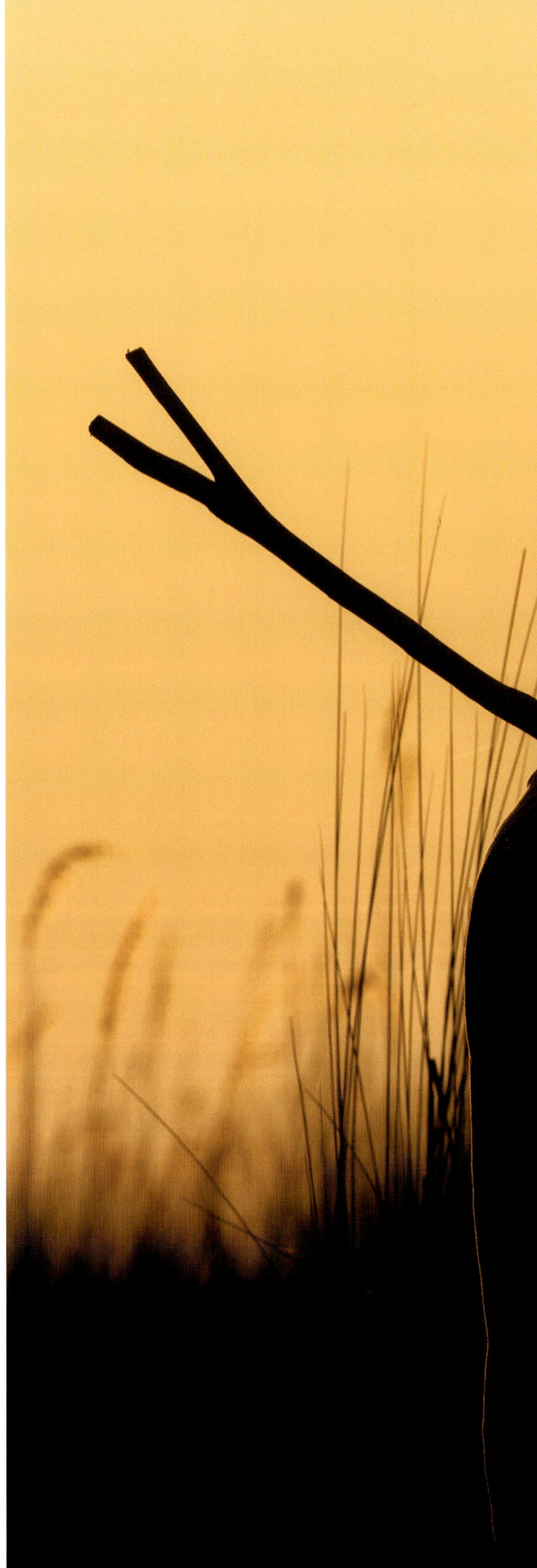

The oral histories passed down between generations in songs and stories, myths and legends, capture the essence of the landscape. **PAGES 30–31:** *The Okavango Delta has captured the imagination of every person who has ever gone there. It's the kind of place that takes you back in time to the way the world was before us—a last glimpse of true wilderness.*

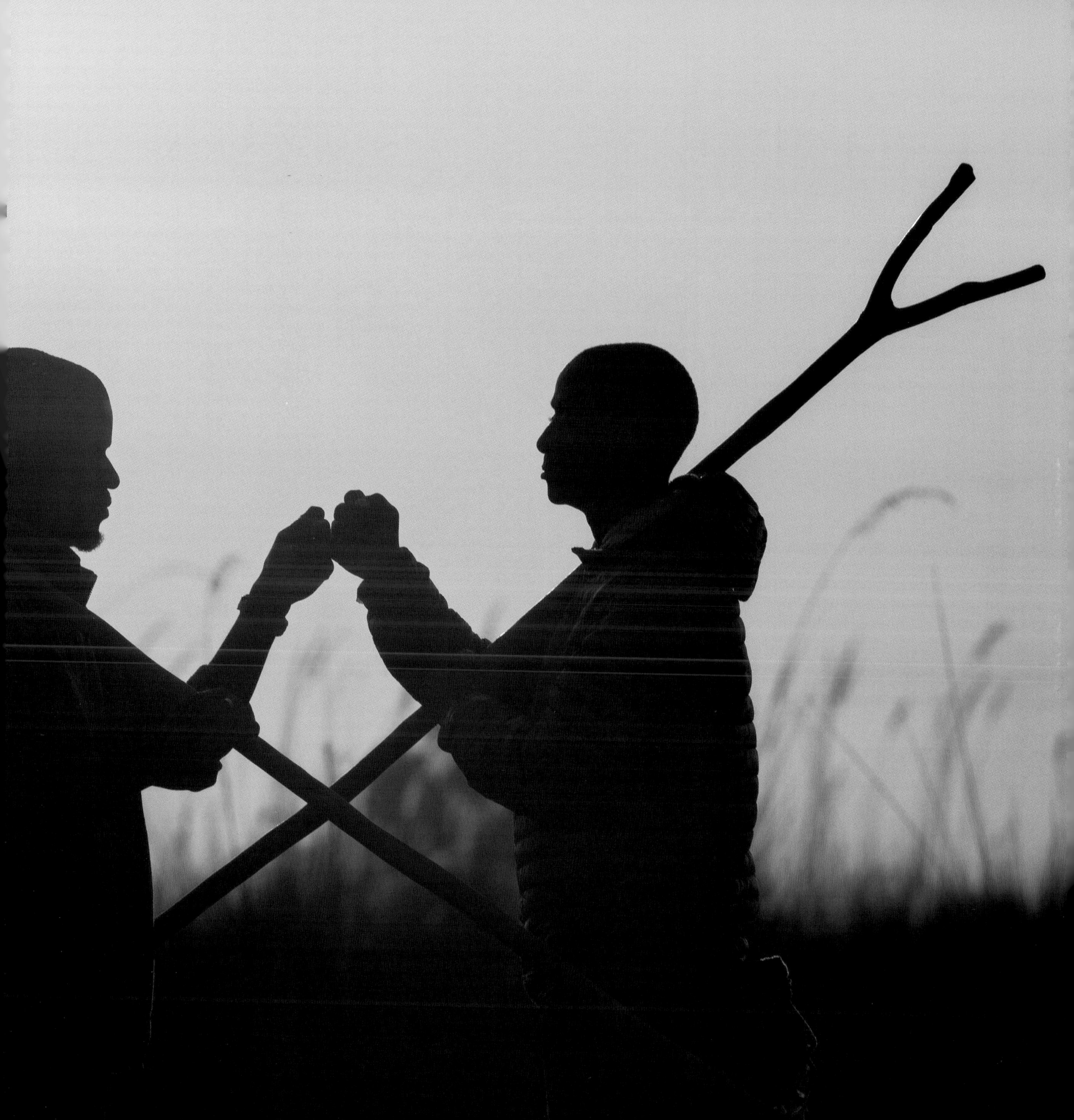

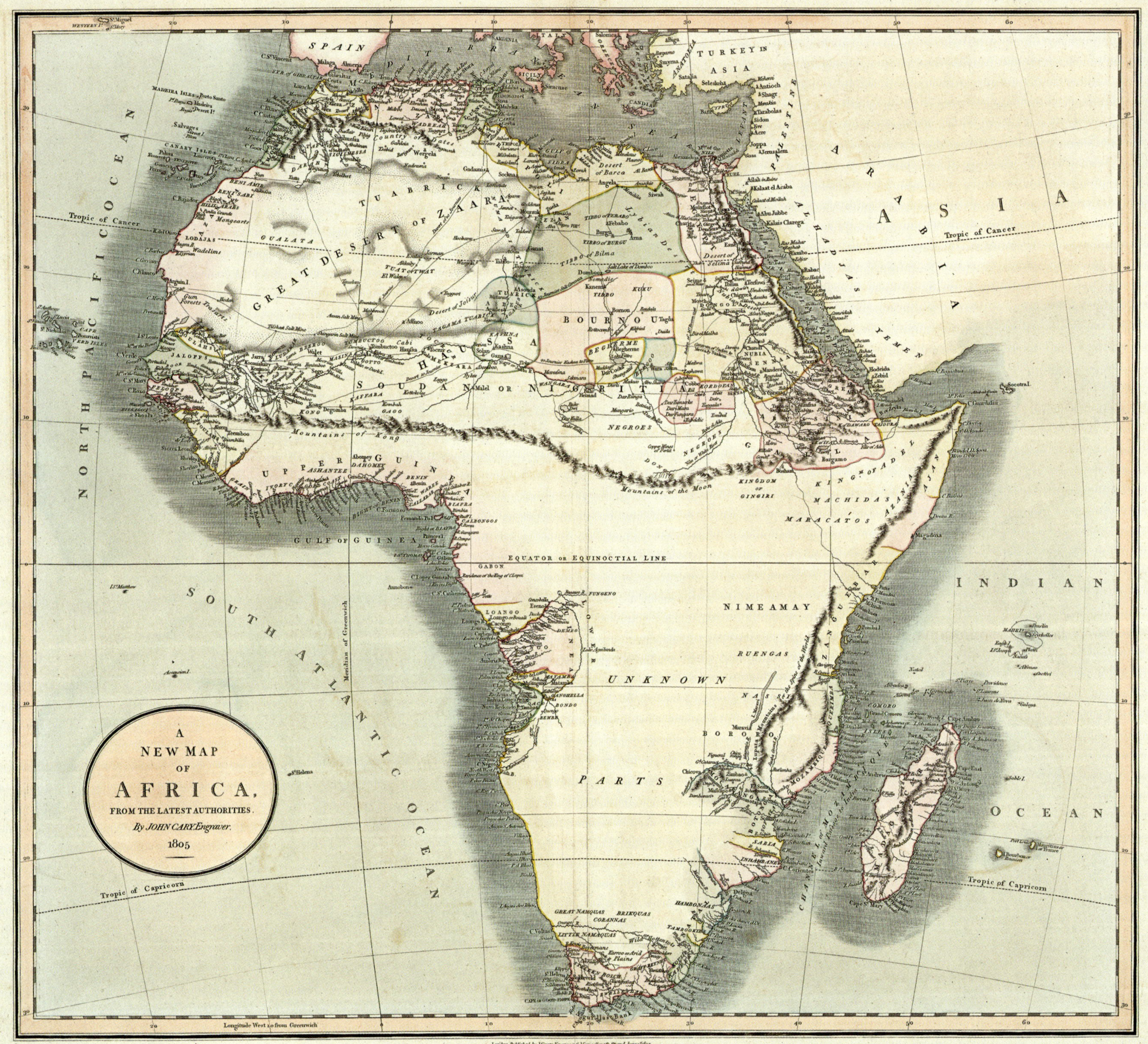

London Published by J. Cary Engraver & Mapseller 181 Strand June 1st 1805.

table, with a change in elevation of just 60 meters (195 ft) over 250 kilometers (155 mi).

The cycle of constant change is key to the ecological functioning of the Okavango Delta. The right amount of disturbance supports diversity, creating a living system at a geological scale, not just an ecological one. A time-lapse of the Okavango Delta over thousands of years would look like the pumping valves of a green heart in the middle of a living desert.

In addition to these broader landscape forces, the Okavango's alluvial fan also has three ecosystem engineers, so to speak: Hippos, elephants, and termites are as important as the flood regime and channel switching. The Okavango Delta is maintained by its biology and would collapse without this web of life that sustains itself. This oasis in the dry Kalahari is home to an estimated 530 bird, 160 mammal, 157 reptile, 38 amphibian, 105 fish, and 3,500-plus plant species. It's home to the world's largest populations of megafauna: a paradise for the world's largest elephant population and keystone populations of hippos, lions, giraffes, African wild dogs, cheetahs, and African buffalo. We humans, too, are part of that web, and have been since the beginning.

When you walk on an island in the delta, you feel life wrap around you. An imperceptible heartbeat, a love and a threat, permeates all experience.

IN EARLY 2003, two years into getting to know the Okavango Delta, I drove my 1972 Series II Land Rover out to Vundumtiki Island to take a role as head of housekeeping. From that point onward, my housekeeping team and the little brown-and-turquoise parrots became my life. Between driving around looking for parrots, checking rooms, washing sheets, drinking free wine with guests, poling my mokoro around the island, and walking with wildlife, I met some of the most interesting people on the planet. At that time, Vundumtiki felt like the busiest camp in the Okavango Delta, with 18 new guests—retired generals, professors, movie stars, librarians, politicians, teachers, pharmacists, ballet dancers, honeymooners, and baseball coaches—arriving every three days, year-round.

I learned how to speak in public by doing what we called a "delta talk" every few days and sharing bush stories around the campfire. With my working and resident permits, I became a true resident of the Okavango Delta, unwilling to leave even for holidays.

By late 2003, at the age of 24, I was made general manager of Vundumtiki Camp. I had arrived. We had a camp staff of just over 50, and I was the only white African. We worked hard, kept each other safe, covered for each other when we made mistakes, and had a lot of fun doing it.

We had some extraordinary adventures on the island: fighting wildfires for days and nights; driving our babies to the clinic through floods and hailstorms to be weighed and vaccinated; drumming and dancing all night under baobab trees; fixing generators struck by lightning by instructions conveyed over two-way radio; and, on one occasion, building a remote camp for a Russian billionaire who stayed in it for only an hour. I lived offline, far away from

Up until recently, the African interior was considered "Unknown Parts," as this 1805 map reveals: uncharted, undocumented, and cut off from the rest of the world by dangerous animals, sleeping sickness, malaria, and the mighty African kingdoms that protected it. Africa's greatest mysteries are still being revealed—the central mission of the Okavango Wilderness Project.

civilization, in my small tin roof house a kilometer from camp in a beautiful ebony forest filled with hundreds of baboons. Just before dawn each morning, I had to walk to the main area with an umbrella as the baboons were doing their morning business.

The entire camp staff became a family—muddled by language but there for each other. I grew from being a naive university student in my early 20s to the game ranger I had always dreamed of becoming as a child. I was now a living stereotype, wearing my khaki shirt and short shorts and talking all day about animals and their antics. I was happily mad and yet as sane as I'll ever be. It was paradise for that time in my life.

IN MY FREE TIME, I tried to learn everything I could about the Okavango, its people, and its history. I went to libraries, bars, old homesteads, makeshift coffee shops, and universities to find published and unpublished manuscripts, books, journals, and maps and to have extraordinary conversations with experts, academics, old hunters, tribal elders, and Wayeyi polers. The story I found may be different from how others tell it; others may have read the old English and bad Portuguese translations differently, cryptic as they are. Many of the old maps were hand-drawn, torn, and stamped with random dates. The place-names on maps either had been long since forgotten or were incorrectly spelled. I could find no online resources at the time, so this became a year-long side project, one that revealed a history I found quite inspiring.

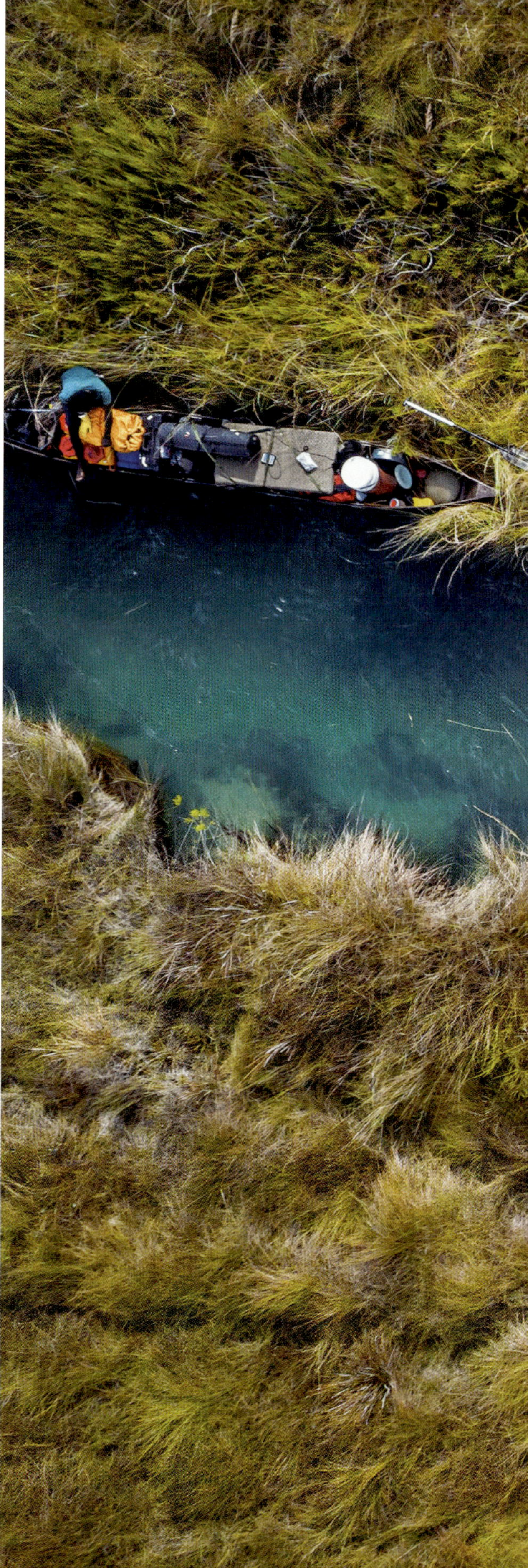

In 2015, escorted by the HALO Trust, we did the impossible: We took seven mokoros and all our gear in armored trucks and Land Rovers to set forth into the undocumented eastern source of the Okavango Basin, the Cuito River. This land of crystal clear blue-green water felt like a place for lost things, like us.

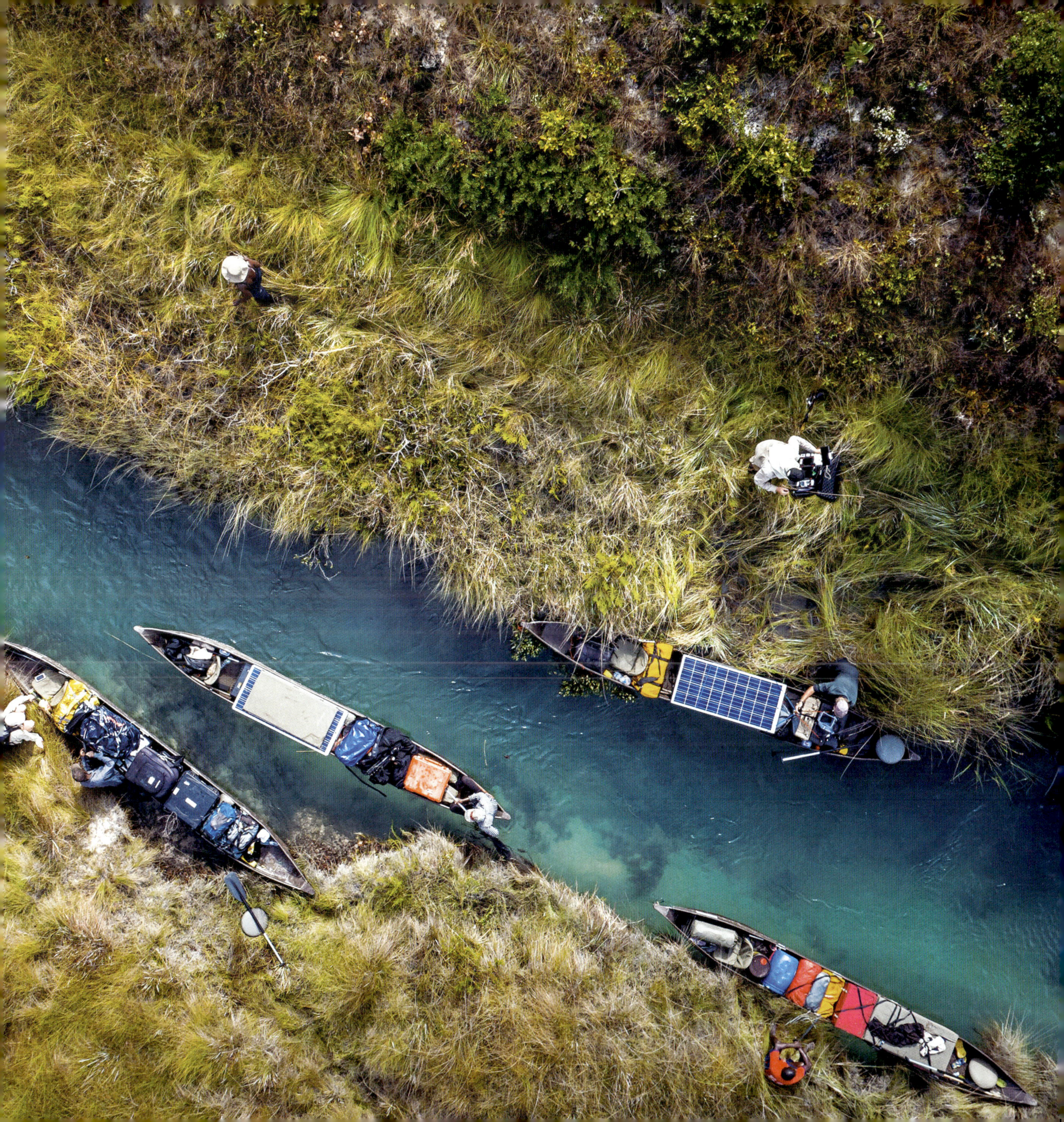

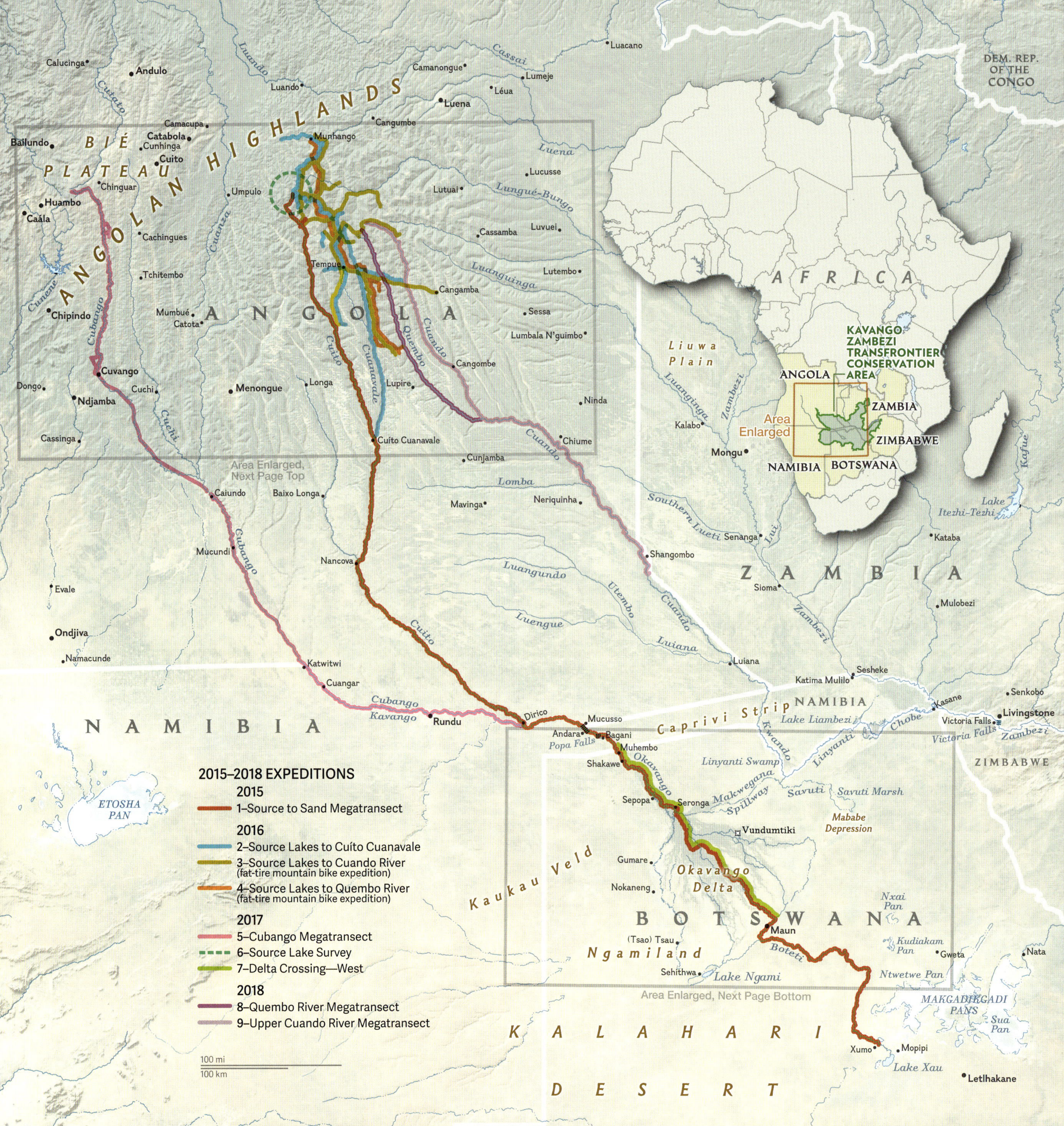

2015–2018 EXPEDITIONS
2015
1–Source to Sand Megatransect
2016
2–Source Lakes to Cuíto Cuanavale
3–Source Lakes to Cuando River
(fat-tire mountain bike expedition)
4–Source Lakes to Quembo River
(fat-tire mountain bike expedition)
2017
5–Cubango Megatransect
6–Source Lake Survey
7–Delta Crossing—West
2018
8–Quembo River Megatransect
9–Upper Cuando River Megatransect
100 mi
100 km
Area Enlarged, Next Page Top
Area Enlarged, Next Page Bottom
AFRICA
KAVANGO ZAMBEZI TRANSFRONTIER CONSERVATION AREA
Area Enlarged
ANGOLA
ZAMBIA
ZIMBABWE
NAMIBIA
BOTSWANA
DEM. REP. OF THE CONGO
ANGOLAN HIGHLANDS
BIÉ PLATEAU
ANGOLA
ZAMBIA
NAMIBIA
BOTSWANA
ZIMBABWE
KALAHARI DESERT
Caprivi Strip
Kaukau Veld
Ngamiland
Okavango Delta
Liuwa Plain
ETOSHA PAN
MAKGADIKGADI PANS
Calucinga
Andulo
Luando
Camanongue
Luacano
Lumeje
Léua
Luena
Cangumbe
Munhango
Camacupa
Catabola
Cunhinga
Cuito
Bailundo
Chinguar
Huambo
Caála
Umpulo
Lucusse
Lutuai
Cachingues
Cassamba
Luvuei
Tchitembo
Tempué
Lutembo
Cangamba
Mumbué
Catota
Sessa
Chipindo
Lumbala N'guimbo
Cangombe
Cuvango
Dongo
Cuchi
Menongue
Longa
Lupire
Ndjamba
Ninda
Cassinga
Cuíto Cuanavale
Chiume
Cunjamba
Caiundo
Baixo Longa
Mavinga
Neriquinha
Mucundi
Shangombo
Nancova
Evale
Ondjiva
Namacunde
Katwitwi
Cuangar
Rundu
Dirico
Mucusso
Andara
Bagani
Popa Falls
Muhembo
Shakawe
Sepopa
Seronga
Vundumtiki
Gumare
Nokaneng
(Tsao) Tsau
Sehithwa
Maun
Xumo
Mopipi
Gweta
Nata
Letlhakane
Kalabo
Mongu
Senanga
Sioma
Kataba
Mulobezi
Luiana
Sesheke
Katima Mulilo
Kasane
Senkobo
Victoria Falls
Livingstone
Cuanza
Luando
Cassai
Luena
Lungué-Bungo
Luanguinga
Cuando
Quembo
Cuito
Cuanavale
Cubango
Cuchi
Cunene
Cuatir
Lomba
Luangundo
Utembo
Luengue
Luiana
Southern Lueti
Luanginga
Zambezi
Lui
Kafue
Kwando
Linyanti
Chobe
Kavango
Okavango
Boteti
Lake Liambezi
Linyanti Swamp
Makwegana Spillway
Savuti
Savuti Marsh
Mababe Depression
Nxai Pan
Kudiakam Pan
Ntwetwe Pan
Sua Pan
Lake Ngami
Lake Xau
Lake Itezhi-Tezhi

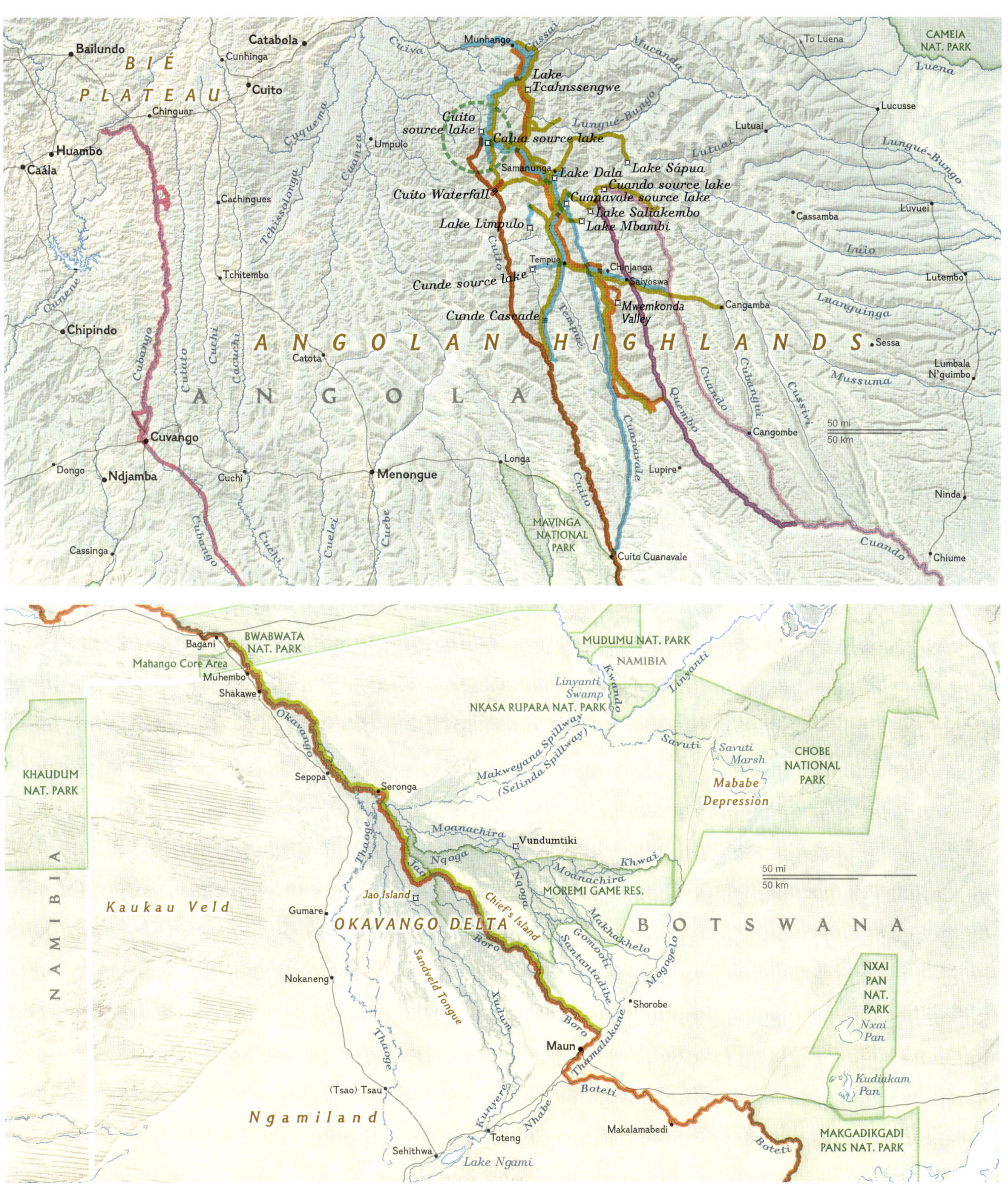

Bailundo
Catabola
Cunhinga
BIÉ PLATEAU
Cuito
Chinguar
Huambo
Caála
Cachingues
Tchitembo
Chipindo
Cuvango
Dongo
Ndjamba
Cassinga
Cuchi
Menongue
Longa
Catota
Umpulo
Munhango
Lake Tcahnssengwe
Cuito source lake
Calua source lake
Samanunga
Lake Dala
Lake Sápua
Cuando source lake
Cuanavale source lake
Lake Saliakembo
Lake Mbambi
Cuito Waterfall
Lake Limpulo
Tempué
Chinjanga
Saiyoswa
Cunde source lake
Mwemkonda Valley
Cunde Cascade
Cangamba
ANGOLAN HIGHLANDS
ANGOLA
Sessa
Lumbala N'guimbo
Cangombe
Lupire
Nindа
Chiume
Cuíto Cuanavale
MAVINGA NATIONAL PARK
To Luena
CAMEIA NAT. PARK
Lucusse
Lutuai
Cassamba
Luvuei
Lutembo
50 mi
50 km
Bagani
BWABWATA NAT. PARK
Mahango Core Area
Muhembo
Shakawe
MUDUMU NAT. PARK
NAMIBIA
Linyanti Swamp
NKASA RUPARA NAT. PARK
Makwegana Spillway (Selinda Spillway)
Savuti Marsh
Mababe Depression
CHOBE NATIONAL PARK
KHAUDUM NAT. PARK
Sepopa
Seronga
Vundumtiki
MOREMI GAME RES.
Jao Island
Chief's Island
Gumare
OKAVANGO DELTA
BOTSWANA
Kaukau Veld
Sandveld Tongue
Nokaneng
Shorobe
Maun
NXAI PAN NAT. PARK
Nxai Pan
Kudiakam Pan
(Tsao) Tsau
Ngamiland
Toteng
Sehithwa
Lake Ngami
Makalamabedi
MAKGADIKGADI PANS NAT. PARK
50 mi
50 km

PAGE 38: *Our expeditions from 2015 to 2018 began in the remote Angolan Highlands in the northwest, where we found the Okavango headwaters and then followed those rivers through Namibia and Botswana. Since we had such limited knowledge of the terrain, all these expeditions were logistically difficult.* **PAGE 39 (TOP):** *We arrived at the Cuito source lake in 2015. By 2018, we had explored most of the Lisima Landscape using mokoros and fat-tire mountain bikes.* **PAGE 39 (BOTTOM):** *By 2018, we had poled across the Okavango Delta nine times, conducting annual hydrological and ecological surveys.*

The great Barotse Kingdom, in present-day Angola and Zambia, was a powerful state with a culture of trade, river exploration, war, and conquest. According to legend, the Wayeyi people, led by Hankuzi the Hippo Hunter, left this kingdom and their ancestral origins along the Zambezi River, in what is now western Zambia and eastern Angola, to find a peaceful home downstream. They found heaven on Earth in the Okavango Delta, a wild paradise where God lived among humans before joining the ancestors, as the Wayeyi narrate.

In my research, I sought to place a timeline to these legends. What I found suggests the Wayeyi people arrived in the delta in the late 1770s, at the height of centuries of Indigenous African river exploration, documented only through oral histories of the local tribes. The Wayeyi arrived with two revolutionary Iron Age technologies: iron axes and 5.5-meter (18 ft) mokoros. Living alongside the Basarwa and N|oakhwe San Bushmen, who had been living in and around the Okavango Delta for thousands of years, and intermarrying with these peaceful people, the Wayeyi became "river Bushmen," forever connected to the fate of the Okavango's waterways. They learned how to hunt and gather on the islands of the delta, and to walk between seasonal flats during the dry season. They learned the medicinal plants of the Kalahari Desert and taught the San, their new brothers and sisters, how to make mokoros and cultivate millet and sorghum.

With the aid of their mokoros, the Wayeyi were the first people to fully explore the waterways of the Okavango Delta. They learned the intricate workings of this alluvial fan better than anyone. There was no written record and limited oral history of the flood dynamics before the arrival of the first European explorers and traders in the 1840s. However, the Wayeyi taught their sons and daughters how to watch wetland birds to predict the water's arrival, to read the floods, and to follow hippos to the best fishing areas and lily beds. They learned to know the delta now and not reflect on the past.

In the early 1800s, the first stories about an almost mythical place called Ngamiland started circulating among Portuguese slave traders operating in the swamps northeast of the Okavango Delta. In 1843, David Livingstone became the first non-African to visit Lake Ngami in Ngamiland, which is seasonally flooded by the Okavango from the north. At the Royal Geographical Society, Livingstone enthused about massive elephant tusks used as doorframes in the impressive settlement at today's Sehithwa, southwest of the delta. Cattle traders and fishermen spoke of a "land of a thousand rivers" filled with elephants and hippos to the north, impossible to traverse and far too dangerous to live in. As was almost always the case, local African tribes told the first European explorers that the place most special to them was inaccessible, too remote, too dangerous.

This land of a thousand rivers was, of course, the Okavango. Malaria and sleeping sickness had kept people with livestock from sustainably settling in this pristine wilderness, leaving it inhabited by hunter-gatherers like the Basarwa and the farming and fishing Wayeyi. Though no doubt a hard place to live in the 19th century, it was far from inaccessible, serving as royal hunting grounds for two kingdoms, the Batawana and Hambukushu.

In his writings, Livingstone incorrectly stated that Lake Ngami's water came from snowmelt in distant, impassable mountains. Yet ivory and a sense of mystery from these early reports sparked the interest of Charles Andersson, a Swedish naturalist, who led the first European expedition into the Okavango Delta, funded by the king of Sweden.

In 1853, Andersson became the first non-African to enter the Okavango Delta, traveling up the Thaoge Channel on the delta's western edge. He described an abundance of water and wildlife, demonstrating that this channel system was receiving most of the Okavango's floodwater in the mid-19th century; today, Thaoge Channel is almost entirely dry, depending on very high flood levels to be inundated.

More European hunters began to arrive in the late 19th and early 20th centuries. Livingstone and Andersson traveled northeast of the Okavango Delta, past Vundumtiki Island, to today's Selinda Spillway and Savuti Channel, two ephemeral watercourses connecting the Okavango Delta to the Linyanti, Chobe, and Zambezi Rivers. At the time, both the spillway and channel were flowing strongly. When the legendary hunter Frederick Courteney Selous passed through this same area just north of Ngamiland a few decades later, he found them dry, demonstrating how dynamic the flows of these river systems have been throughout recorded history. Tectonic activity was never reported, but we can safely assume that this was one of the sources of the seemingly random switching and diversion of channels reported through history.

Picking through these original accounts was amazing. To me, Andersson's descriptions of the Thaoge Channel reflected the delta I knew now. Perhaps it hadn't really changed—or maybe it had recovered. Reading these dusty historical accounts, I felt as if I were becoming part of history.

As Europeans continued to arrive, they shaped the landscape in new and unexpected ways. After the Berlin Conference of 1885 formally claimed the continent for European powers, the "Scramble for Africa" set off waves of arrivals and created fresh conflict. It remains a source of Pan-African pride that Emperor

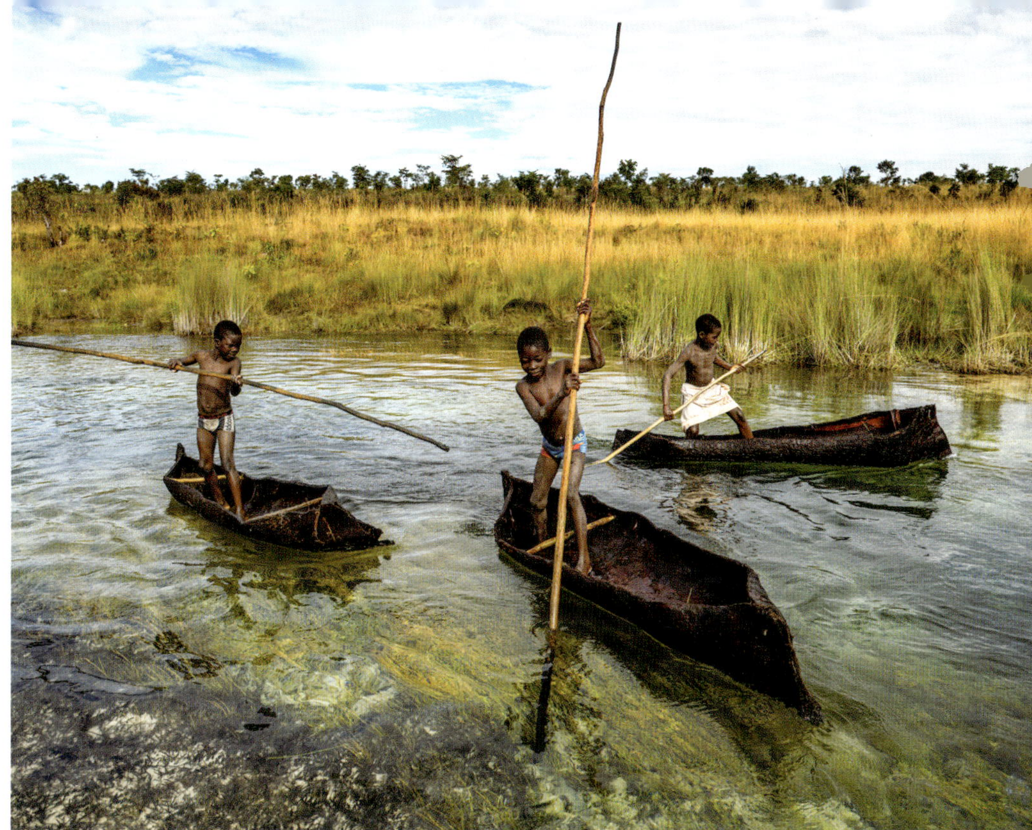

Around the Okavango's sources in the Angolan Highlands, young boys and girls learn how to pole from a very young age, steering traditional canoes made from tree bark.

Menelik II and his wife, Empress Taytu, led more than 100,000 Ethiopians to defeat the Kingdom of Italy in the Battle of Adwa in 1896. Yet unbeknownst to them, the Italian forces unleashed a devastating bioweapon: rinderpest, or cattle plague, which devastated most of Africa. By 1899, it had reached the continent's southern tip. In most areas, 90 percent of the cattle had died, wrecking socioeconomic and political stability and severely weakening any resistance to subsequent colonial expansion. Millions of Africans died of famine within a few years, although numbers are impossible to determine. The aftermath of the rinderpest outbreak shaped Africa's historical trajectory for generations.

The virus spread rapidly throughout the Horn of Africa, eventually crossing the Zambezi River in early 1896. In the Okavango Delta, the impact on wildlife populations, especially African buffalo, was catastrophic. By 1900, tsetse flies, which spread sleeping sickness to people and cattle, had gone locally extinct—a strong contrast to what happened in eastern Africa, where former cattle fields returning to bush allowed tsetse fly habitat to grow and an epidemic of sleeping sickness to ravage the region.

Yet the outcome in the Okavango was no less devastating. The temporary retreat of sleeping sickness encouraged ambitious ivory hunters to operate permanently in the heart of the Okavango Delta, decimating wildlife populations. In 1905, one trading post in Ngamiland reported more than 10,000 elephant tusks in a single year. It must have been a time of great distress.

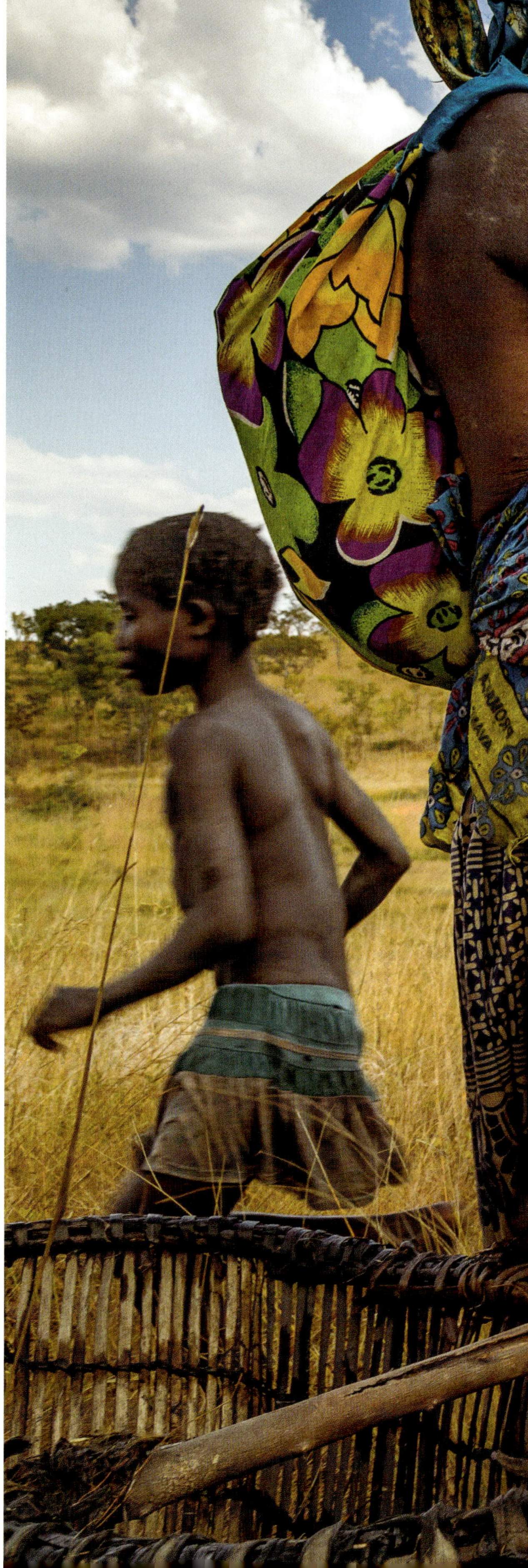

In 2015, these Luchazi women were the first people we met while exploring the remote Cuito River in the eastern Angolan Highlands. They were living with their children far from their village to work cassava fields and fish with traditional traps. They lit a bushfire as they left, as if to make sure we didn't follow.

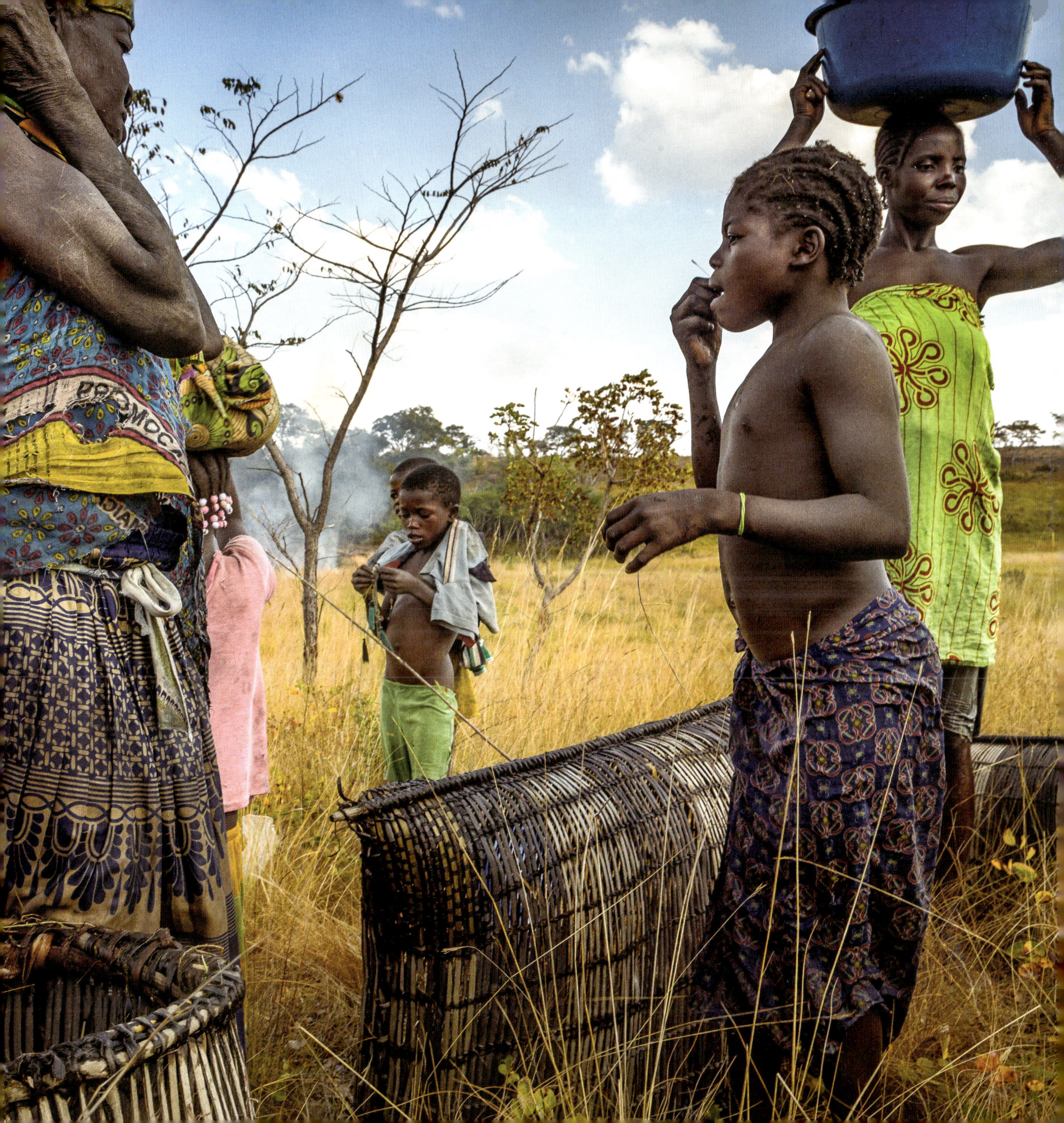

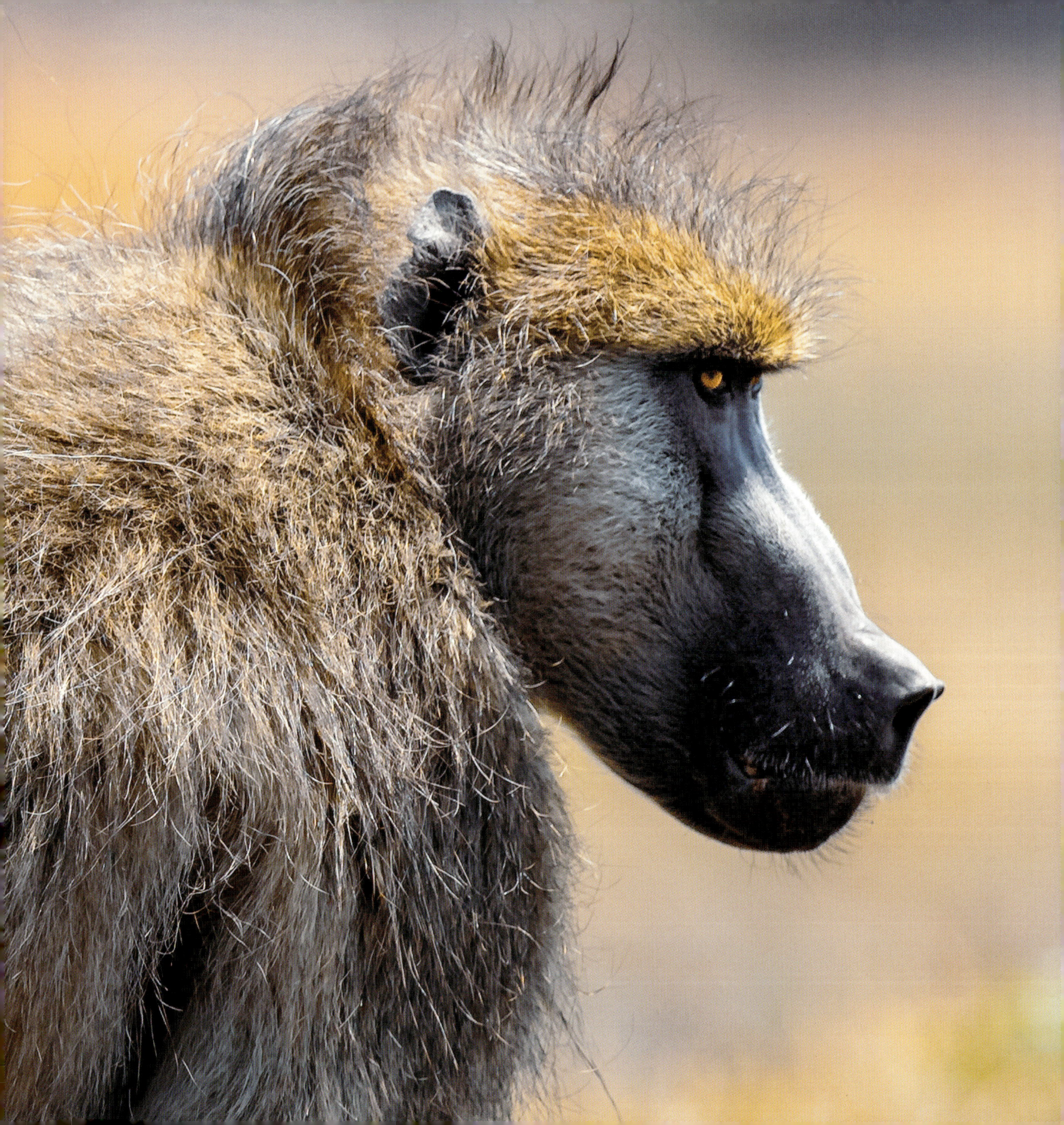

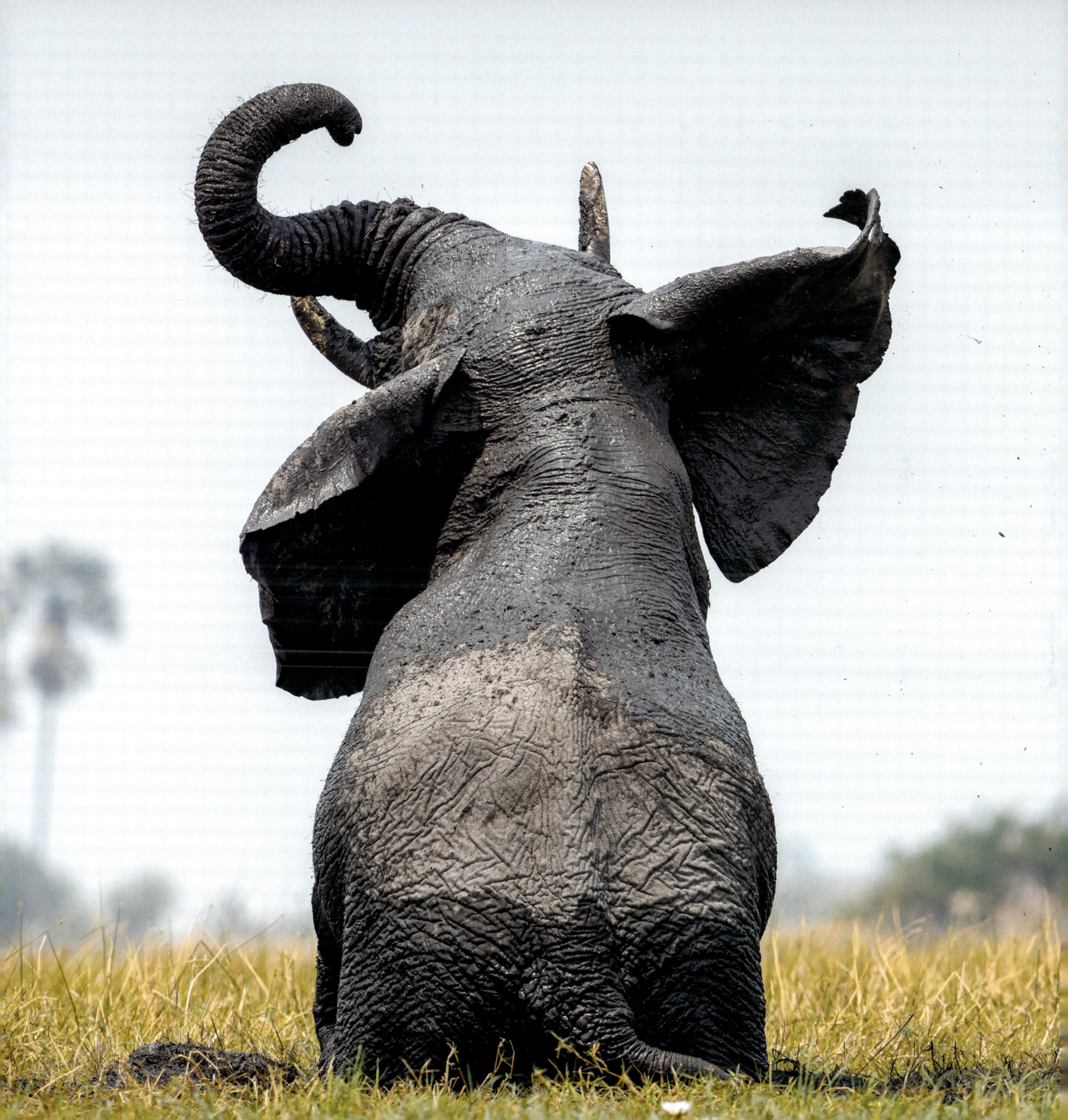

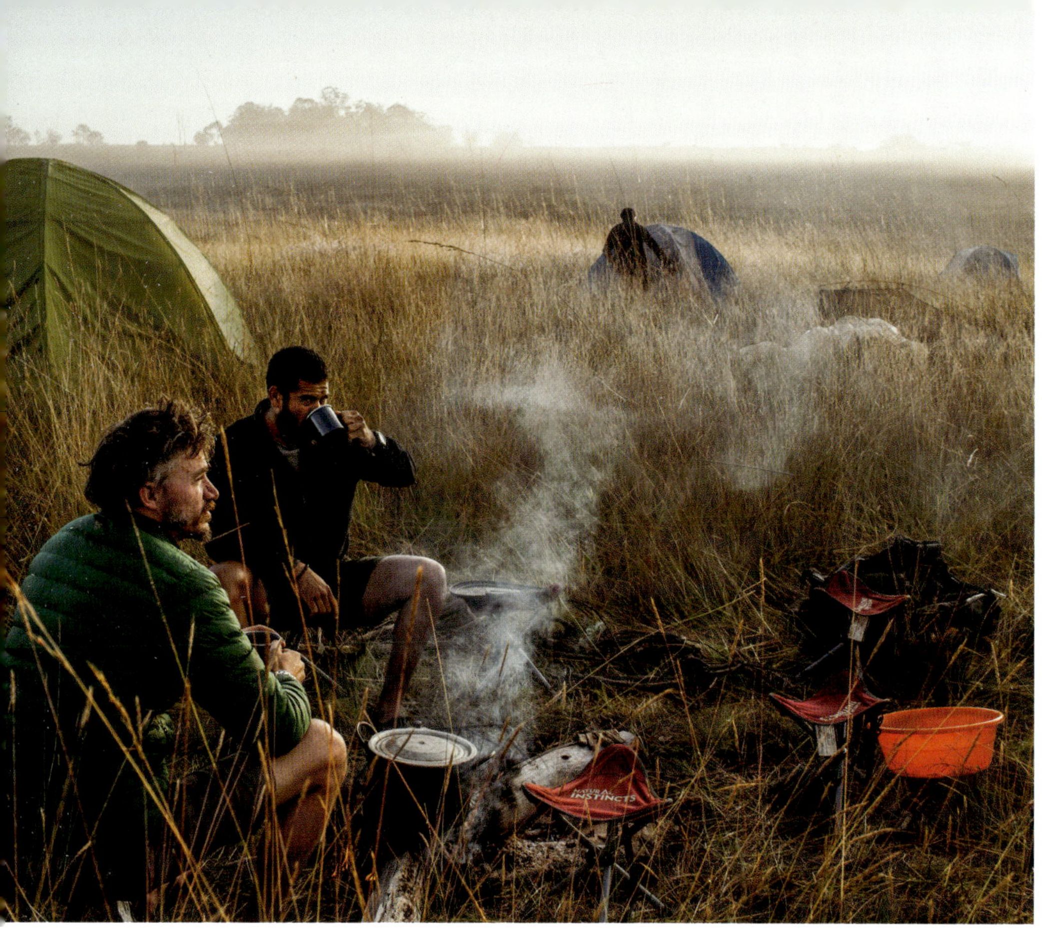

There's something extraordinary in the feel and smell of a campfire on a cold, misty morning on expedition. You've survived another day, you've slept another night under the stars, and the coffee tastes wonderful. **PAGES 44–45:** *Baboons are powerful primates. They move around in large troops. Yet, like the most powerful animals of them all, the delta's elephants, they have a playful side to them. Wherever you find savanna elephants, you find baboons.*

THIS WAS THE DELTA of the grandfathers of elders that I talked to. I met old men and women who lived childhoods similar to what was described in these dusty history books. One of the old men born in the central wilderness said that there were no tall palm trees when his father was a child; the elephants had knocked them all down. A tall palm tree takes 80 to 100 years to grow, and after European hunters arrived and killed all the elephants, the fan palms grew again. Today, elephants fight over palm nuts, and all over the delta in August and September, you hear palms shaking violently and nuts falling to the ground like bags of golf balls. There are now tens of thousands of tall palms, but I have never seen middle-age palms. Tens of thousands of elephants sought refuge in Botswana over the last half century to avoid war in Angola, and as a result, the young palms are all being disturbed so much by elephants that they never grow up. One day they will be gone again.

Over the past 175 years, we have seen significant channel-switching events, shifting the floodwaters of the Okavango Delta from west to north to central to north to central on a cycle of between 50 and 100 years. My own observations poling around the delta showed me that localized changes in water distribution were linked to changes in dense reedbeds and papyrus, often driven by fires. Channel switching was happening all the time at different scales, year after year, and the importance of the annual floodwaters was key to wetland ecology.

Over time, I began to understand another essential self-renewal strategy in this dynamic wetland system. For most of the 20th century, elephants and hippos disappeared. Now, they are a dominant force in the Okavango Delta again, reclaiming their status as ecosystem engineers alongside the mighty termites, island builders responsible for all permanent land above the floodwaters. Trees eventually colonize the termite mounds, and hippos, feeding on the lush grasses growing on them, deepen major channels and lagoons around these proto-islands. Hippos can't really swim, so they walk along the bottom, moving along narrow paths between lagoons and distant pastures, kicking up sand and sediment as they porpoise below the surface. These hippo paths are the channels of the future, the arteries of the alluvial fan.

RIVER GUARDIANS

Comet Masupa

A Life Beyond Time

THERE ARE PEOPLE on Earth who still live without time. They wake up with the sun and the morning chorus. They live with the rain and stars, connected to the ebb and flow of living. They are present and alive. Comet Masupa was born in the Okavango Delta and lives for the rain and floodwaters. He lives with change in the present moment.

Comet taught me to sit quietly, alone in the wilderness, and wait, connected. A mouse scurries between your feet and stops. A leopard pauses meters away and looks at you. There you are, present and aware. Nothing is happening. Nothing is moving. There is no sound. Just existence. Revel in the vitality of wildness and experience the fiction of time and space. You are alive. This is life in the wilderness, in the Okavango Delta. This is the world Comet Masupa knows.

You don't schedule a meeting with Comet. You find him in the present moment. You can't arrange a time to meet him. Comet follows the floods, rains, and wildlife movements. He is connected to place, rhythm, change, and movement—not time. I have only ever experienced timelessness with Comet, conscious experiences I can't explain, moments without time. Comet lives in an eternal and objective reality. Lost without measure, he seems unreal, even supernatural when he appears out of nowhere as soon as you wonder where he is. He has showed me that as wildness emerges, past, present, and future cease to exist. We are a story that has no beginning, Comet once told me, because we live only in the present moment.

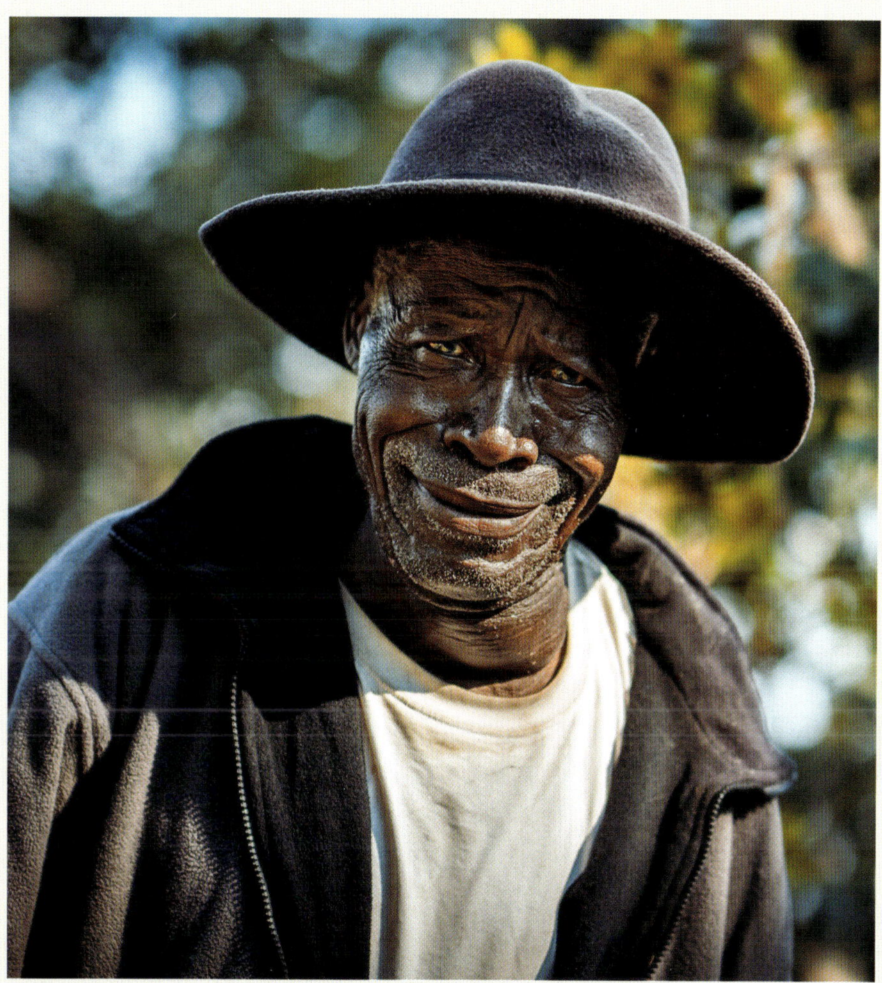

Comet is a Wayeyi elder from Jao Island who knows the central wilderness of the Okavango Delta around the northern peninsula of Chief's Island better than anyone.

Like hippos, elephants create well-trodden trails that guide floodwaters into new areas. Without elephants, the delta would be overwhelmed by dense woodland thickets and papyrus rafts that restrict access and clog waterways. The sheer number of elephants in the delta is a powerful force of nature. Hippos and termites help create the islands, while elephants garden the woodlands, holding the islands together.

IN 2001, David Quammen famously published a series of *National Geographic* stories about J. Michael Fay's 465-day, 3,219-kilometer (2,000 mi) Megatransect expedition across the Congo Basin, hiking by foot to undertake ecological surveys. At the time of its publication, I was doing coursework and planning my master's dissertation. This series of *National Geographic* stories inspired me so much that I based both my master's and Ph.D. dissertations on large-scale spatial and temporal research transects.

A research transect is a repeatable scientific survey methodology undertaken along a standardized route, like a section of river, trail, or road. The first transect is the all-important baseline, against which all future transects will be compared. After this baseline is set, repetition over long periods of time is key. We don't know what is there until we measure, count, or document it, and we don't know how it works until we measure, count, and document it again and again. The more I reviewed the scientific literature, the more I became convinced that securing early 21st-century scientific baselines for Africa's rivers, wildlife populations,

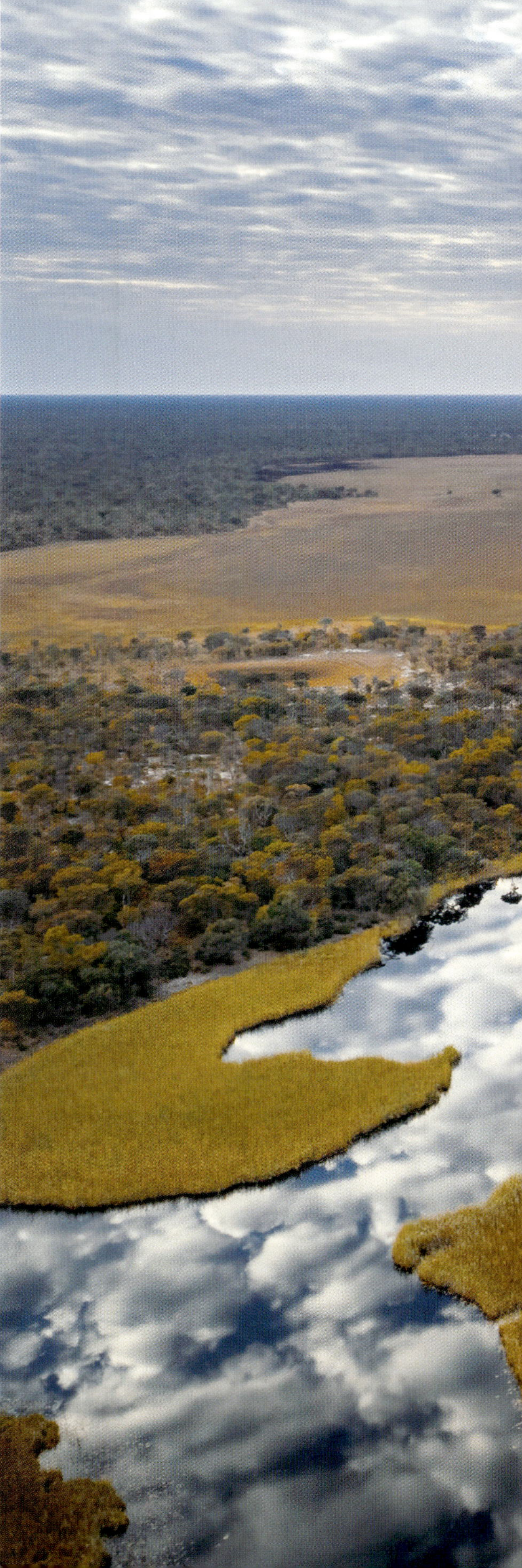

Here is the Lisima lya Mwono—the Source of Life—revealed in an aerial view of the Cuando River, which we explored in 2018. Our exploration of the Okavango Basin led us to all the rivers with sources in the eastern Angolan Highlands: the Kwando, Zambezi, Congo, and Cuanza, Africa's wildest rivers.

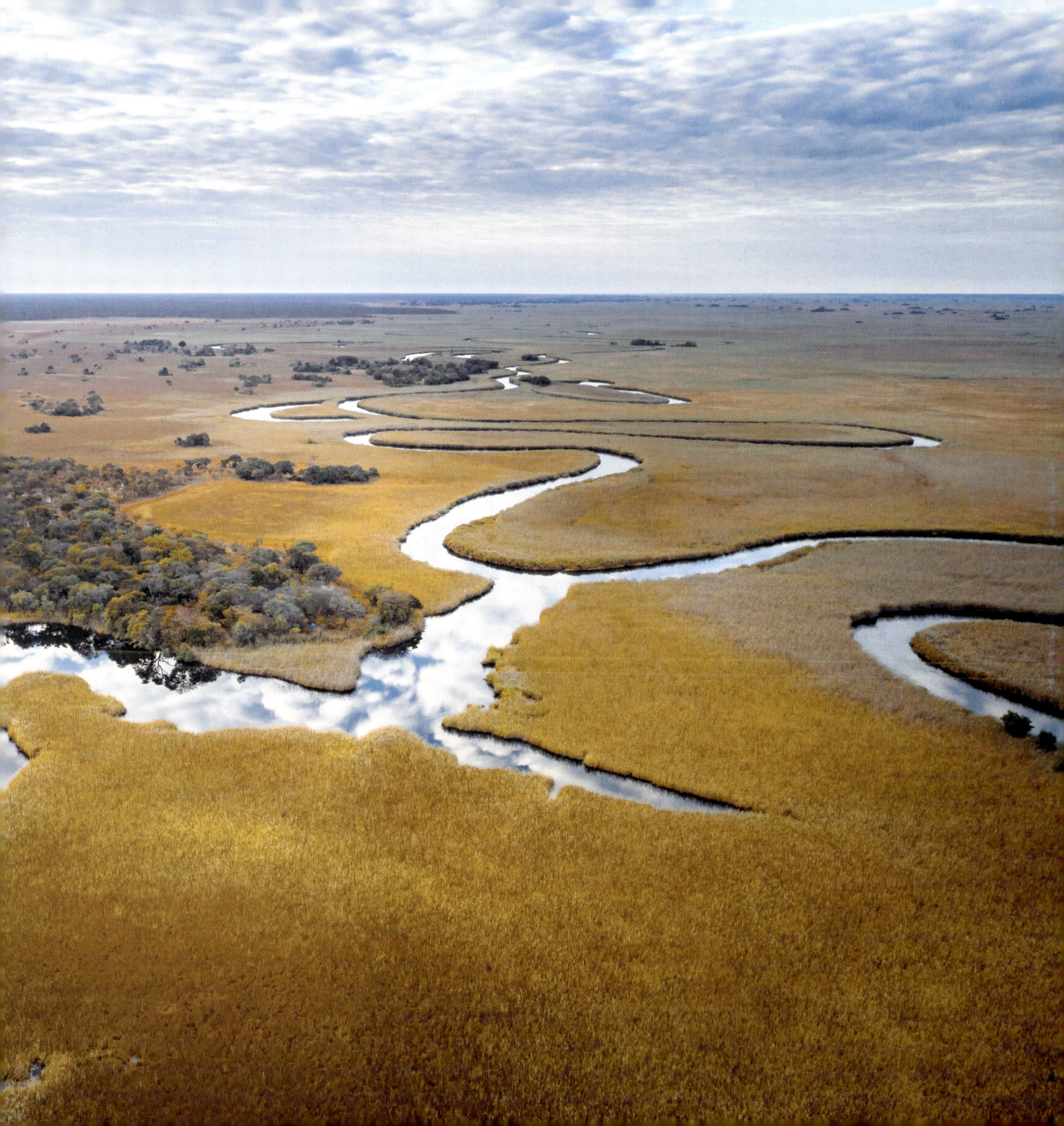

It felt like the zenith: David Quammen, one of the best science writers on the planet, writing about our exploration of previously undocumented rivers in the Okavango Basin.

biodiversity, human impacts like deforestation and fires, the behavior of keystone species like elephants, and ecosystems was going to be the focus of my work.

In 2009, I took some of my first steps at turning this research into conservation action: With my wife, Kirsten, we founded the Wild Bird Trust, which focused on protecting critical bird habitats. That year, just after getting my Ph.D., I also got a contract from the Bronx Zoo to work with none other than Mike Fay, ecologist and National Geographic Explorer in Residence, whose Megatransect had so inspired me. We were to conduct biodiversity surveys and wildlife transects in the Maputo Special Reserve in Mozambique: a dream come true. I arranged all logistics for the expedition and did the bird surveys, while Mike and his team walked baseline transects across the reserve.

In June 2013, my life changed significantly when I became a National Geographic Emerging Explorer, a prestigious award, alongside an amazing cohort of scientists, storytellers, academics, innovators, risk-takers, and thought leaders. I used the $10,000 Emerging Explorer grant to fund the 2013 Okavango Wetland Bird Survey, which was becoming a series of detailed hydrological and ecological baselines.

In July 2017, four years later, David Quammen joined us on the Cubango River, a major tributary of the Okavango Delta, for our final hydrological and ecological river transect or baseline, in a series of three expeditions we called the Okavango Megatransect. Our megatransect covered more than 5,000 kilometers (310 mi). We counted all wetland birds, taking fish samples every day, measuring water quality and flow, using 360-degree photography to document habitat types along the river, and enumerating all people, structures, boats, farm fields, and impacts from fires along the river. We were conducting the most comprehensive river transects ever undertaken, measuring the very heartbeat of the Okavango Basin in the early 21st century.

It was an extraordinary feeling when, later that year, Quammen's 22-page story, called "The Mission to Save Africa's Okavango Delta," was published in *National Geographic* magazine. In his words, "a fervent South African conservation biologist named Steve Boyes, with the support of the National Geographic Society, has embarked on a grand effort of exploration, data gathering, and conservation advocacy called the Okavango Wilderness Project." At the time, it felt like the zenith: David Quammen, one of the best science writers on the planet, writing about our exploration of previously undocumented rivers in the Okavango Basin. I was living out the inspiration of my life. Looking back, I know it was just the beginning. We had just begun to understand the Source of Life—the land and waters feeding the Okavango Delta.

ON JUNE 22, 2014, the Okavango Delta was inscribed as our planet's 1,000th UNESCO World Heritage site—a long overdue accolade for a wilderness beyond comparison. Within weeks, our attention at the Okavango Wilderness Project shifted from achieving this listing and monitoring ecosystem health to finding the source of the floodwaters that sustain the Okavango Delta's crucial flood regime. We knew the waters came from the little-charted Angolan Highlands, a place all Angolans knew. In the 15th century, Portuguese explorers named the Angolan Highlands *terra do fim do mundo*, meaning "land at the end of the earth." To make sure the Okavango Delta, key to the livelihood and sustainability of so many, would be there for future generations, we needed to know where its water came from. Who lived up there? What threats did these life-giving rivers face? And how could we support the Angolans who hold the future of the Okavango Delta in their hands?

All the pieces seemed to be falling into place. In years prior, these questions could not have been answered. Angola had been snarled in a bloody, decades-long civil war, making exploration untenable. In the 1950s, Angola was known as the "living room of Africa," legendary for its giant sable antelope, enormous elephants, and herds of wildlife. In 1961, just as the world was becoming enthralled with this wild country, *National Geographic* published a feature article called "Angola, Unknown Africa" in the September issue. But that same year, Angola's war of independence broke out, ushering in decades of armed conflict and isolating this extraordinary country in south-central Africa.

THE ANGOLAN BUSH WAR had made any work there impossible. All the rivers were heavily fought over. The conflict began as a civil war between three military movements, each seeking to control Angola after it became independent from Portugal. Yet it also became a proxy war in the Cold War between the United States and the Soviet Union, as each country supplied training and aid to the groups with their preferred ideology. The fighting decimated wildlife populations and depopulated southeastern Angola. Almost all the elephants, perhaps as many as 100,000, were either killed or driven to become refugees in northern Botswana. Several other large game species, including zebras and giraffes, were believed to be extirpated locally. By 1990, the southern African wildlife tourism industry was booming everywhere but Angola.

The war displaced some 4.28 million people, an estimated one-third of the population, separating children from families and plunging thousands into famine, disease, and poverty. Rural areas, in particular, were abandoned as military groups set up strongholds in the wilderness, planting dangerous land mines as they

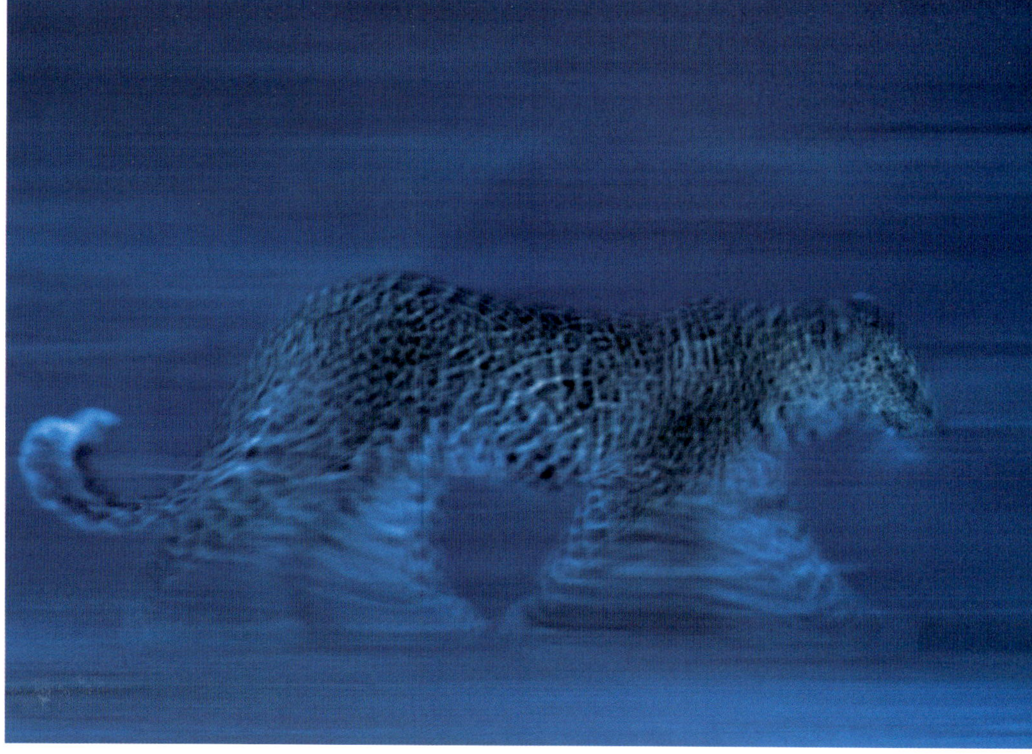

Leopards are ghosts in the darkness, ever present yet never seen, captured here in motion thanks to a camera trap. Their footprints often appear in camp in the morning. A symbol of power and intelligence, leopard skins are worn by the kings of Africa.

traveled. I met many Angolan refugees in Ngamiland, living around the delta, but they didn't speak of their homeland. Many friends, typically older guides and rangers, had fought in the Angolan Bush War. After a few whiskeys or a bottle of rum, they would tell their stories of a war that nobody would ever win, of heroics and good people dying for no apparent reason. None of them wanted to go back.

The war finally ended in 2002. Thirteen years later, we would become the first explorers to attempt to enter that "land at the end of the earth." But what we actually found was, in the local Luchazi language, Lisima lya Mwono—the Source of Life, a lost wilderness that would give hope to a 21st-century world where everyone presumed we had explored everything. This wilderness holds the undocumented source lakes that feed the Okavango Delta and the surrounding wilderness. It truly is the source of life for millions of people, as well as for the wildlife, biodiversity, and ecosystems downstream. This Lisima Landscape functions as a highland water tower—in this context, not a human-built water storage made of wood or metal but a stretch of high-altitude, high-rainfall forested watersheds with incalculable water-storage capacity, delivering fresh water to distant river systems in arid regions susceptible to drought and famine. In this era of climate change, the region has global importance as one of our planet's largest, though largely unaccounted for, subtropical carbon sinks.

Following four decades of relative isolation, central and southeastern Angola are slowly opening up. Resource exploration and investment potential are growing, exposing this ecologically sensitive, and largely misunderstood, region to new development opportunities and new economic drivers: ecotourism, mining, agriculture, and other pressures that could adversely impact the vast intact woodlands, grasslands, source lakes, wetlands, and rivers that sustain these uniquely biodiverse Angolan Highlands and downstream ecosystems.

Sheltered from the European colonial conquest by secrecy and inaccessibility and cut off from modern exploitation by terrible war and the lingering threat from land mines, the Lisima Landscape now presents us with an opportunity: to protect this unique, globally important hydrological structure, the largest of its kind on Earth, and to do so in partnership with Indigenous communities. This is the only way to save the Okavango Delta.

And so, these became the goals of the Okavango Wilderness Project: to document the secrets of the Source of Life landscape to secure the long-term future of the Okavango Delta and to work to protect Lisima lya Mwono and its residents—whether tribes isolated from the world for a generation or "ghost elephants" that represent the last of their kind in Angola. This book tells the story of how we left the Okavango Delta to enter the unknown and how we documented the previously untold riches of upstream landscapes. My hope is to help the world understand the importance of the Lisima Landscape—to encourage the world to explore this wilderness with us, knowing that explorers search for these last wild places not to change them but to learn from them.

This book tells the story of how we left the Okavango Delta to enter the unknown and how we documented the previously untold riches of upstream landscapes.

RIVER GUARDIANS

Koketso "Koki" Mookodi

A Champion for Africa's Women

In 2002, I met Koki at a camp on Vundumtiki, a remote island in the Okavango Delta. It was the beginning of a friendship for the ages. I was head of housekeeping and Koki ran the front of house, greeting guests with bubbling exuberance. I gave the "delta talks" and she brought the passion and fire. Guests from around the world loved it.

In 2019, Koki became Botswana's country director of the National Geographic Okavango Wilderness Project, and in 2020 she became a National Geographic Explorer for her role as a conservation educator. She is a Mandela Washington Fellow, a community leader, a tireless guardian of national pride, and a mother dedicated to the next generation.

Like Koki, the women of Mother Africa are a force of nature. As mothers and caregivers, they are pillars in their communities, keepers of knowledge and wisdom, and custodians of their cultural heritage. As farmers and gatherers, doctors and healers, spiritual leaders, dancers, artists, and creators, they embody resilience and innovation in urban and rural communities across the continent. All Africans live between their roots in rural areas and their aspirations in big cities. From the bustling rural and urban markets that feed millions to cutting-edge entrepreneurial ventures in African megacities, women are at the forefront of an African renaissance as the continent emerges as a world power. Africa's women are rising like sparks from a fire.

Despite lingering gender inequalities, limited access to education and health care services, and the impacts of conflict and poverty, African women are at the forefront of social movements, advocating for human rights, equality, and justice. Koki embodies their powerful voice for peace and reconciliation. She has become a champion for better education and health care, advocating for the rights of women and girls across Botswana.

Koketso, known to her friends as "Koki," is a passionate educator and community leader who works tirelessly to make sure that Botswana's Ngamiland has a bright future.

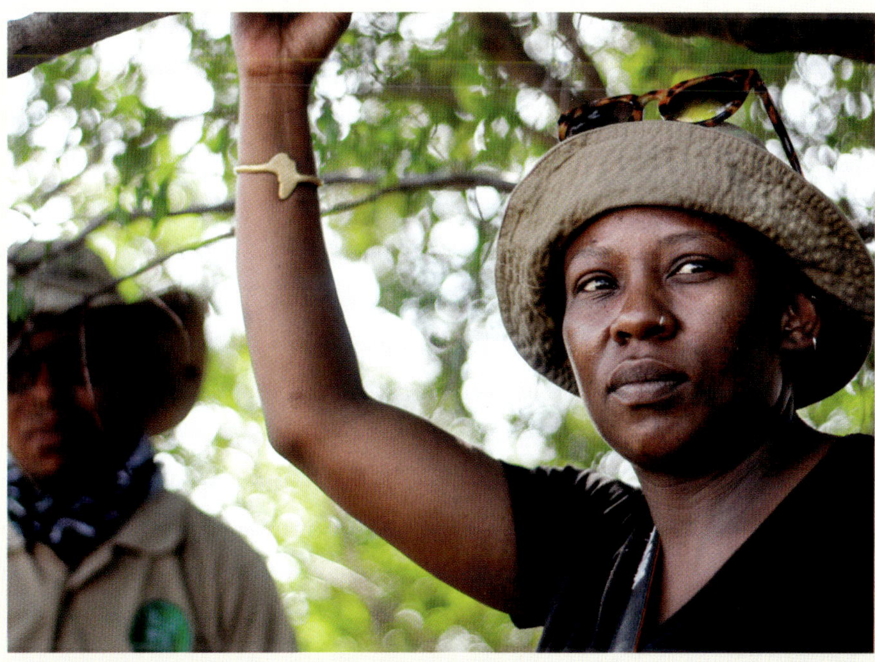

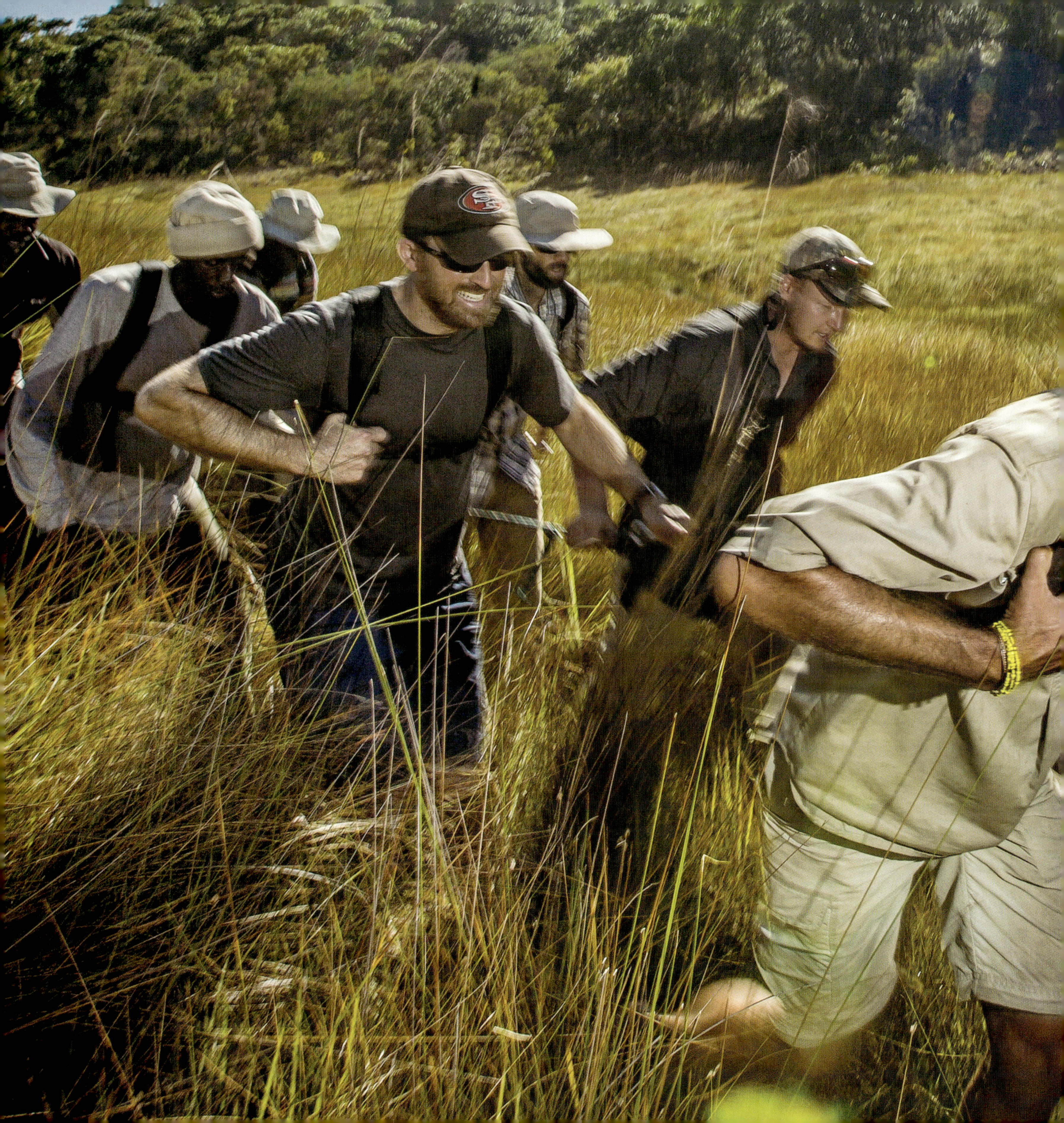

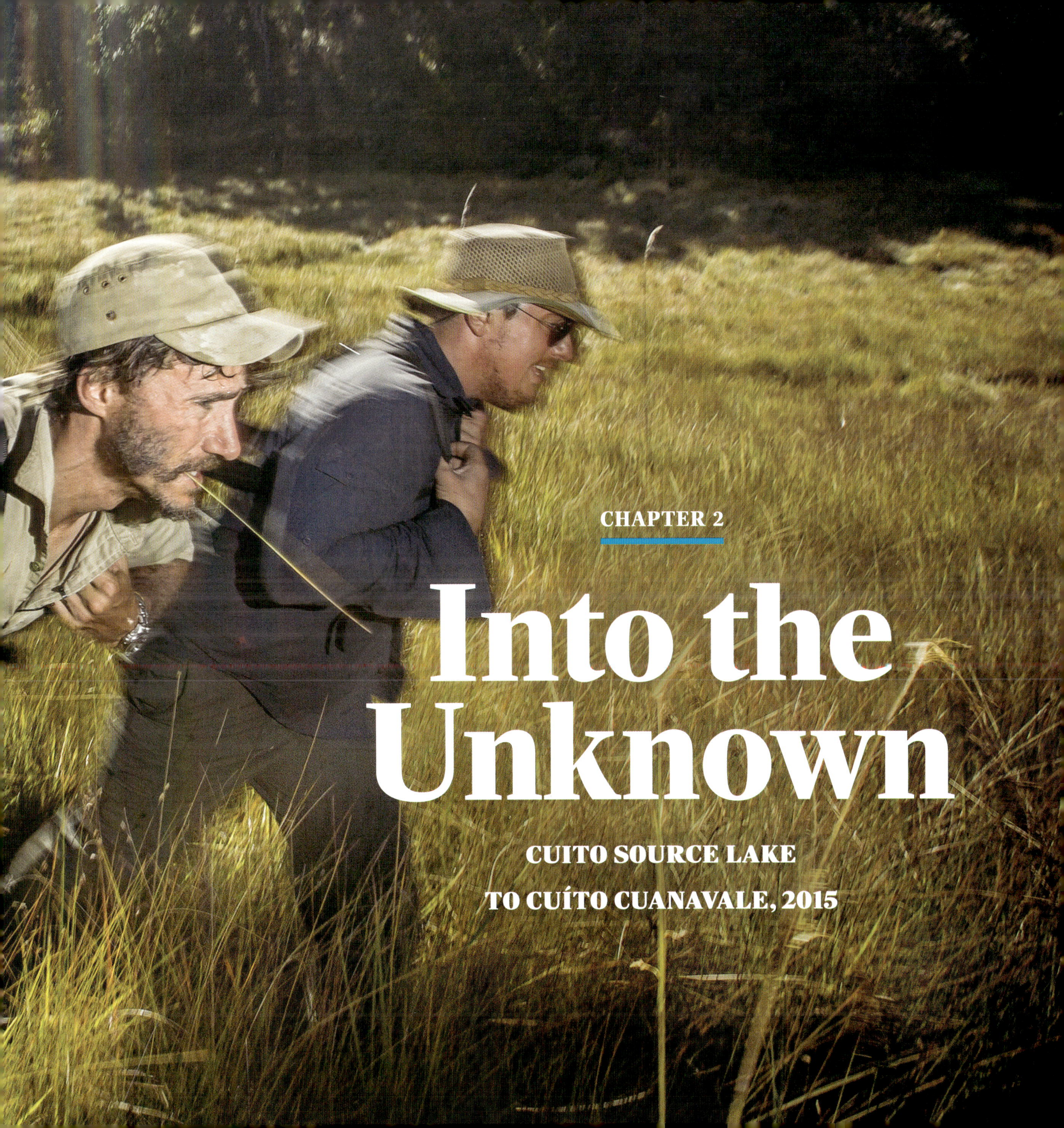

CHAPTER 2

Into the Unknown

CUITO SOURCE LAKE TO CUÍTO CUANAVALE, 2015

ON MAY 25, 2015, European Space Agency astronaut Samantha Cristoforetti, the first female commander of the International Space Station, tweeted down a photograph of the Okavango Delta, wishing my team good luck as we launched the Okavango Megatransect. After holding a broken cable in place just long enough for the photo to download, I showed the tweet to Tumeletso "Water" Setlabosha, my close friend and mentor, a Wayeyi elder from the Okavango Delta. He marveled at the photograph, asking where it had come from. I said a lady in the sky, living in a bus in space, took this photo with her camera. He looked up at the sky contemplatively, paused, and looked at me before saying, "But we aren't in the delta!" He didn't believe my story about where the photo came from, though he looked up at the sky again, as if to check that nothing was up there.

Mr. Water was right: We weren't yet in the delta. Instead, we were dragging fully loaded mokoros alongside the narrow and constricted Cuito River. Deep in the little-known eastern Angolan Highlands, we were five days into an expedition that would change our lives.

It had taken us about six months, using satellite imagery and exploratory forays into the highlands, to find what we thought might be a drivable route to the undocumented eastern source of the Okavango Delta, the Cuito River. In late 2014, we had tried to access this impossibly remote location from the southwest but failed due to the inescapable abundance of land mines. We had decided that approaching from the north might be viable instead.

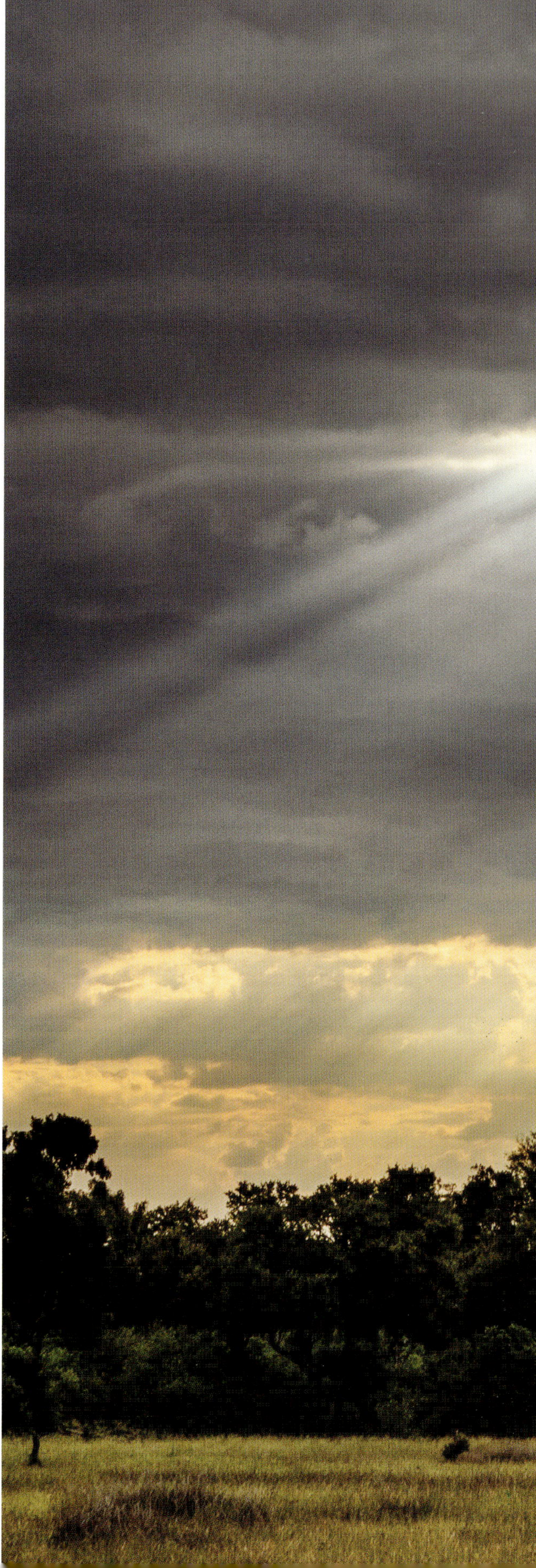

Hope shines through: In 2014, the Okavango Delta became the 1,000th UNESCO World Heritage site, long overdue recognition for a wilderness beyond comparison. This was our call to action. **PAGES 54–55:** *In 2015, as we embarked on our first expedition, we really had no clue what we would find. Only the Luchazi people knew anything about this so-called land at the end of the earth—and we hadn't yet met them. We dragged our fully loaded mokoros for two weeks before finding navigable water.*

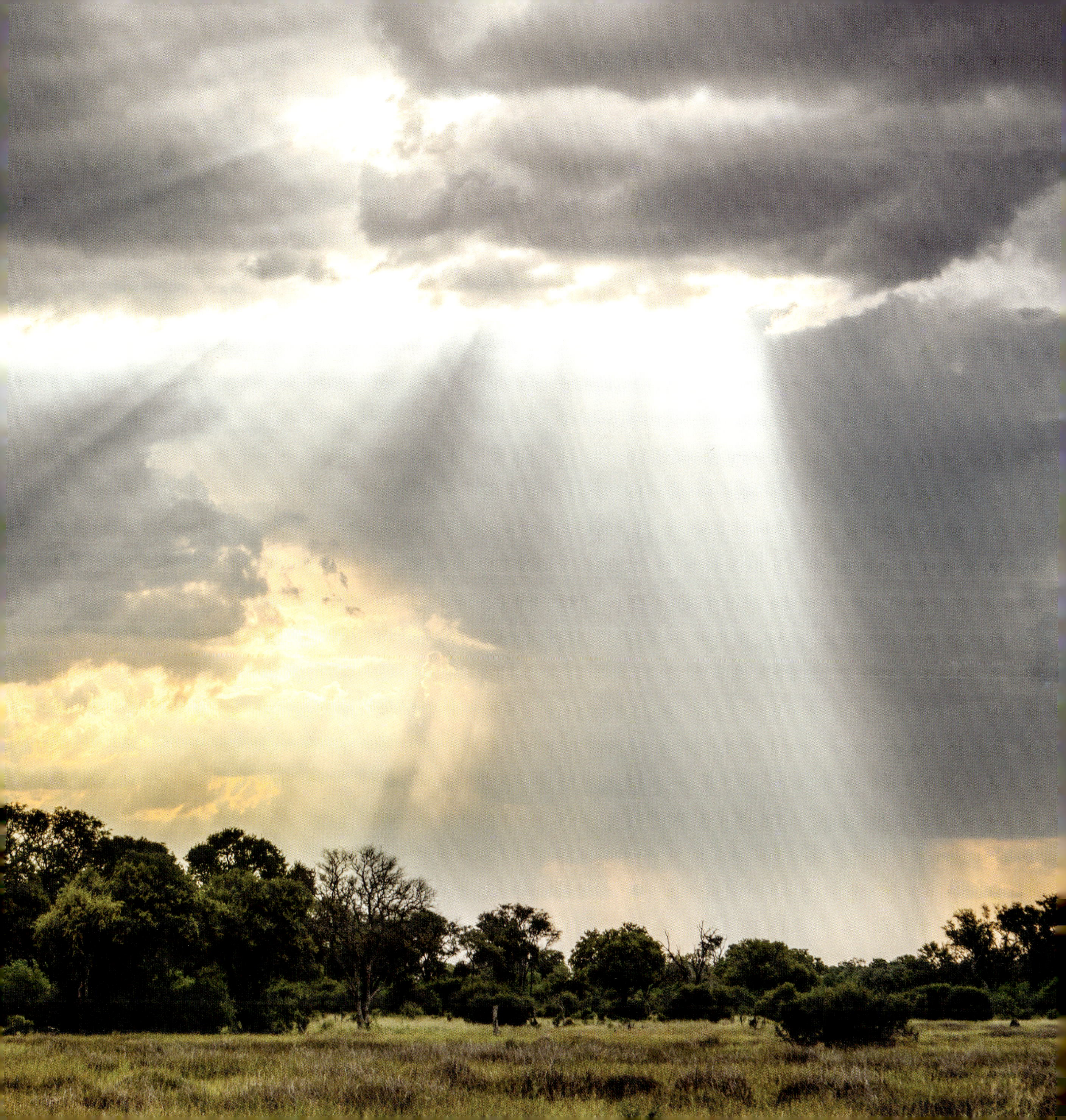

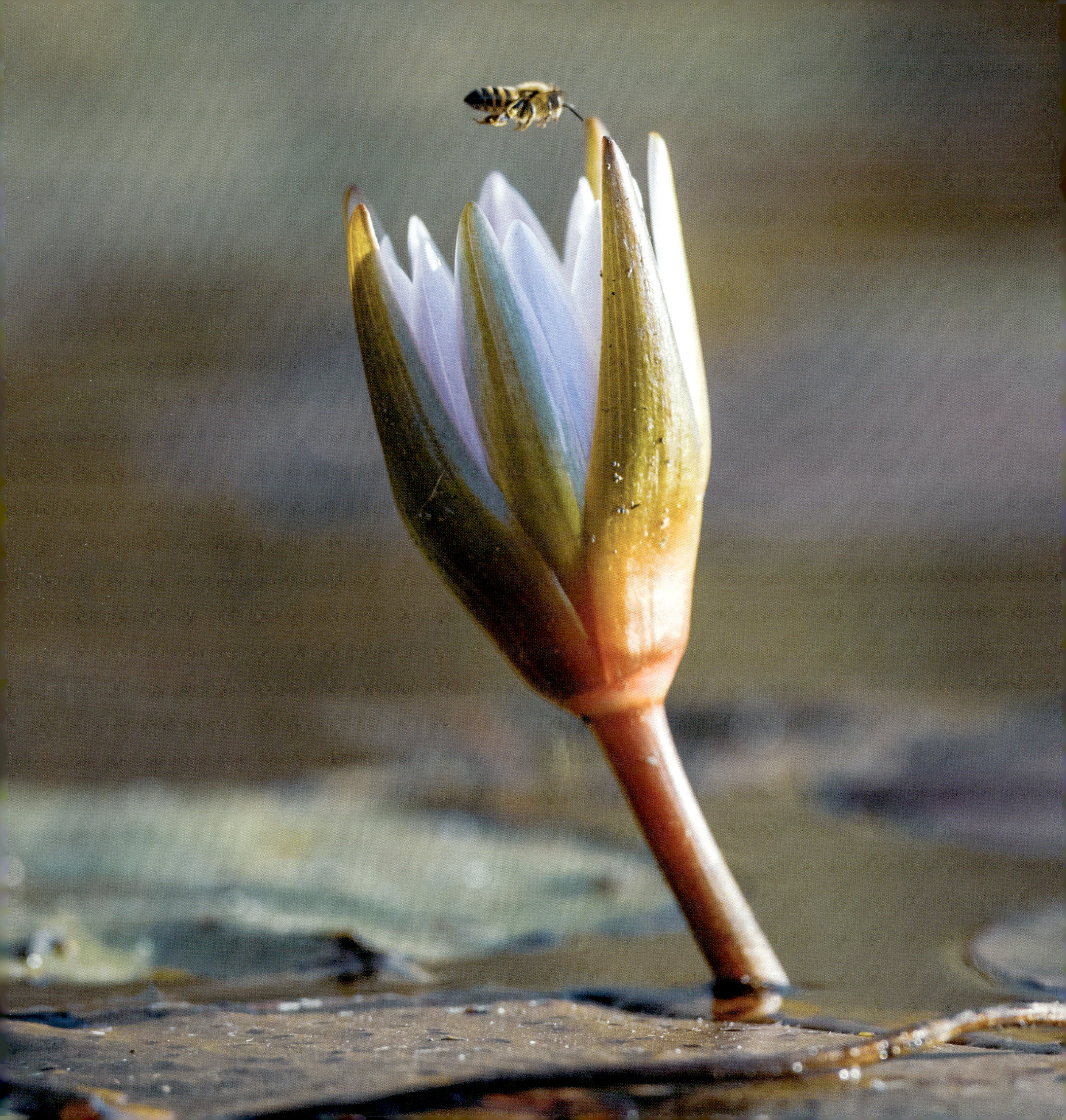

We crossed the border with the help of local government officials and barely made it with our fully loaded trailers to the town of Menongue, down the barely existent and treacherous Katwitwi Road, which links southeast Angola with Namibia. This broken track was intentionally left in ruin to isolate this part of Angola from outside influence. Though it had been more than a decade since the end of fighting, small towns were still traumatized by the civil war. It was a reality check.

A few days before departing Menongue for the source, I met with the governor of Cuando Cubango. He was an imposing man, a general from the war years. After offering to put two police helicopters on standby for medical evacuation, if needed, he asked me one favor: "Please find out who lives there." He called the region *terra do futuro*—"land of the future"—a sharp change from the 27 years of civil war, when it was called *terra da fome*—"land of hunger." The governor considered where we were going to be the most dangerous, unforgiving part of the country. Yet he was trying to break the stigma attached to these lands, looking instead toward what they could be in his still healing nation.

ON MAY 17, I HAD woken up at 4 a.m. to supervise packing up the vehicles and get our bustling convoy under way and off into the unknown. Loaded in our trucks were eight traditional mokoros from the Okavango Delta and supplies for about a month on the water. Our Land Cruisers were escorted by a fleet of armored Land Rovers and trucks from the HALO Trust, a humanitarian land mine clearance NGO made famous in 1997, when Princess Diana walked across an active minefield in the Angolan Highlands, wearing a protective visor and flak jacket labeled with the trust's name. HALO's presence at our side reflected yet another remnant of the decades of war.

At my side was a crew handpicked for the expertise they could bring to whatever we would find out in the wilderness. One part of this team was made up of biodiversity experts: Paul Skelton, a famous ichthyologist and director emeritus of the Southern African Institute for Aquatic Biodiversity, would survey the aquatic life in the upper reaches of the Cuito River; Adjany Costa, an Angolan scientist, would assist him as an expedition ichthyologist; Bill Branch, a famous herpetologist, would be on the lookout for lizards; and David Goyder, a botanist from London's Kew Gardens, would try to describe the plants and ecosystems surrounding us. Götz Neef, a young Namibian scientist from Rhodes University, joined to support the biodiversity surveys.

Making up the rest of the expedition specialists, we had Pete Hugo, expedition cook and an award-winning poet; Giles Trevethick, an expedition expert and data recorder in my mokoro; Neil Gelinas, a filmmaker working on his first feature documentary; James Kydd, a photographer and wilderness expert I knew from university; and Cory Richards, a National Geographic photographer on assignment with his assistant, Mark Stone.

The core of the expedition team were five Wayeyi fishermen and polers from the

Water lilies abound throughout the Okavango Basin, from source to delta, and bees are the secret agents of beauty and abundance, offering us flowering plants and life-giving honey. The honey harvested around the Angolan source lakes was once world-famous for its medicinal properties.

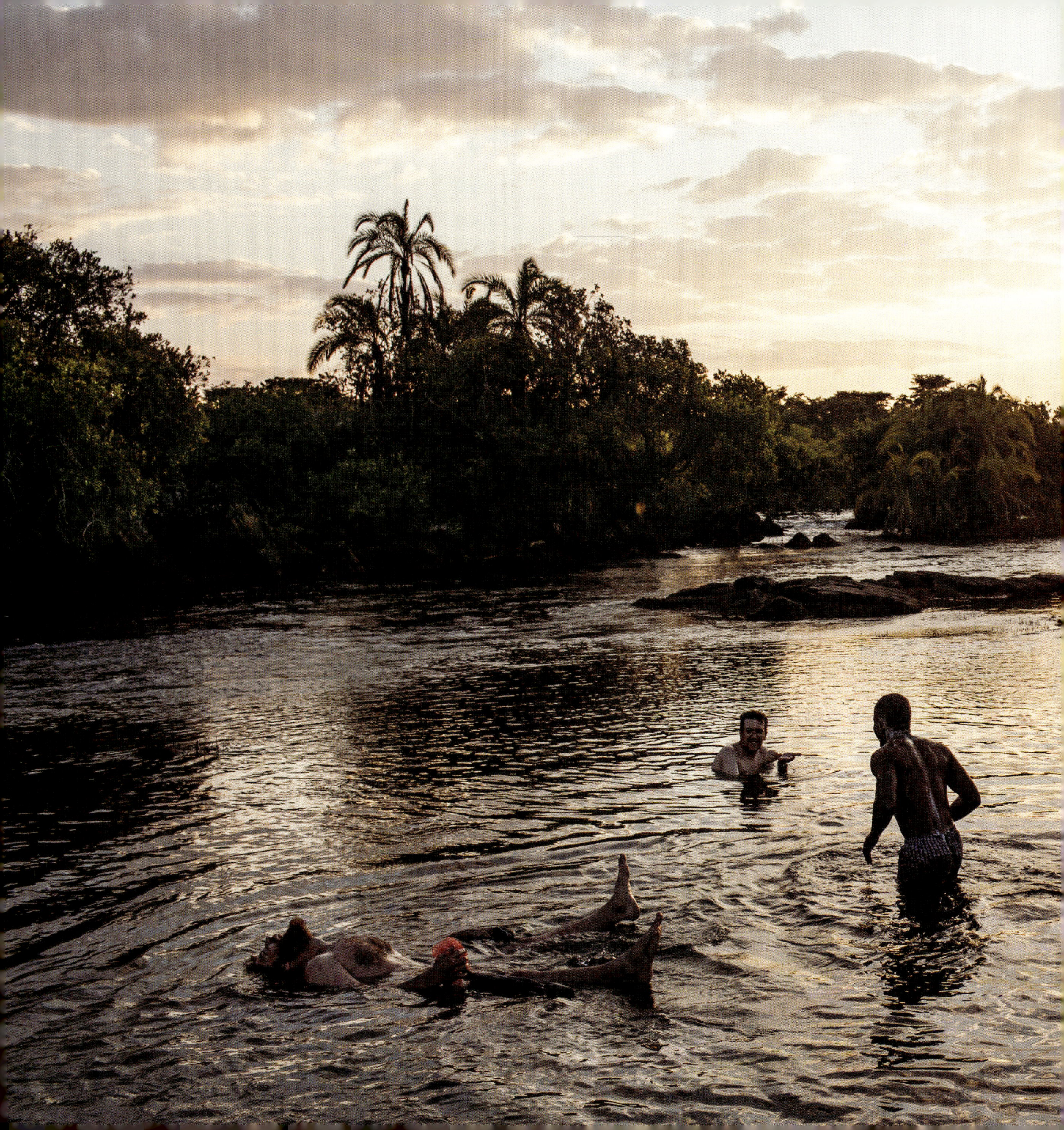

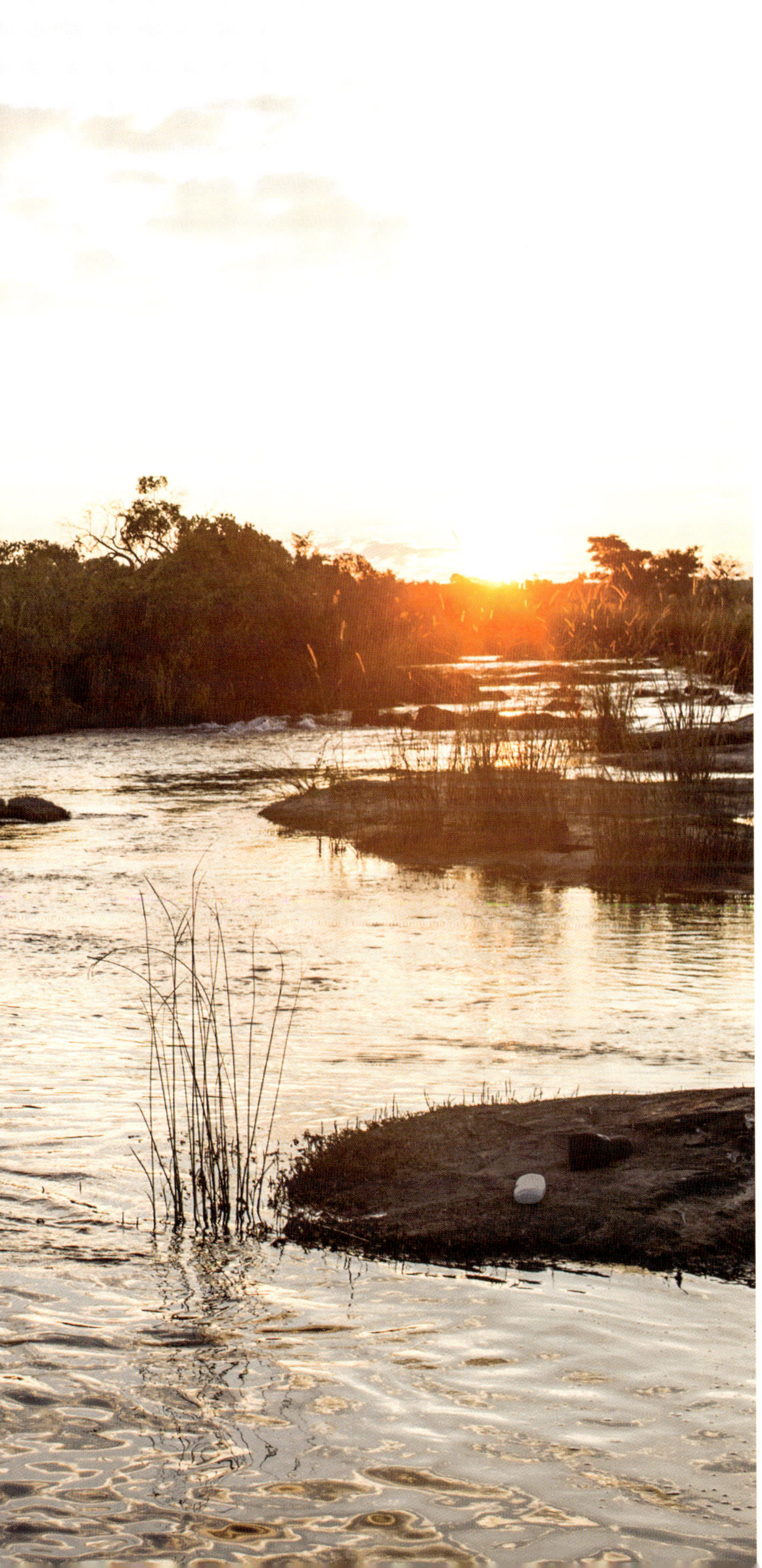

Okavango Delta: Gobonamang "GB" Kgetho, his brother Leilamang "Snaps" Kgetho, Topho "Tom" Retiyo, Tumeletso "Water" Setlabosha, and Kgalalelo "KG" Mpitsang, all of whom we had known for many years and with whom we had poled across the delta several times since 2010. They were all close friends, brothers to Chris and me. They were eager to meet the people living along the river. They were born polers from the delta. They had been exploring that wilderness all their lives, but they had never imagined they would visit Angola.

We drove onto the main road to Luena, capital of Moxico, the province hiding the Okavango's eastern sources. As in most of Angola at the time, the main road wasn't really a road. Either it had washed away, or it had been diverted through adjacent woodlands due to fear of land mines—or it was never there to begin with. Driving into the unknown was a strange feeling. When setting out on a trip, one usually has a destination in mind. But we had no idea where we were going or how we would get there. No one would be able to help us out there. We had only satellite imagery to guide us.

I always say that the hardest part of river exploration in Africa is getting to the river. Police roadblocks in remote parts of Africa are usually very tricky, but we had signed letters of credential from the governors and, most important, the president of Angola. We drove past endless signs—DANGER, DANGER, DANGER—warning about active minefields; skull and crossed bones everywhere; do not pass on all roads. It felt like a one-way ticket.

The last town we passed was Munhango, a heavily mined outpost, where we were restricted to camping at government buildings. Shortly after leaving Munhango, we joined the railway line, passing more active minefields. Finally, we turned south onto a narrow, sandy track in dense miombo forests, a type of African

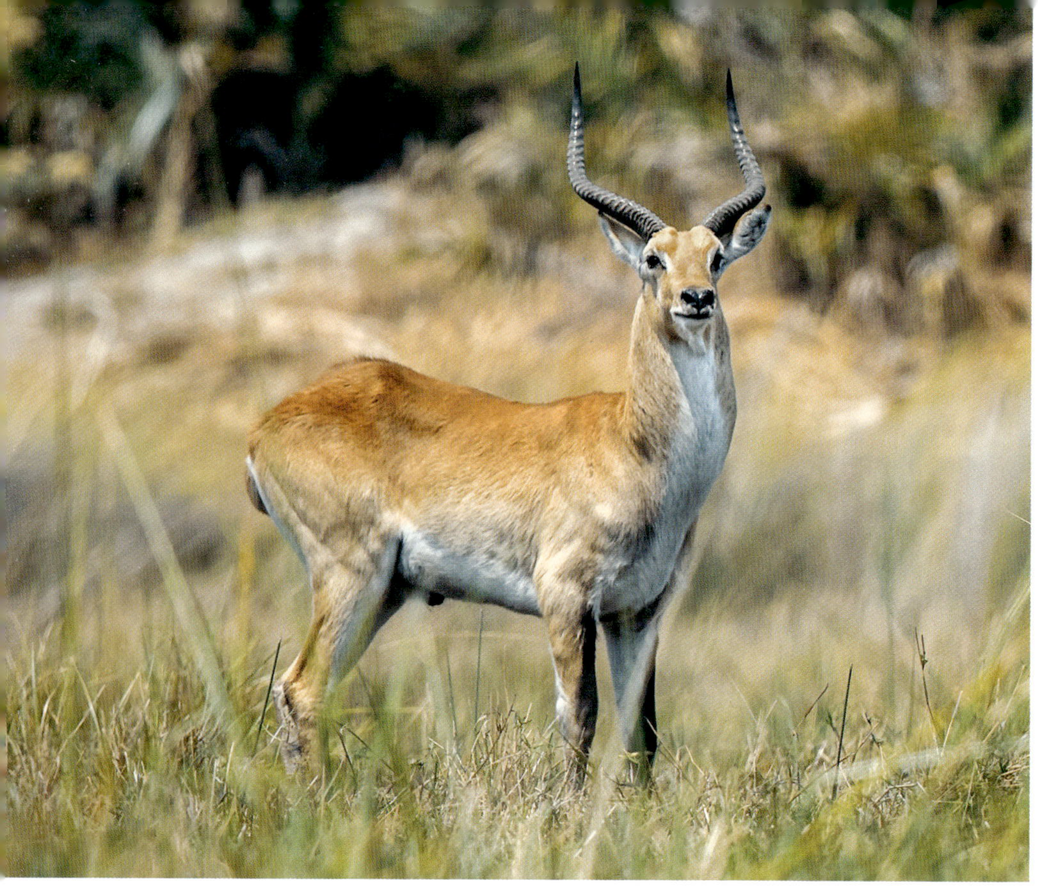

While we confirmed that red lechwe, a medium-size antelope, had gone locally extinct along the Cuito River, we did find a small relict population along the remote Cuanavale River. **PAGES 60–61:** *Despite the abundant crocodiles, a swim in these highland rivers is often tempting.*

biome dominated by oak-like miombo trees. We missed the track entrance—just a gap in the forest canopy—several times, only finding it using aerial photographs taken by our drone. We began slowly crashing through thick forest, following vehicle tracks last used during the war. It was unnerving, to say the least, to drive past blown-up convoys of military vehicles and supply trucks—some bombed from the air but most destroyed by land mines.

We had now entered the unexplored, uncharted, and undocumented *terra do fim do mundo*. No scientists had ever studied this landscape. Everything we knew was based on extrapolation, assumptions from satellite imagery, and remote sensing datasets.

We drove from dawn to dusk through two previously undocumented minefields and woodlands we couldn't describe, full of tree species we didn't recognize. It was hot and humid, light only occasionally breaking through the trees. Within hours, the track was gone. We were forced instead to follow gaps in the forest canopy, footpaths and motorbike tracks used by bushmeat hunters, and cut marks indicating hunters had used tree bark as dinner plates. The coaxial cable had been ripped out of our satellite modem, so the unit only worked while someone held the cable in place. Our GPS units were constantly recalculating our location. We were effectively lost.

As we got closer and closer to what satellite imagery suggested was the source, the vehicle track reappeared in clearings and then disappeared again in dense overgrown forest. It was slow going as we studied the topography and looked for signs of human activity to minimize the risk of encountering an undocumented minefield or a hunter's pitfall trap. Ahead of us, a Kamaz truck, originally built to traverse the Russian tundra, bashed through trees and undergrowth using its armor plating and extraordinary horsepower. At one point, one of our Land Cruisers literally disappeared into a bog and had to be winched out by the Kamaz.

Every time an obstacle delayed the expedition, we were plagued by millions of stingless sweat bees, named for their love of landing on humans to drink perspiration. They choked the air around us, their harassment worsening a growing feeling of hopelessness. We held our collective breath as we felt the forest folding in behind us.

IN MEMORIAM

Topho "Tom" Retiyo

A Song for the Delta

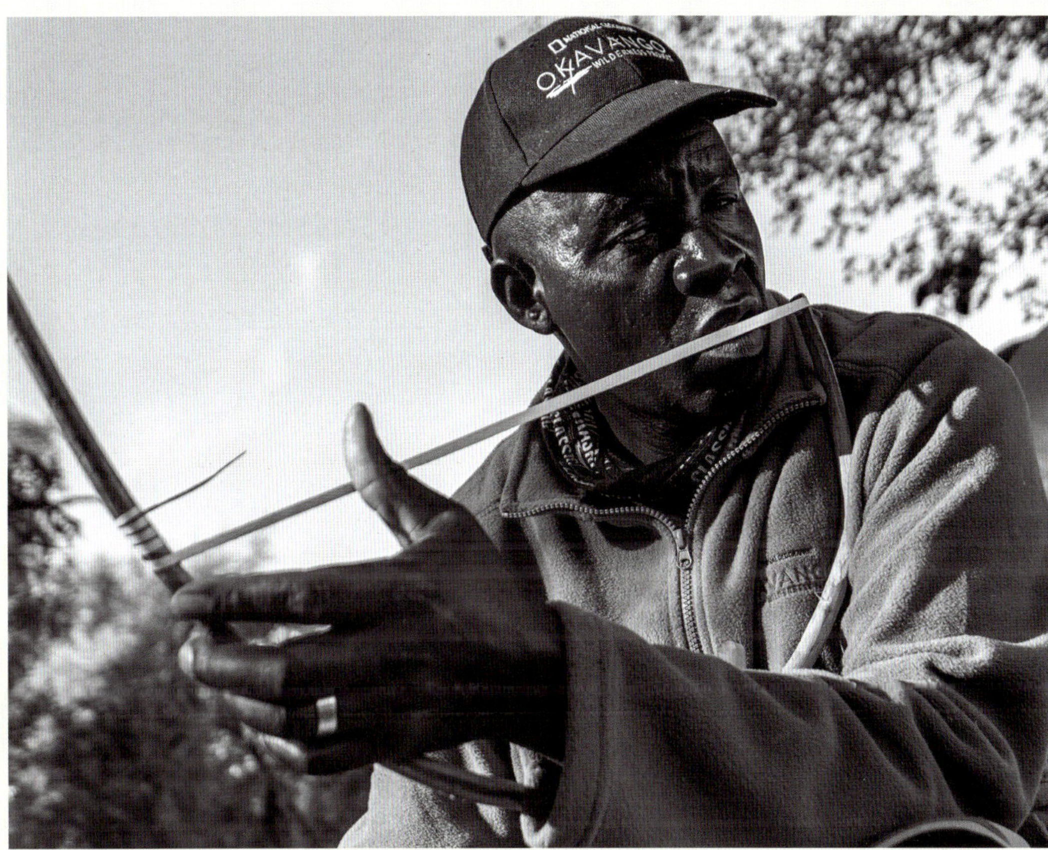

Topho, known to his friends as "Tom," was a poler, farmer, and musician. On every expedition, he made a new tswororo mouth bow to play around the fire at night.

MR. TOPHO "TOM" RETIYO passed away tragically in Seronga in 2021. On every expedition, Tom "the Music Man" found a notched brandy bush bow and a palm frond to make a new *tswororo*—an African version of a kazoo. He used a carved stick to vibrate the instrument as he made it sing with his lips. His music danced like flickering flames, echoing primordial rhythms, whispering old stories to the stars, the water, and the earth below.

When Tom played, time stopped. As if in a trance, we all joined him as he lost himself in the music that became the air the delta breathed. Often Tom would stop playing, look around the campfire, smile broadly in the sudden silence, and laugh in joyous celebration. We joined him in a raucous moment of laughter and applause. His music was always a spontaneous gift that uplifted and balanced us.

Topho was a gentle man who loved his wife and daughters. He was a father, a farmer, a born poler, and my friend. He was killed by an elephant that a neighbor had chased onto his farm. His smiling, unbreakable spirit still calls to us, in fish eagles at sunset and in the undulating chorus of a million bell frogs each night. Tom is now the flames and warmth of the fire, both given without thought of getting anything back. He played the music of the wilderness, and though he is no longer with us, his melodies will forever float across the delta's floodplains he loved so much, a timeless symphony.

The sun was beginning to set as we approached the ridgeline above what we thought must be the source, which had appeared as a bright green clearing, perhaps a wetland, in the satellite imagery. Too tired to be happy or excited, we hoped for first sight before last light. Eventually, the trees parted to reveal a sparkling lake at the bottom of a burned valley. This was totally unexpected. We had found it: the source we had been seeking for so many months, the place no one had ever documented.

Severely dehydrated and overwhelmed, we unpacked the Kamaz before darkness fell. I didn't eat that night and woke up an hour before dawn. At sunrise, I walked to the edge of the water. The autumn morning was crisp, and mist hung over the lake. I tasted the water. It was sharp and crisp—pure. It was the most remote place I had ever been, yet something about it was strangely familiar. I had dreamed about this place, imagined it countless times in the weeks prior.

As the sun rose, more people joined me at the water's edge. Bill Branch was running around with elastic bands, shooting them to stun some of the hundreds of sand lizards scurrying around the shore, which he believed to be an unknown species. He couldn't fathom why they were here in such great numbers. Paul Skelton was surveying the catch from a seine net deployed by his team at dawn. Bill and Paul had written the field guides I had pored over as a young scientist; now they were now running around like excited students themselves at a lake they had never imagined existed.

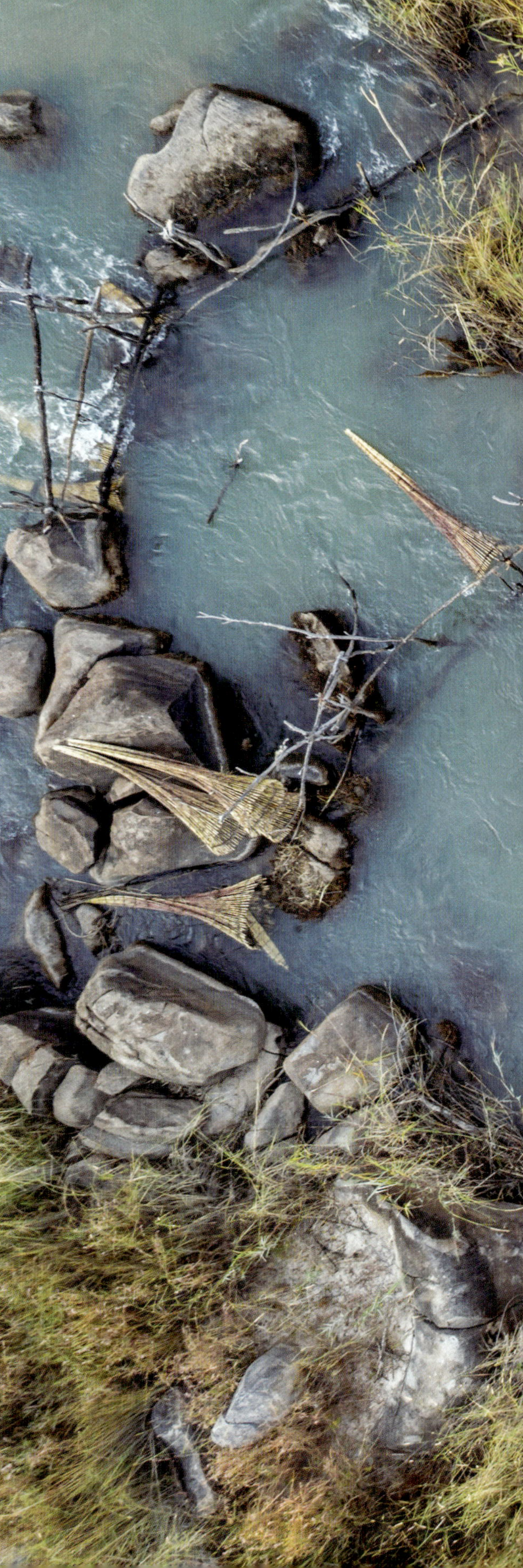

Along the meandering Cuito River, local fishermen have blocked off the outflows of most of the oxbow lakes to channel the fish breeding in them into their traditional handmade traps.

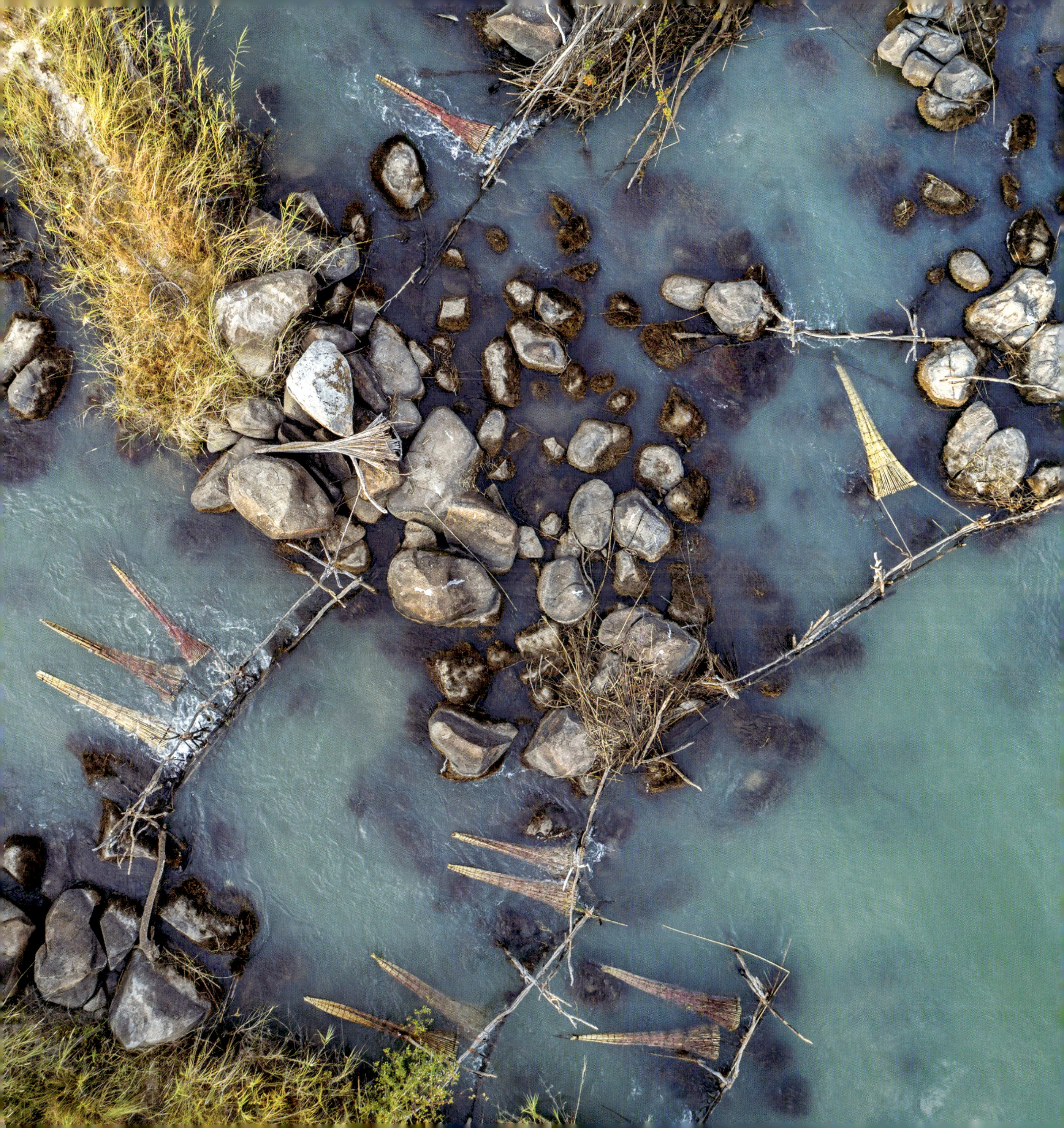

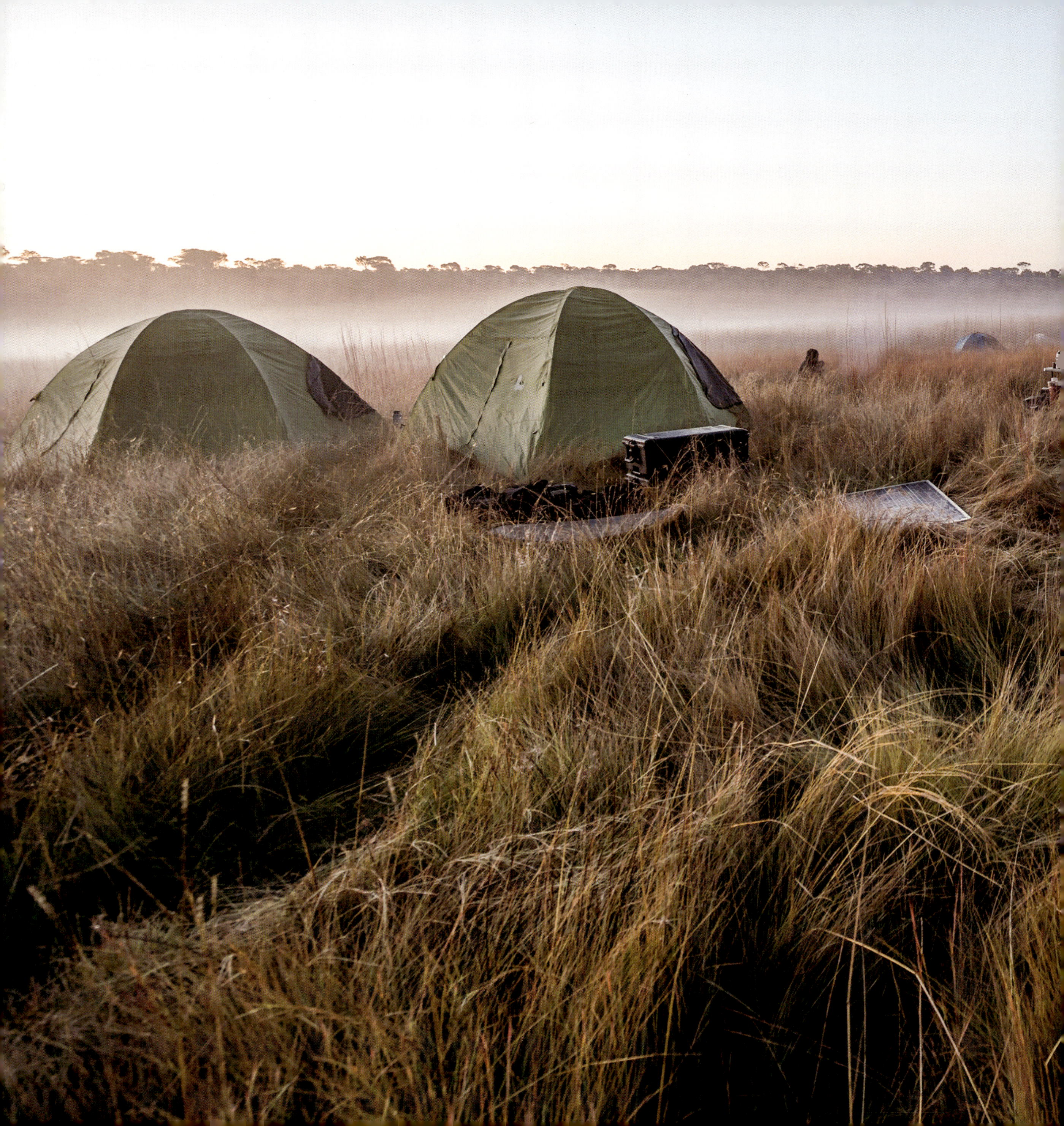

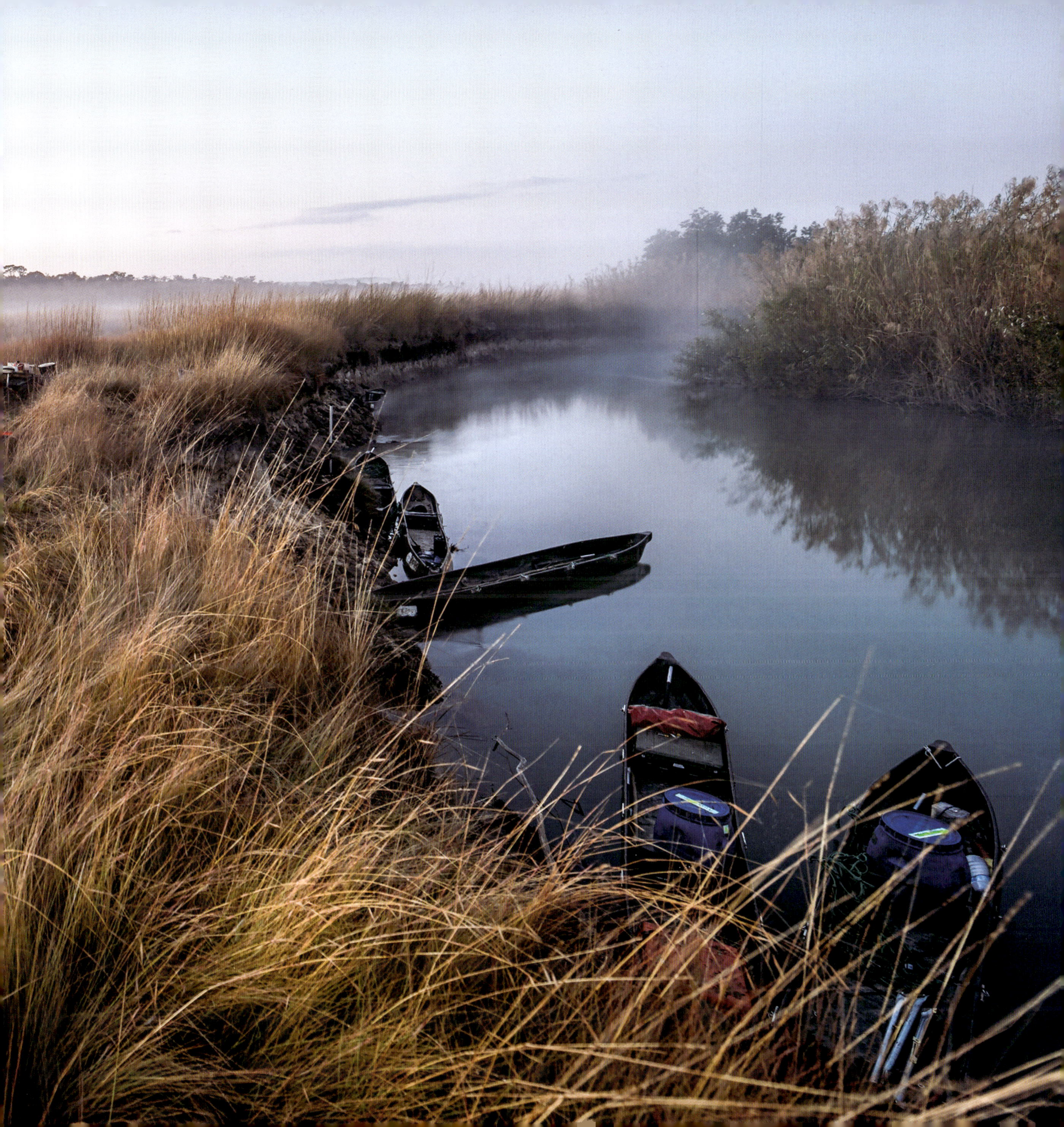

Paul showed me a beautiful striped fish that he exclaimed was from the genus *Microctenopoma*. (Paul always speaks in Latin names, and I hardly ever know what he is talking about; I just nod and smile to keep up appearances.) He believed it to be a new species: a climbing perch with a striking pattern. He also documented a family of fish he called "barbs," a colorful ray-finned fish species that was surely new to science or at least a new species for the Okavango Basin. Within an hour, Bill had captured a few sand lizards. After closer inspection, he told us a long story about how special they were in the evolutionary tale of their genus. We were beginning to fill an immense gap in our knowledge of local biodiversity.

Paul also observed that the lake was surrounded by peatlands, an important carbon sink, which an excited David Goyder noted were poorly known in tropical Africa. This was only the second discovery of this nature.

We began carrying all our gear and supplies from the trucks down to our mokoros, waiting on the edge of the extraordinary source lake. Our intention was to begin at this source lake and follow the water as it traveled down toward the delta. The team was tired from the long drive. We had far too much gear, and we found ourselves packing and repacking the boats. As the morning got warmer, we became more and more nervous.

We decided that we had to add our backup mokoro, but we didn't have enough polers to captain this extra canoe. I asked Götz Neef what sports he'd played at school. He said he played handball and kept fit by mountain biking. I told him that he was now going to figure out how to captain a mokoro across one of the world's largest undeveloped river basins. Without flinching, he accepted the challenge. This was the Götz I came to know: intelligently fearless.

Eventually, at about 11 a.m., we were ready to launch. The biodiversity experts and their teams had been up since dawn sampling and checking traps and were still behaving like kids in a candy store. The river team looked nervous. We didn't like that there were no local people to be seen—nobody to warn us about sacred sites or unseen hazards. We were going to pole and paddle ourselves all the way home through the unknown, carrying with us a story for a generation, a journey that would become folklore.

ON MAY 19, 2015, we had launched the eight overloaded mokoros into the source lake only to find that the lake's outlet was completely blocked by an unfamiliar species of water berry. Though the valley was relatively straight in the satellite imagery, the small, crystal clear stream that emerged from the tree blockage was too narrow, turning upon itself, constricted by high ridges on either side. There were switchbacks every few meters, making it impossible for us to navigate in our six-meter (20 ft) dugouts. We'd decided to haul the mokoros—each 300 kilograms (660 lb)—out of the stream and to drag them down the marshy river margin.

We had brought ropes and harnesses, expecting to portage around obstacles. Yet we had not expected to drag the mokoros, six to eight of us at a time, for almost two weeks, about a mile a day. The conditions were extremely

The most striking thing about exploring the life-giving peatland-sustained floodwaters of the Okavango Basin is the crystal clear water. It's like paddling or poling through a freshwater aquarium, with water lilies and fish sparkling beneath you. **PAGES 66–67:** *There's a rhythm to expedition life. You set a new camp every afternoon, break it down and pack it into the mokoros every morning, never to return. Life is perfectly uncomfortable.*

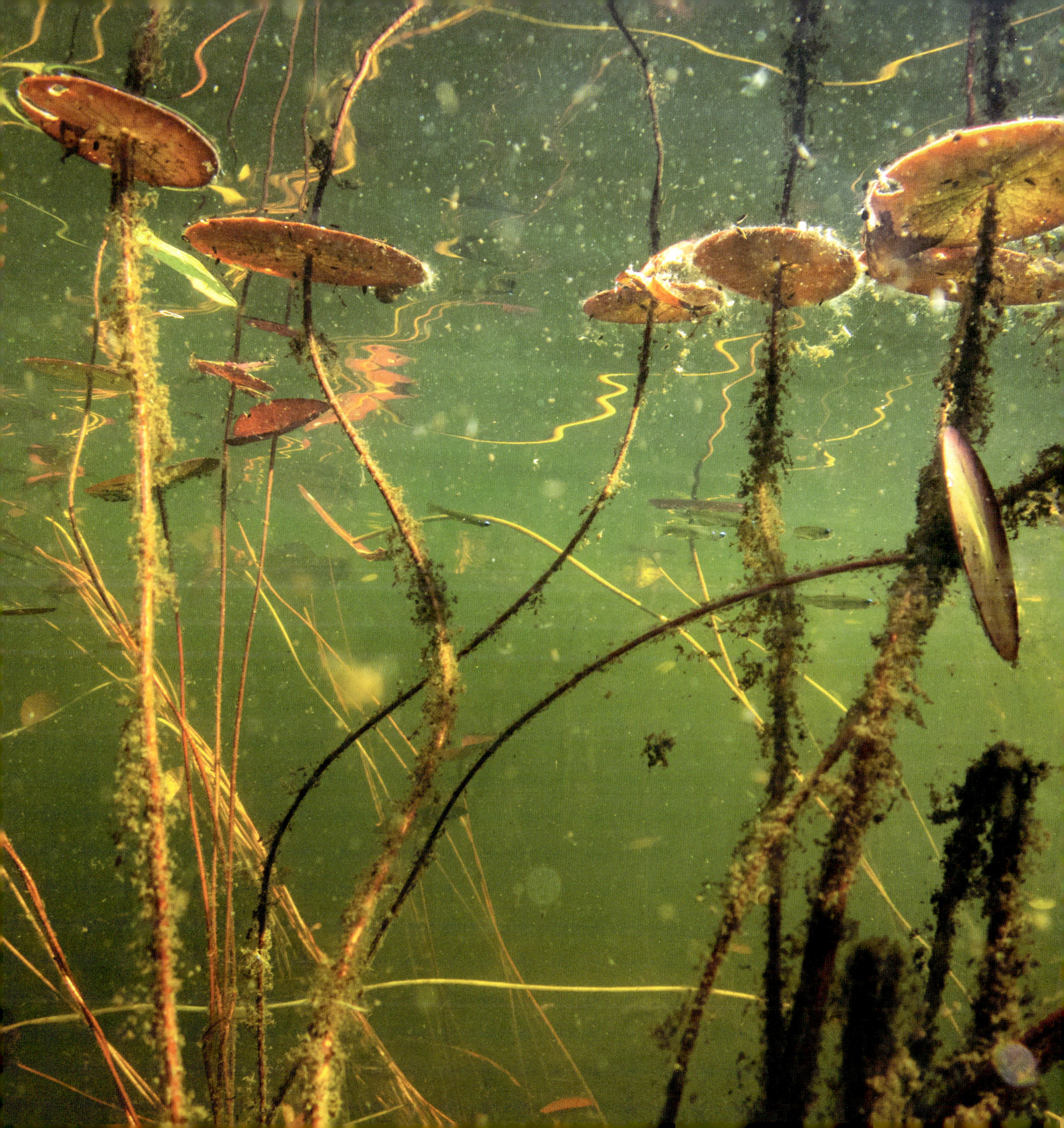

The residents told us that during the civil war, they had isolated themselves from the fighting. We were the first visitors from the outside world in 42 years.

difficult. Cory Richards, the National Geographic photographer, had climbed Everest several times and said our travel felt like alpine training. My biometrics watch told me that my average heart rate over 12 hours was 135 beats a minute, and that I should rest for 143 hours before exercising again. None of us were performance athletes. The days ended up being very hot and humid, while the nights were freezing cold, misty, and damp. We weren't prepared for the cold and wet, so we slept in all our clothes and huddled around the fire each morning. We still saw no sign of local people in the eerie silence of this lost wilderness. We saw many signs of life, but no animals.

Every day felt like we were disappearing deeper and deeper into a place from which there was no return. Our satellite connection wasn't working, our solar power system was failing in the mist, and the GPS units were confused as to where we were. I had never felt so remote and isolated. We couldn't turn around, as the support vehicles had left; it was just us. We had entered a green, lush, unending desert made of forest, marsh, sand, and water. I had traveled all over Africa—to almost all the countries in sub-Saharan Africa—and never seen anything like this. This land was a paradox—a vast, thriving yet seemingly empty wilderness, one that left me both overwhelmed by its vastness and inspired by its potential. I resolved to focus our future efforts on forgotten postwar wildernesses like this.

At sunset each day, I'd run up to the highest ridge to whistle loudly and call for help. I desperately wanted to meet local people who might be able to help us move the mokoros or warn us about the next danger around the bend. We would only be safe with them.

On day nine, I'd told the team to rest up and dry their gear while I ran about 10 kilometers (6 mi) downstream to find the first possible launch site. I wanted to give the team some hope that night and tell them how many more days of pulling we had to go—without satellite imagery, I'd been unable to answer this question for more than a week. It was disheartening to move so slowly that we could look back when we stopped every afternoon and see where we'd camped the night before.

On my run, I flushed an oribi, a small wetland antelope, out of deep grass, giving me the fright of my life. I also found diggings by bushpigs; there was life in these forests after all. Then, to my astonishment, I found human footprints: barefoot tracks of a family of four, walking away from the river. I followed them.

I had been told about a long-forgotten San community, known as "forest Bushmen," living in this remote part of the eastern Angolan Highlands. Many thought they had disappeared during the war, had been absorbed into local cultures, or had been extinguished altogether. It was wonderful to learn that these people might remain—and might be nearby. I suspected, however, that they had seen us and moved away. A marching army of explorers dragging loaded canoes, creating what looked like the trail of a giant snake, would have appeared as a threat. Giant pythons were known to have formed the great river valleys in San mythology. I'm sure they wanted nothing to do with whatever made that track.

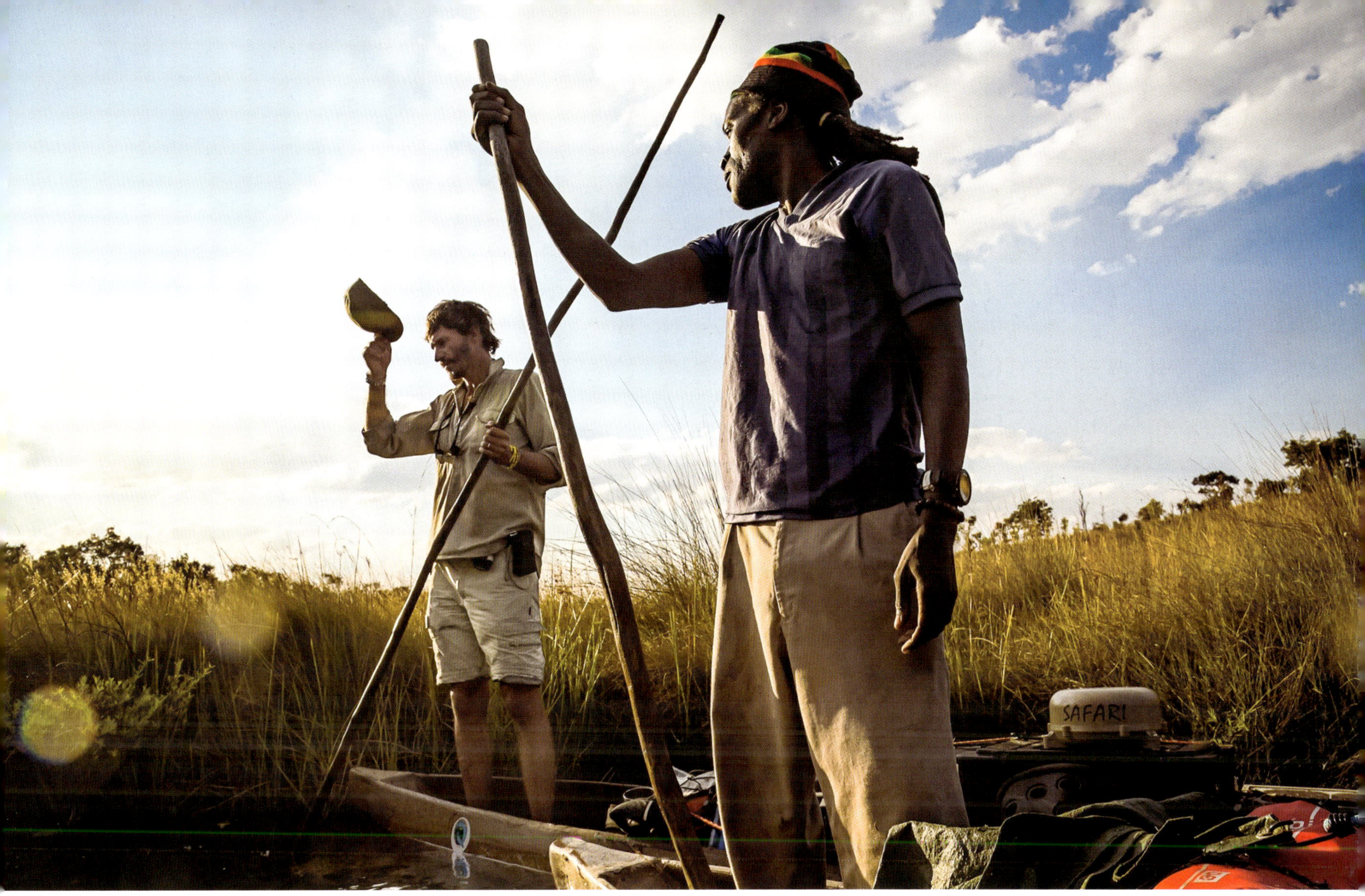

About a kilometer (0.6 mi) away, I found a smoldering fire below a tree, where the family appeared to have been honey hunting. The footprints disappeared deeper into a dense forest patch, where the ground was compacted, and their tracks became harder to see in the low light. I started to get nervous as the trees wrapped around me. It was eerily silent. I felt like I was being watched. Though I couldn't get lost, as there was only one footpath, it was getting late, and the thought of darkness frightened me. At some point I turned around and headed back to camp.

BY EARLY JUNE, after 14 days of hauling, we had finally made our way along the riverbank to open water, launching into the near-unnavigable upper reaches of the Cuito River. Every day the feeling became stronger; nothing was familiar. It was an otherworldly place, the remains of fires along the river the only signs of people. The crystal clear waters switched

In June 2015, we had been exploring the Cuito River for a month and were already three weeks behind schedule. The river was winding upon itself in the upper reaches. Sometimes we would pole for four kilometers (2.5 mi) to progress one kilometer (0.6 mi) in a straight line. We still had not met any local people. We felt lost.

from blue to green to blue as inlets from adjacent peatlands flushed acidic water into the mainstream. Thousands of small fish darted about in the clear water, chased by large African pike. Every night, Paul, Götz, and Adjany set fishing nets, and every morning Paul shouted out Latin names and enthused about the discoveries made in the preceding 24 hours.

The farther we went, the more burned the sedgy peatlands and adjacent grasslands became. Lit deliberately by the local people, whom we hadn't yet met, these fires didn't seem to harm the forests. Yet I had never heard such deep, eerie silence. Why was this landscape seemingly abandoned? We silently feared for our lives. After all, no one had ever made it out of here to tell the tale.

The steep ridges on either side of the upper Cuito were held together by what we now know are underground forests: trees hidden underground by ancient sand dunes that buried their car-size, centuries-old trunks. Only their canopies now poke above the surface, leaving slopes covered with prickly, hard twigs impossible to camp on. Navigating this landscape made for very, very tough going. The river was wild and fast-flowing, bending and turning upon itself. We could travel four kilometers (2.5 mi) on the river and progress just one kilometer (0.6 mi) in a straight line.

One morning, in frustration, I sent Pete Hugo and Götz out to cut a path through dense water berries around a blocked corner in the river while the team packed the mokoros. We poled downstream to meet them three hours later, finding Pete and Götz neck-deep in the river, cutting submerged branches. Not more than three meters (10 ft) upstream from them was a three-meter crocodile, watching them sawing and bobbing. We had seen several large crocodiles, probably all females, moving slowly upstream in those upper reaches near the source lake. I touched the waiting croc with my nkashi pole. She turned and slowly moved upstream under us. I don't think she was interested in eating us; more likely, she had never seen people foolish enough to stand in her river for such a long time.

Within 10 days of our setting out, the river was about 12 meters (40 ft) wide and flowing nicely. We were at last enjoying ourselves, making progress, laughing at one another as we surfed the fast corners. Then, it happened: Voices of women and children laughing and talking broke the usual eerie silence. We came around a river bend, where the group froze at the sight of us, dropped the cassava they were washing, and ran into the bush.

Adjany called after them and they returned. One old lady could speak a little Portuguese. The women, some with babies on their backs, were wearing blankets, traditional woven cloths called *kikoi,* and old dresses. Though they were clearly very scared, they told us about a waterfall ahead, and that their village was up a tributary below the waterfall. They retreated into a tree-lined clearing behind us, where they lit a bushfire that raged through the night—perhaps to make sure we didn't follow them. It was a beautiful scene, watching them walk through the smoke and flames. They were extraordinary

Driving to the sources of the Okavango Delta in the Angolan Highlands is almost impossible. We had to use either old military roads plagued by land mines or motorbike tracks used by hunters. For years, the HALO Trust—an NGO dedicated to clearing land mines—escorted us on week-long drives using armored trucks and Land Rovers.

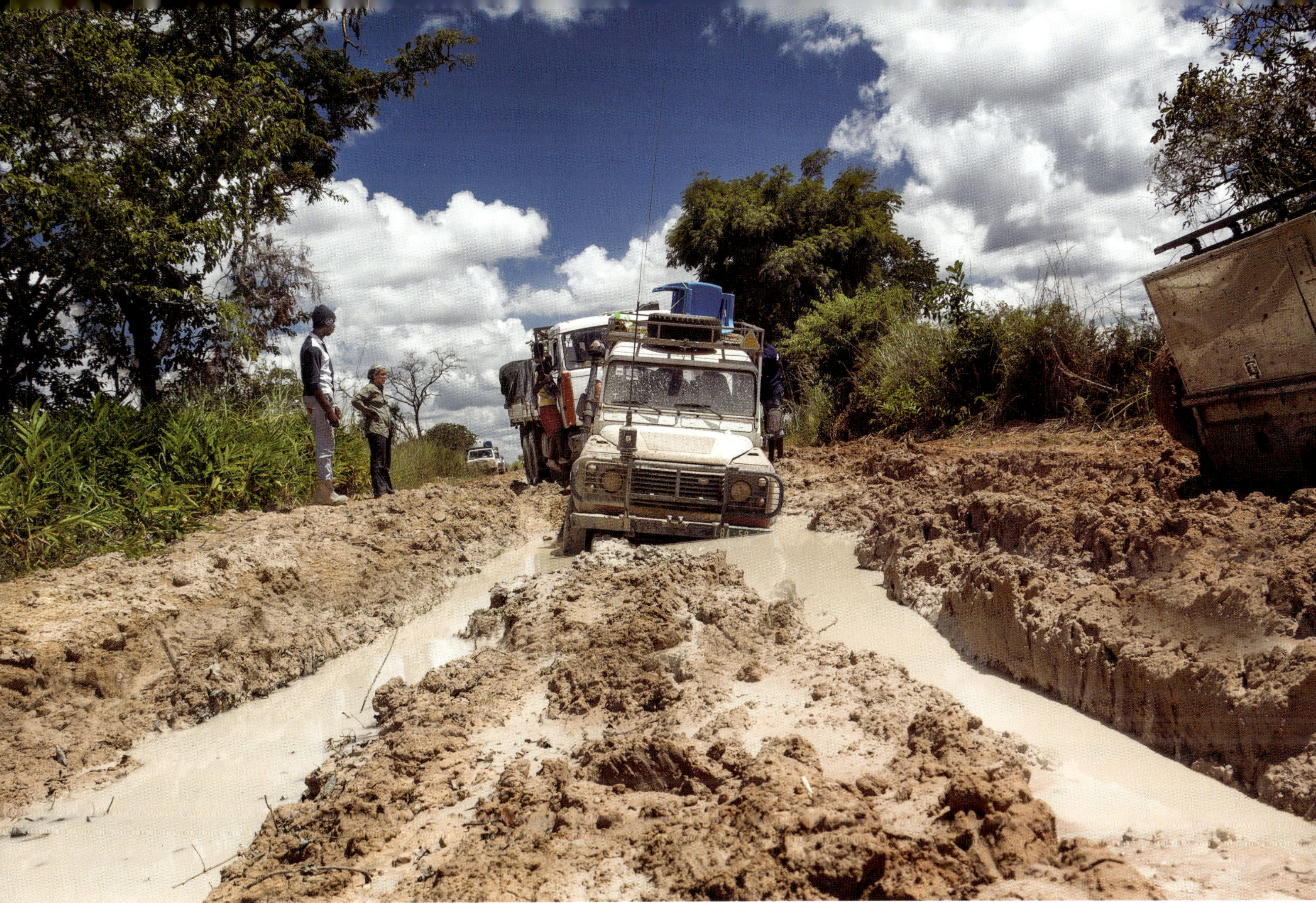

women, powerful in body and spirit: as remote and unbreakable as the wilderness they inhabited.

A day later, we stopped to portage around the waterfall, and a small group of us left to visit the village. We had no idea what to expect. When we arrived unannounced and uninvited to Soba Nhangu's village (*soba* meaning chief or leader), residents disappeared into their huts before gathering in the middle of the village dressed in old suits and dresses from the 1970s. It was like going back in time.

The residents, who called themselves Luchazi, told us in Portuguese that, at some point during the civil war, they had destroyed a bridge and isolated themselves from the fighting. We were the first visitors from the outside world in 42 years.

I could see that Mr. Water, my Wayeyi mentor, felt conflicted: He couldn't decide whether we

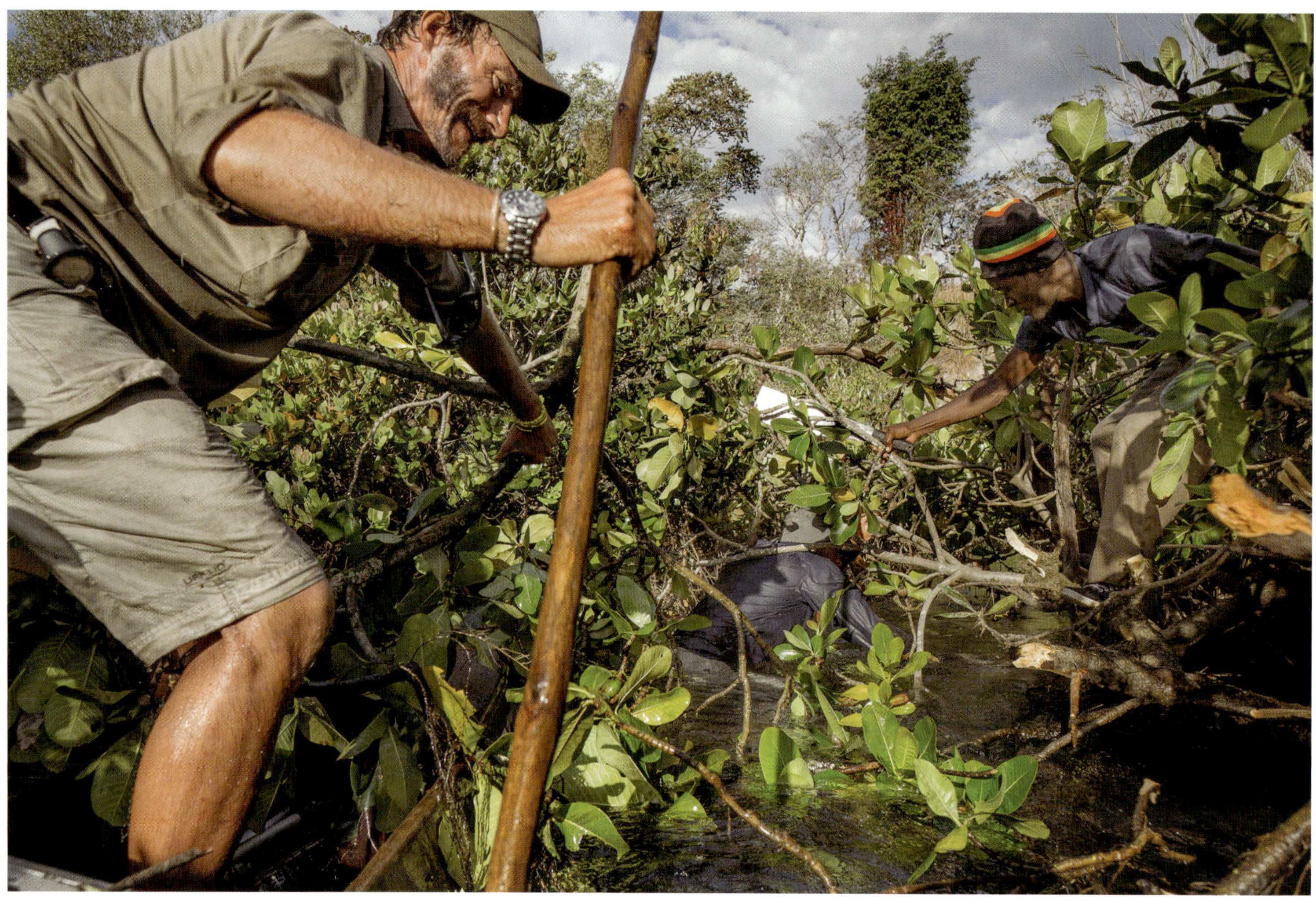

Below a waterfall on the Cuito River, the Okavango's eastern source, we encountered impenetrable tree blockages. It took a week to cut our way through. The upper reaches of this river had never before been navigated. As scientists, we were the first to try.

should help the Luchazi recover after four decades of war or leave them to live free. Their lives were hard, but this paradise at the end of the world had preserved their freedom for four decades. Being lost in this place showed me that when the whole world was wilderness, we were all free. We died young, lived hard, and believed in magic. This is the way of the wild.

SHORTLY AFTER WE LEFT the Luchazi, Paul Skelton, 72 years old at the time, dislocated his shoulder while setting a fish trap in a submerged tree. It was the final straw in a string of mishaps. We had capsized a mokoro that afternoon and lost all our plates and spoons. We were a month over schedule and had already run out of most of our food rations.

I requested a resupply and evacuation for Paul using police helicopters. Project director John Hilton, doing expedition support from Menongue, was preparing the food resupply from local markets and arranging emergency medical evacuation when the governor of Cuando Cubango, almost 500 kilometers (310 mi) away, heard about the accident. The governor offered a resupply from his palace pantry. When the helicopter arrived early the next morning with the governor's private medical team, it held rice, beans, tins of peas, and some vegetables, but more than half of the cargo was expensive red wines and cognac. We used the pea tins as plates and our bush knives as spoons and gave back the alcohol with a letter of thanks, to the dismay of some team members.

A few days later, there was a cartoon in the national newspaper, *Jornal de Angola*, of a professor sitting on a toilet in a private jet with a broken arm. The article told the story: how Paul was evacuated in a police helicopter, and then flown to Luanda in the toilet cubicle of a Learjet before being flown to South Africa for medical treatment. The governor's entourage was larger than expected, and the toilet was the only place Paul could fit. This was one of many times it became clear to us that the Angolan government was incredibly excited to have a National Geographic expedition in their country. Over the coming months, every time a helicopter arrived for police checks or resupply, they brought journalists from the newspaper and national broadcasters. Lost in a remote wilderness, we were famous in Angola.

In our planning, we had estimated that it would take us 19 days to travel on the water from the source to the bridge at the confluence of the Cuito and Cuanavale Rivers, where our team could resupply us. It ended up taking six weeks. When we finally got to Cuíto Cuanavale, the first small town, the relief was palpable. We were nearly starving by then and could finally restock our food and medical supplies, as well as pick up a new satellite unit, plates and spoons, batteries, and everything that had been broken in the tree blockages or drowned and lost in the river. We added an additional mokoro to help carry it all and an additional poler: my brother, Chris, who joined us for the rest of the journey.

We were met in Cuíto Cuanavale by the governor and vice-governor, as well as more than 20 high-ranking generals from the Angolan armed forces. They were there to show us the site of the infamous Battle of Cuíto Cuanavale, Africa's largest tank battle since the Second World War. Fought from late 1987 into early 1988, the fighting included armed forces from Cuba, South Africa, and the Soviet Union, making it a significant Cold War confrontation in Africa. Each side claimed victory for their own strategic reasons. An undisclosed number of Angolan infantry died, perhaps as many as 15,000, and are remembered as war heroes at the impressive memorial constructed on the edge of the battlefield.

The trees parted to reveal a sparkling lake. We had found it: the source we had been seeking for so many months, the place no one had ever documented.

The battle remains a symbol of resistance and a turning point in the history of southern Africa, and it showcased the interconnectedness of regional conflicts and their global implications. It altered the course of the Angolan Civil War, contributed to the independence of Namibia, and helped to catalyze the dismantling of South Africa's apartheid regime.

We visited captured South African tanks that had been made accessible by the HALO

We had become the first conservation project to sign a partnership agreement with the Angolan government in more than four decades—a watershed moment.

Trust's decades of land mine clearance, and viewed classified maps of the massive defensive and offensive minefields the South African forces laid. I acted as translator, as all the text was in Afrikaans, an Africanized form of Dutch I had learned at school. The reality that these maps were the plans drawn up to kill thousands of people sunk in only later.

The governor then hosted us as honored guests at a lavish lunch near the battlefield, complete with white tablecloths, napkins, and silverware. The whiteness accentuated how filthy I was. My fingernails and skin were caked with grime. My clothes looked like I had been washing them once a week with my bath soap, which I had.

I was seated at the main table, with the most important generals and the minister. Feeling quite self-conscious, I went to the bathroom and again was confronted with untouchable whiteness. I ruined the bar of soap, and then succeeded in completely muddying the basin and white towels. I threw my socks in the bin, straightened my hair, and escaped when no one was looking. I actually looked worse when I came out, though none of our hosts seemed to care. They were fascinated by these scientists obsessed with the rivers they had fought so hard over. We laughed and drank wine that went straight to my head.

While in Cuíto Cuanavale, the governor asked me to give a lecture about the findings of the expedition so far. He gave me and Adjany rooms in his hotel to clean up. Our biodiversity experts provided information on what they had found, and Cory Richards and James Kydd gave me all their amazing photos. I didn't sleep that night and didn't wash until 6 a.m. The next morning, they drove us two hours from the hotel to the yet-to-be-opened Cuíto Cuanavale airport, the largest venue in this remote little town.

Standing in my expedition clothes, giving a three-hour translated lecture for at least 150 attendees in smart suits, I felt like a teenager raised by hyenas and brought into civilization for the first time. Adjany, demonstrating at the age of 25 the same bravery she had shown when asked to paddle the full length of a legendary river filled with hippos and crocodiles, ran the whole show.

We were then marched off with the minister of environment to sign a memorandum of understanding between the National Geographic Society and government of Angola. Approved by the president and Council of Ministers, we had become the first conservation project to sign a partnership agreement with the Angolan government in more than four decades; this was a watershed moment for conservation in the Okavango Basin, and one that really launched the National Geographic Okavango Wilderness Project. This conservation success was a first significant step to protect one of our planet's last wild places and the heritage—natural and cultural—that it shelters. We now had a clear path to continue our research expeditions and engage with local communities to establish community-driven systems of protection for the Source of Life. Yet we had a long way to go to reach the Okavango Delta. The big, winding Cuito River lay ahead.

Paramount Chief Kgosi Kgolo Tawana II Moremi

Family Heritage in Conservation

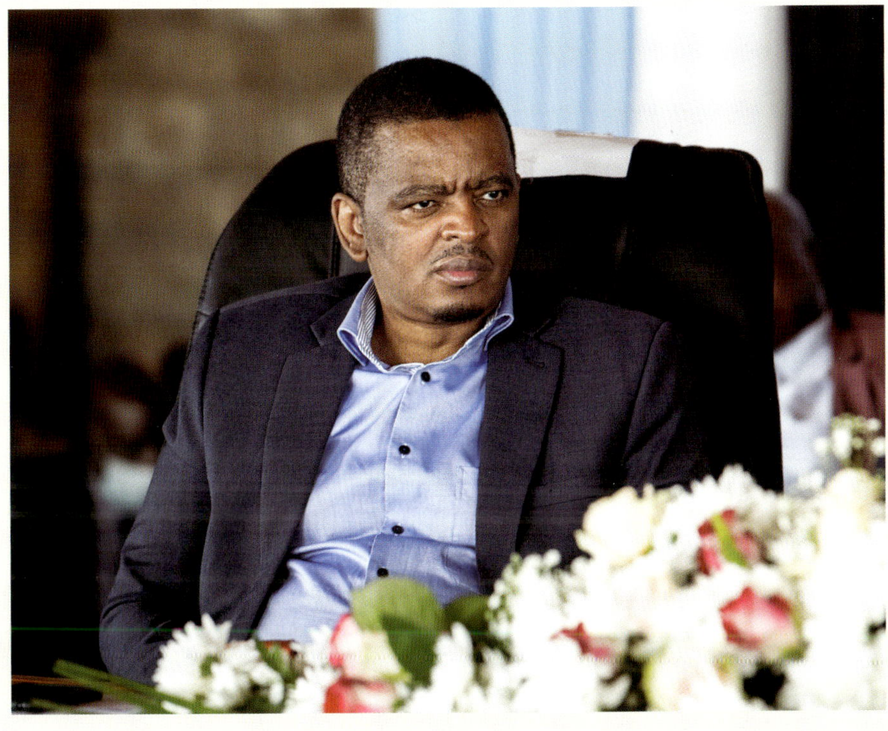

Paramount Chief Tawana II Moremi is the true custodian of Chief's Island, in the heart of the Okavango Delta, giving a voice to local people living in Ngamiland.

KGOSI KGOLO TAWANA II is the son of Chief Letsholathebe II, descendant of the royal Moremi dynasty, who have governed the Okavango Delta for centuries. Deeply committed to preserving and promoting the cultural and environmental heritage of his people, he roots his leadership in a vision that balances traditional knowledge with modern governance.

Kgosi Kgolo embodies a conservation legacy. In 1963, his grandparents Paramount Chief Moremi III and Elizabeth Pulane Moremi established the Moremi Game Reserve in the heart of the delta. By the 1970s, Chief Moremi's Royal Hunting Grounds became part of the reserve. Elizabeth, concerned about the devastating impacts of unregulated commercial hunting on native wildlife, established the Fauna Conservation Society of Ngamiland.

In 2013, driven by his desire to focus on advocating for the rights of Indigenous communities, Kgosi Kgolo resigned from his chieftaincy to pursue a political career and serve as a member of parliament. His leadership blends a deep respect for his people's customs with a forward-thinking approach to development, making him a key figure in Botswana's sociopolitical landscape. Kgosi Kgolo's sister, Kealetile Moremi, served as queen regent of the Batawana people, the Tswana-speaking peoples of the Okavango Delta. In 2022, she stepped down for her brother's re-enthronement as chief.

Mindful of his family legacy, and to champion the harmonious coexistence of people and nature in the Okavango Delta, Kgosi Kgolo Tawana II Moremi has recently reestablished his grandmother's conservation society, now called the Fauna Conservation Trust of Ngamiland. Each action he takes affirms his understanding that the sustainable development of the Okavango Delta must be driven by pride of ownership, community, and a shared purpose.

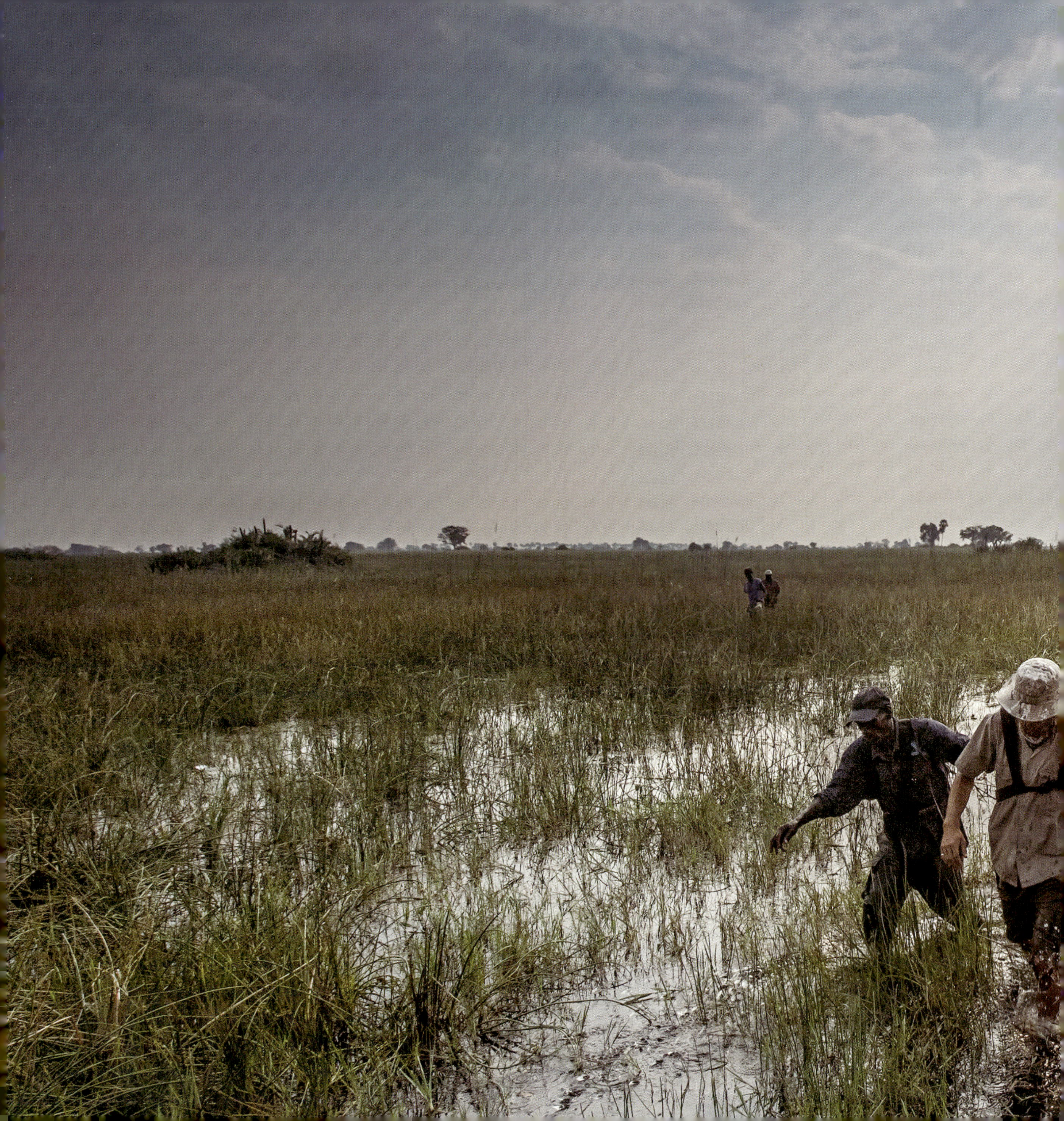

CHAPTER 3

Through Doubt and Fire

CUÍTO CUANAVALE TO THE OKAVANGO DELTA, 2015

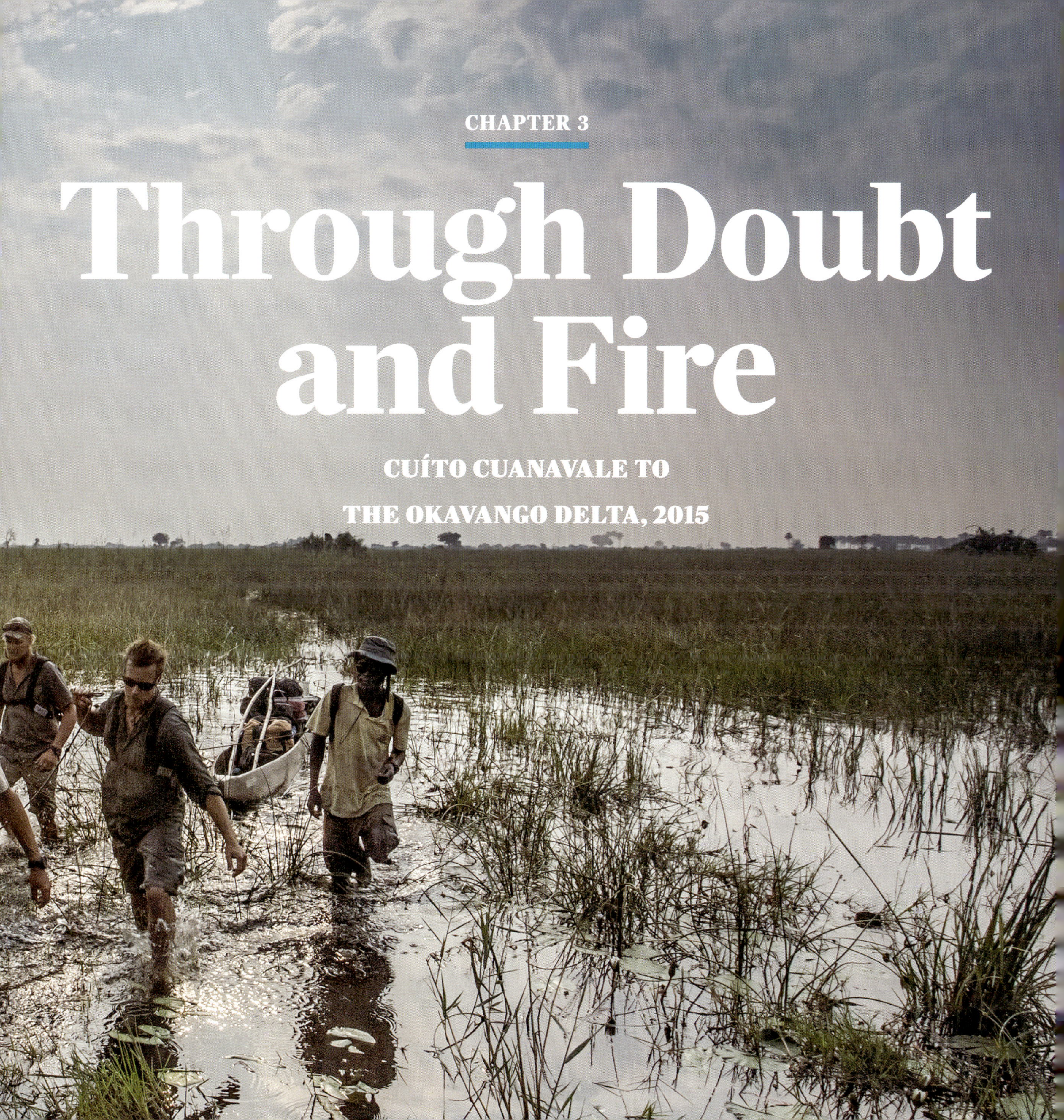

For the next two weeks, as we traveled downstream, wildfires raged around us on the banks. Day after day, the skies were darkened by thick clouds of smoke as we passed burning riverbanks and floodplains. Strange whirlwinds whipped down the valleys, picking up ash and turning black. They terrified the Wayeyi, who believed these were bad spirits that could steal your soul if the whirlwind hit you. The river became sooty and the water from it tasted like smoke. At night we camped in ashes.

Though the smoldering riverbanks and burned forest felt, at times, postapocalyptic, I remembered what the local people had told us about the role fire played in this landscape. At his village, Soba Nhangu had told us that the Luchazi light fires to hunt, clear thick grass, and optimize forest health. Old trees that had survived many fires burned for months in their heartwood, but never died. This had clearly been happening for a very long time. These forests and grasslands were fire-adapted, and human-set fires were part of the ecology.

As the Cuito River dropped out of the highlands, we started to feel that we weren't alone. We were descending into vast open floodplains and reedbeds, with signs of hippos everywhere; it looked as if they had moved downstream just a few weeks before, perhaps because of the fires. We felt an energy around us that made us feel we were being watched. The forest was squeaking and shifting, murmuring and growling; shadows disappeared behind trees. Each day, as we paddled farther, we got fitter, and the river's flow grew stronger.

The river was growing into a large, meandering snake coiled upon itself, strong and unpredictable. At times, we would travel four kilometers (2.5 mi) in the wrong direction as the river wound backward. We would plan to camp at a forested landing we could see in the distance, thinking we would be there soon, only to get there hours later, near sunset. We saw more birdlife and monkeys; when we stopped, we found tracks of roan antelope, reedbuck, porcupines, and wildcats. A week before reaching the confluence with the Cuanavale River, we camped in a beautiful valley with tracks and signs of wildlife everywhere—and for the first time since arriving at Soba Nhangu's village, we found human footprints as well as shotgun shells.

That afternoon, I went for a walk up to the highest ridge. I found a well-used motorbike track and, on it, lion tracks: A pride had walked this trail just a day or two before me. I followed them, the tracks fresh enough for me to be wary. A massive bushfire was raging downstream, billowing smoke that blocked out the sun.

I found old leopard tracks mingling with the lions' fresh prints. At one point, the leaf litter nearby began to crackle and shift. I held my breath, and a flock of strange-looking crested guineafowl—bizarre ground birds that look like a child's drawing of a chicken—appeared around me. The wildness of this valley in the twilight made me feel alive—and terrified me.

By 9 p.m., the fire was towering, flames crackling just across the river from us, nearly coming up to our mokoros and the science tent where the research team was uploading data.

Winter in the Angolan Highlands is fire season, when hunters, fishermen, and farmers light massive bushfires that blaze for days. For two weeks, we traveled through burned floodplains. **PAGES 78–79:** *By the time we got to the Okavango Delta, almost two months later than we had planned, it was the end of the flood season, so dry we had to portage our mokoros.*

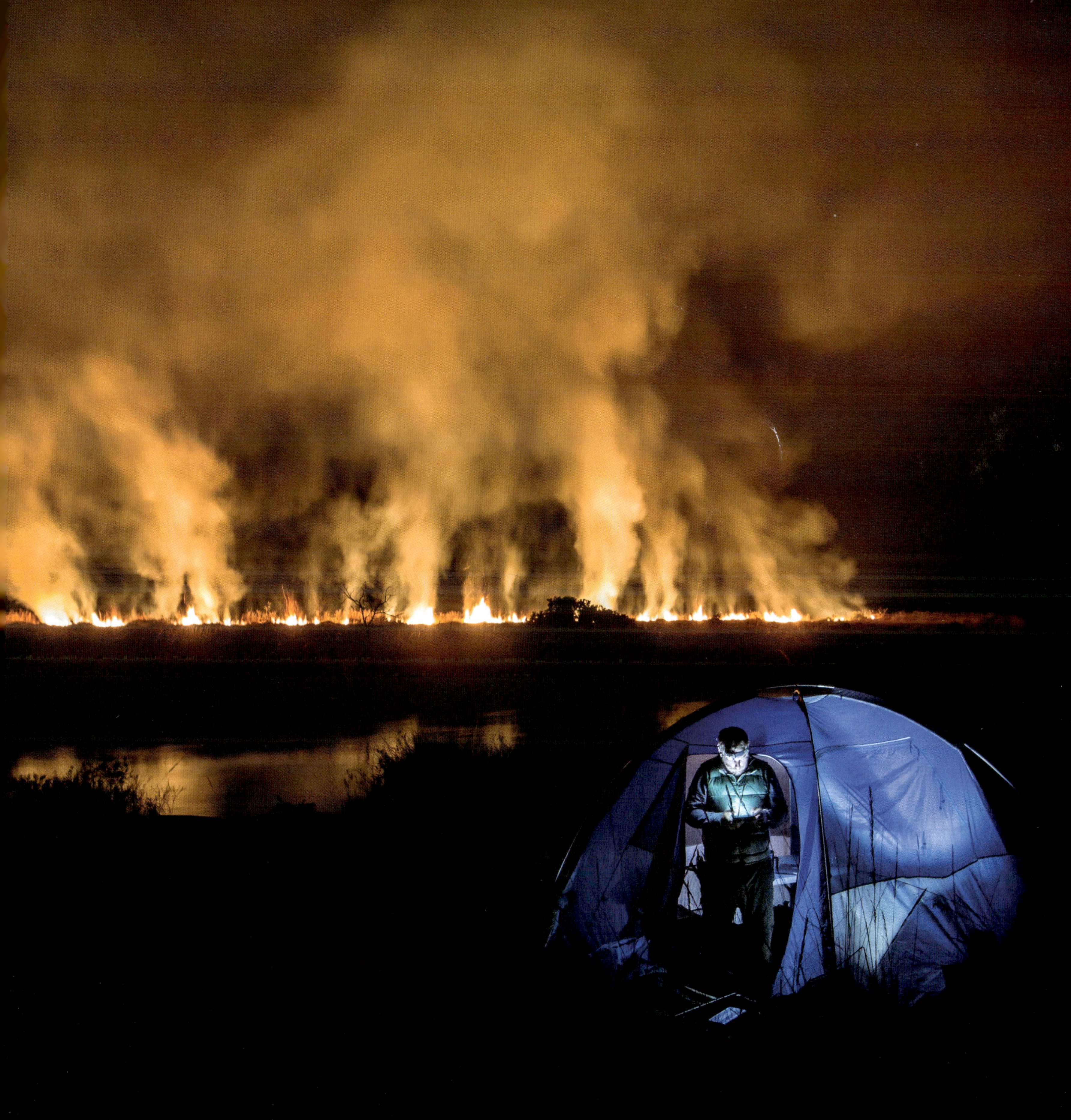

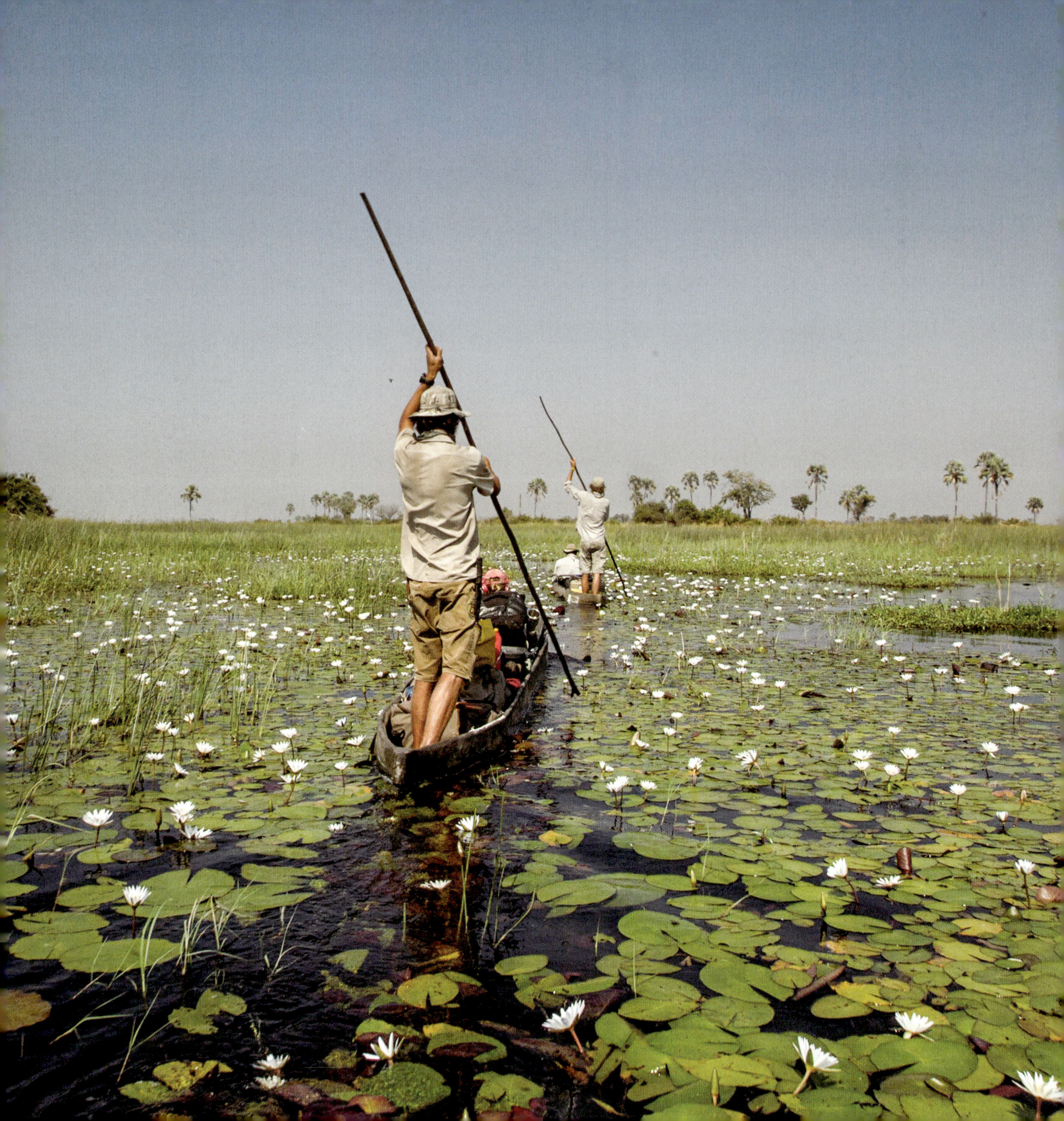

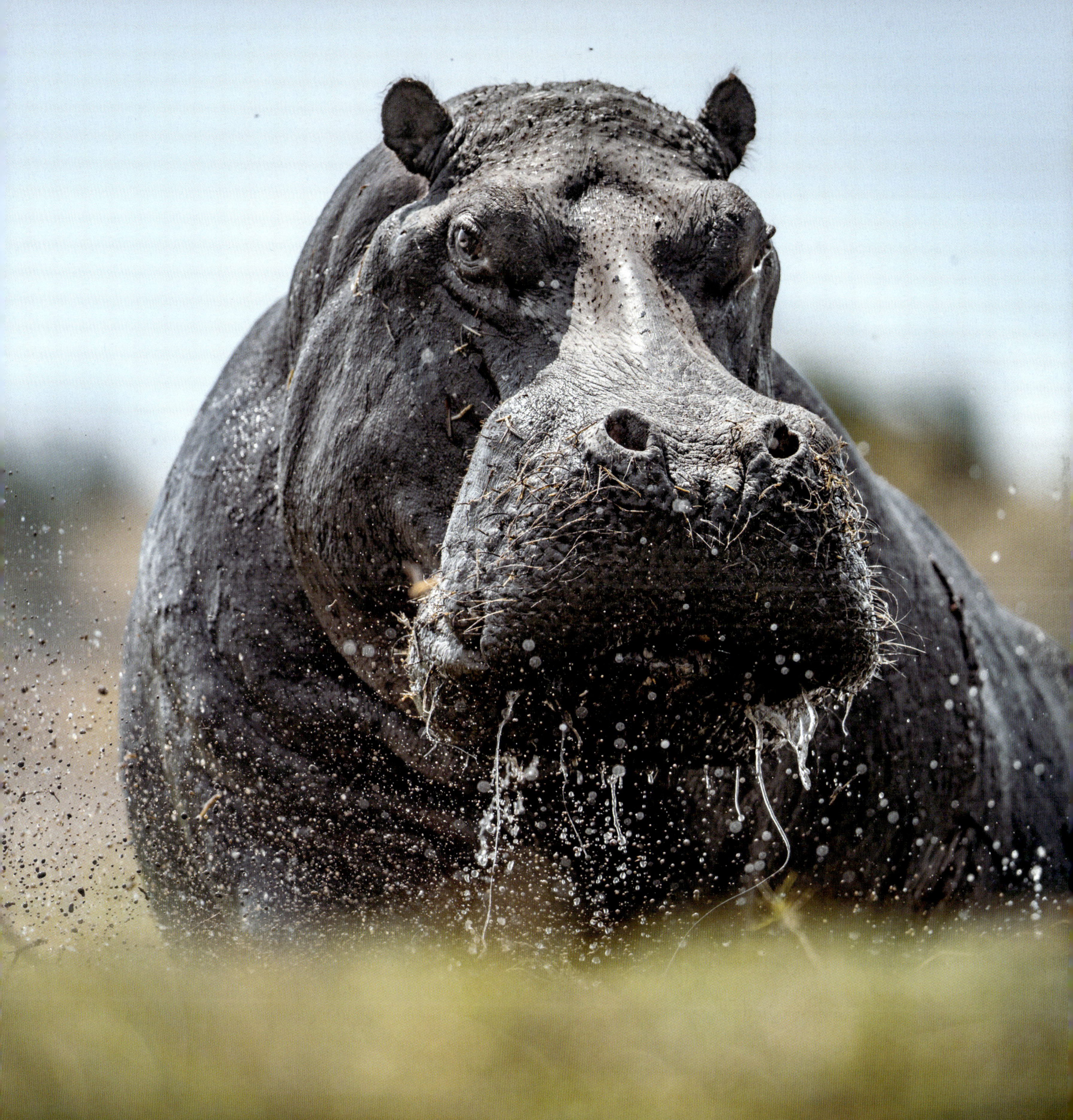

When you arrive on an island in the Okavango and hear or see baboons, you know it's big enough for lions, so you make a big campfire to keep them away. **PAGE 82:** *My brother, Chris (closest in this photo), and I have been poling across the Okavango Delta since 2010, when I completed my first crossing.* **PAGE 83:** *Aggressive bull hippos are a constant threat in this wetland wilderness.*

The heat was intense. All night, we could see fork-tailed drongos—keenly adaptable, insect-eating birds—hawking for insects displaced by the flames. Nightjars, smaller nocturnal birds, flitted in and out between them like angels. We stayed up late that night to watch whether the fire would jump across the river into camp. The wind was on our side, slowly blowing into the fire, then shifting briefly to waft smoke into camp.

Around midnight, I heard motorbikes coming along the track. Obsessed with meeting the elusive local people, I ran up to them in my underpants. Three hunters were carrying goods—batteries, bullets, whiskey, maize meal, and chickens—to remote hunting camps. One of the riders had 14 chickens stuffed into his massive jacket, a cord tied at the bottom to keep them in. We could communicate only with hand gestures, and they indicated they were going downstream. I gave them my business card, which probably seemed a bit odd. We said our goodbyes in the darkness of a moonless night. My flashlight ran out, so I walked back slowly to my tent by the light of the fire still raging across the river.

THE REST OF THE CUITO RIVER, traversing the southeastern corner of Angola, was a big, winding river lost in wide open floodplains and vast miombo woodlands to either side. We saw many hunting camps yet met hunters in very few of them. Extraordinary sandy cliffs, up to 50 meters (164 ft) high, arched around long shady corners. Each night, I would hike up with my gear to camp on these cliffs, with amazing views of the river and the night skies. It was an extraordinary six weeks, and we were paddling up to 60 kilometers (37 mi) a day. We had truly become an expedition, and everyone had fallen into a rhythm.

Until one morning when we departed early in a rush. It was the first time that I, as the expedition leader, felt we could push hard and dominate the river, travel farther, move faster. Now we could start making up time for the month we were behind schedule. Looking back, I realize that decision was reckless.

We knew the river, the landings, and the fires, but we hadn't seen any hippos—until that fateful day. I heard Chris shout from behind me: "Reeds are moving ... Reeds are moving!" I could see from the boat something big entering the water in front of us on the right. We had seen some very big crocodiles sunning themselves on the riverbank, so my first thought was that this disturbance in the reeds was one of them. But the movement was bigger, without the characteristic scampering. Something wasn't right. I stopped paddling, allowing us to drift so I could look for the animal in the clear water below us. I saw the sandy trail of something much bigger, with no distinctive tail drag.

"Dirty water!" I shouted—which, in the Okavango Delta, signals hippos entering the water. I started to think I had made a terrible mistake and gestured to the expedition team to pull over to the left bank. As I was doing this, the water in front of us took a breath, as the Wayeyi would say. It lifted. Something big was rising under the surface. I shouted, *"Kubu!* Hippo!" and the next thing I saw was the face of a hippo, jaws open, hurtling from the murky depths toward my left foot.

Time stopped as the hippo smashed into the mokoro's hull, throwing Giles Trevethick, the data recorder, and me several feet into the air and into the water. In the chaos, I remember hauling myself onto the overturned hull, Giles somewhere beside me. Then I heard my brother shout, "Swim! Swim!" He could see the hippo and knew we could make it to the bank. It was about 30 meters (100 ft), but it felt like 100. When Giles touched my leg on the way there, I thought it was the hippo in pursuit and nearly walked on water for the last few meters.

Chris managed to save the mokoro from sinking. There were two big puncture holes in the hull from the hippo's tusks. Giles and I hugged on the riverbank, and James Kydd, the expedition photographer, held us. We had been very lucky.

The Cuito River taught us many lessons. One that we learned that day is to be humble and stay present at all times. There's no room for ego or ambition on wild rivers guarded by hippos and crocodiles: just respect and a resolute plodding attitude to progress. Within 30 minutes of the accident, the team had made a small fire with reeds and twigs to boil water for sweet tea meant to calm everyone's nerves, while Pete, the expedition cook, repaired the broken hull with fiberglass and resin.

I wanted to talk to my wife, Kirsten, about the accident, so I used a satellite phone to call her. The phone seemed to ring forever. I had stayed fairly stoic up until that point, but as I heard her voice crackling through the connection, I collapsed to my knees, crying, nearly

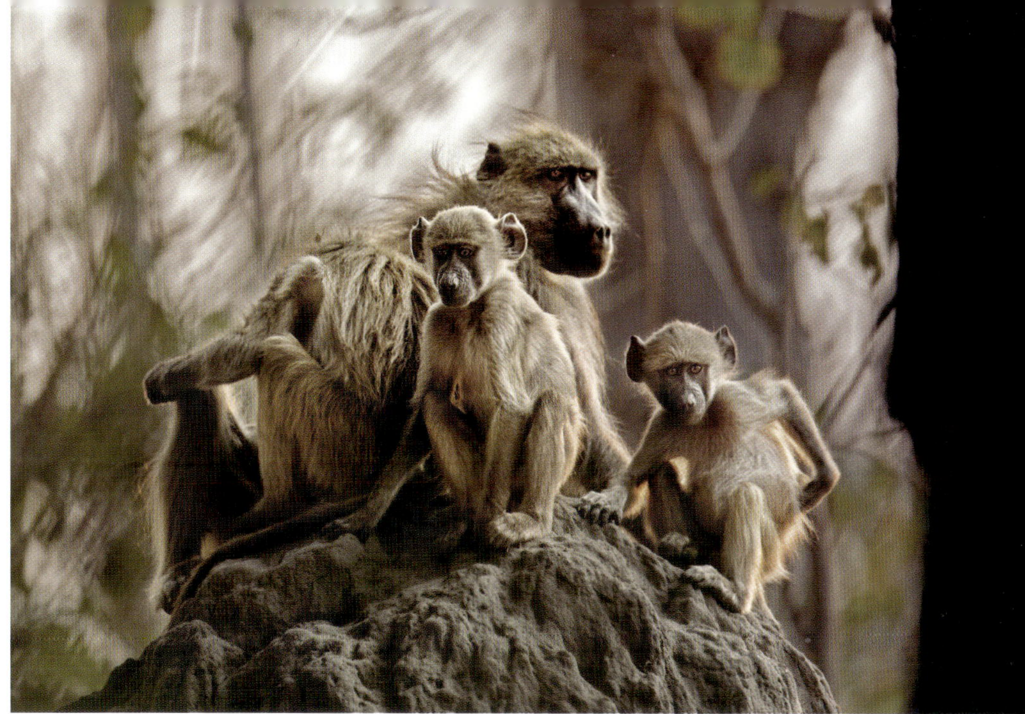

hyperventilating. She couldn't make out what I was saying through the sobbing. The reality of what had happened had dawned on me. The thought of never again seeing Kirst and my son, Jack, then two years old, was terrifying.

After two hours, we were back on the water. We had lost only the solar batteries from my mokoro and Giles's shoes.

We would see hippos most days from that point onward, but most were scared of us. We learned to stick to the shallow water. Even so, we had many more close calls. Around blind corners, Giles would shout, "*Bom dia!*—Good morning!" and tap the hull with his paddle. This alerted resting or sleeping hippos, but it didn't work on the older ones. I will never forget the many times I locked eyes with one: Lilies draped over an eye, half asleep, it would process what I was for a moment and then act with fury and fire. Though hippos can't swim, they are so strong that water vaporizes around them as they storm toward you.

The water in front of us took a breath, as the Wayeyi would say. The next thing I saw was a hippo, jaws open, hurtling toward my left foot.

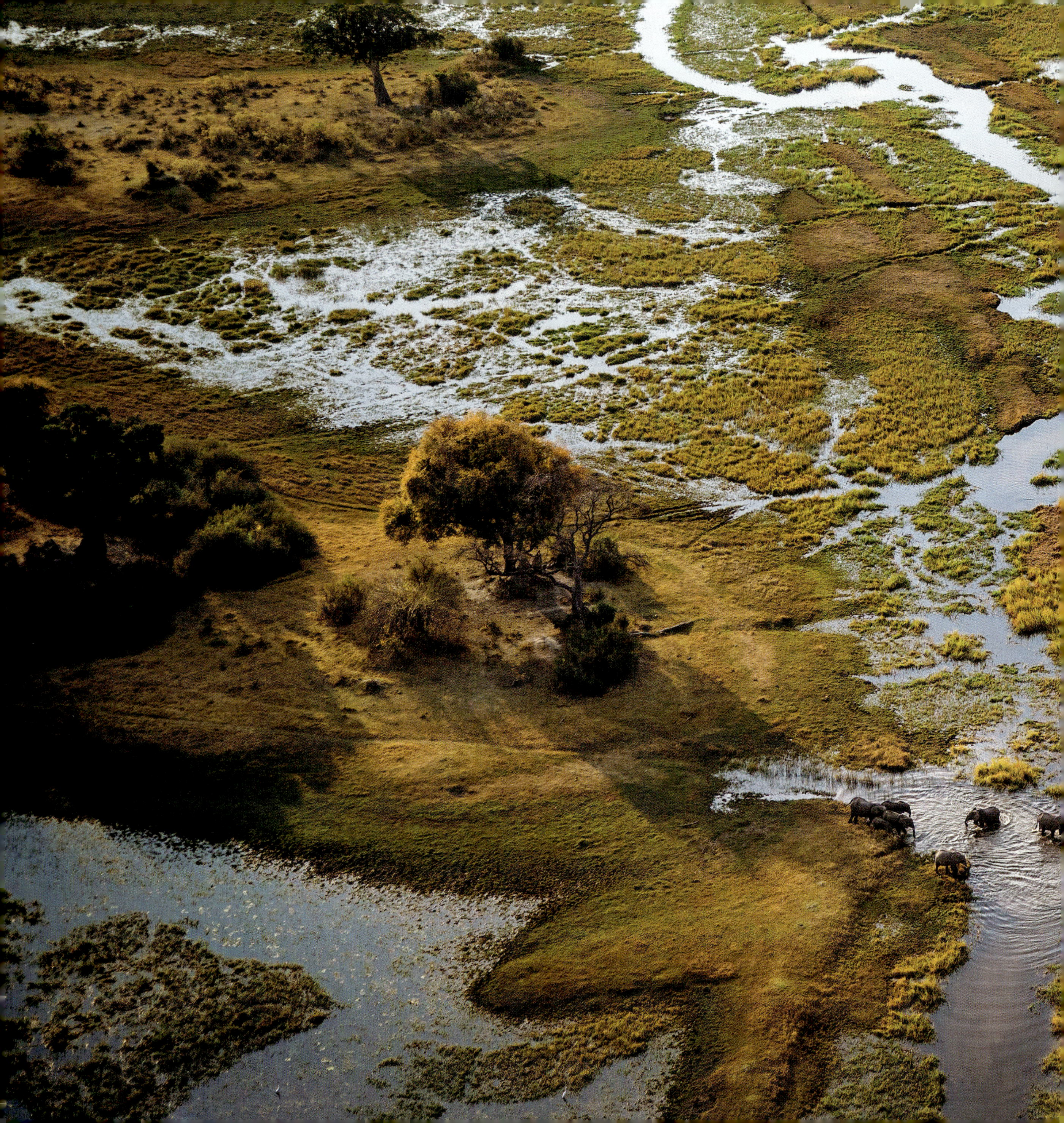

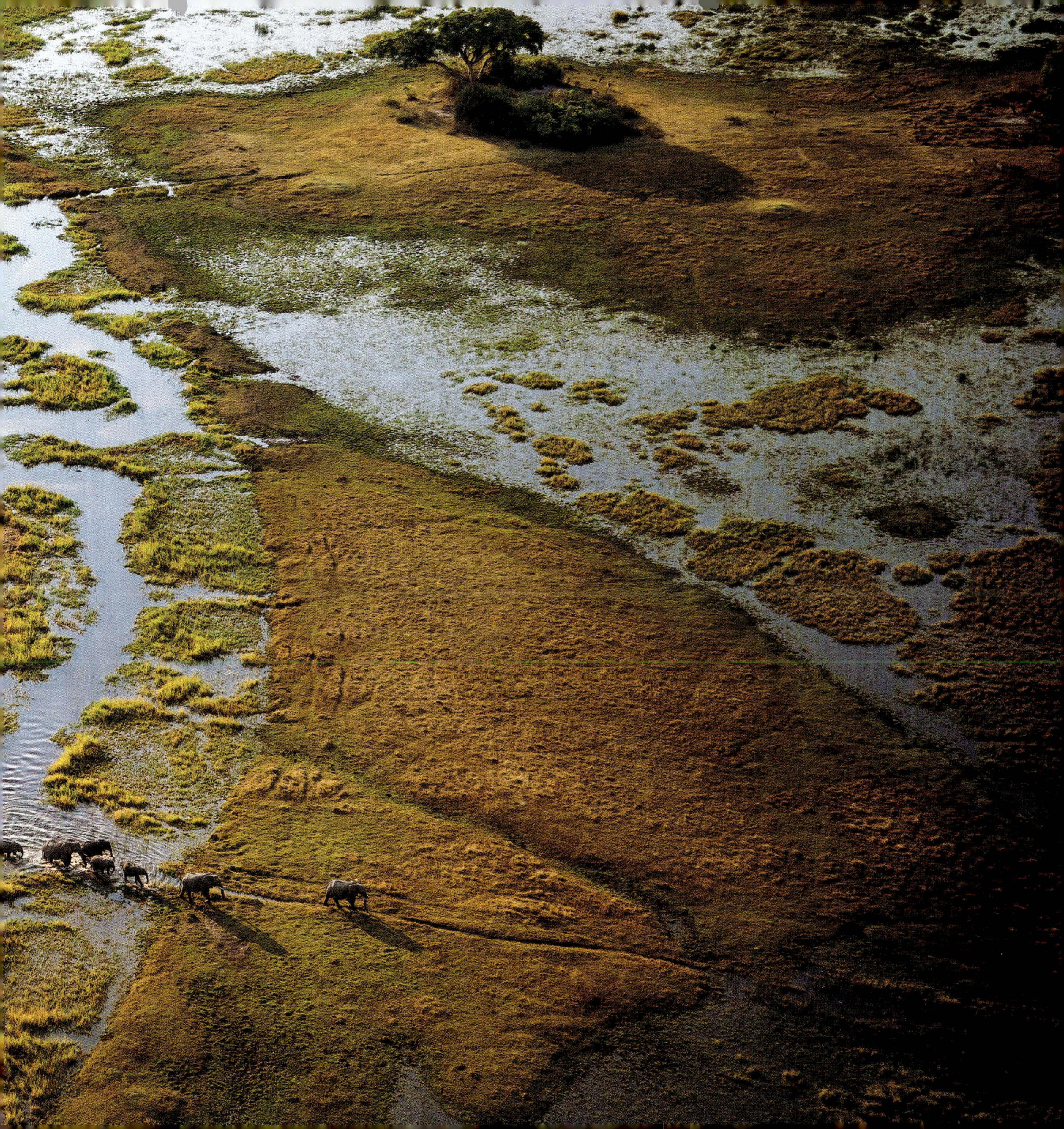

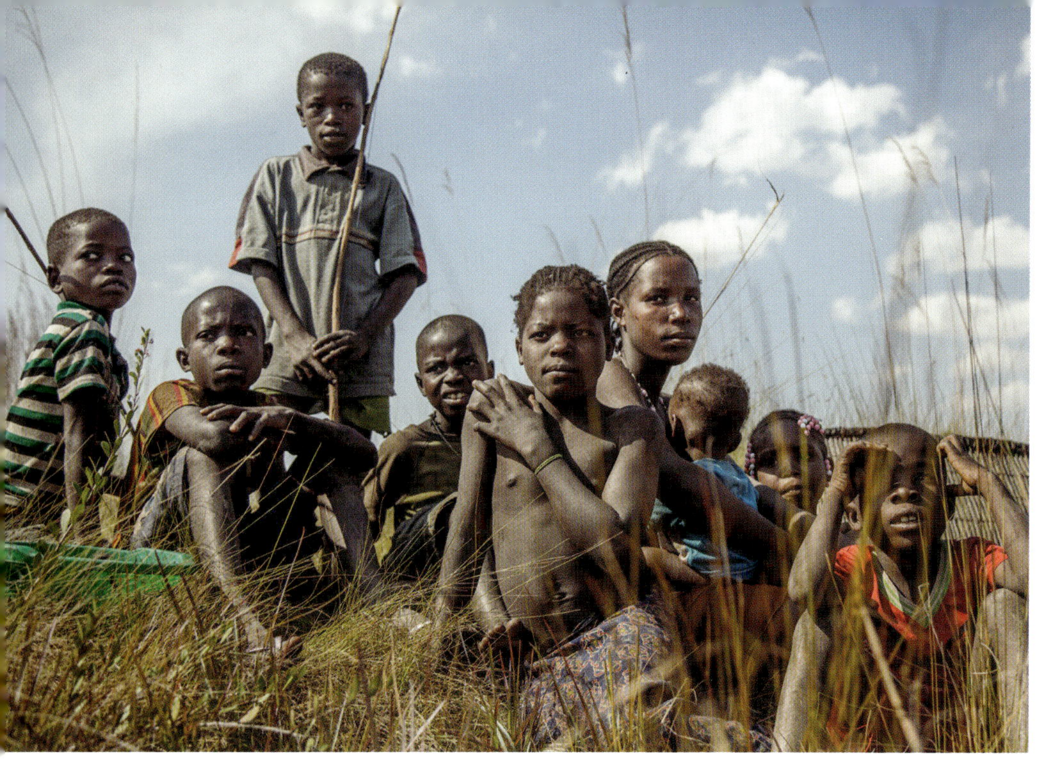

I have never met a person living along a river in rural Africa who isn't welcoming, generous, and friendly. Water has a strange way of calming us, uniting us, and making us whole again. **PAGES 86–87:** *The Okavango Delta is home to the largest population of elephants on Earth, but Wayeyi elders say that there were hardly any elephants in the delta 30 years ago. Where did they all come from?*

THE WILD CUITO RIVER seemed more and more traumatized by war the closer we got to the Namibian border. We found blown-up bridges and sunken supply trucks and tanks in the river. Every small town we passed had land mines and stories of famous battles. This part of Angola is incredibly remote, with limited law enforcement. At one point, we found a family fishing with more than 40 monofilament gill nets and some 1,500 large fish spread over several small sandbars. This wasn't sustainable, even by the standards of this mighty river.

We were desperate to find elephants returning to this part of Angola, which was once famous for legendary hunting safaris. Though many believed they had been hunted or driven out by war, we had seen signs of elephants upstream, and the villagers along the river told us about elephants visiting during summer months.

Two days before we crossed into Namibia, we heard it: the rumblings of elephants. I stood up in my mokoro and saw them, about 60 males on a walkabout in southeastern Angola, using the Cuito River as a corridor into the fertile floodplains we had just traversed. They were clearly nervous, as ivory hunters still targeted them. They were in what should be an elephant paradise, though one with a painful past. Walking between impressive African rosewoods and Zambezi teaks, trees that smugglers also target, the elephants were on a mission to reclaim their homeland.

Adjany, a native Angolan, cried out, "There are elephants in Angola!" We marveled at how they must have survived in this war-torn landscape. Elephants can smell and avoid land mines. The oldest bulls lead the rest in search of food, remembering the war, helicopters, and AK-47s. One day, the displaced breeding herds, groups of related females, would follow.

When we crossed into Namibia, the beautiful clear water of the Cuito River mixed with the cloudy, silty waters of the Cubango, the delta's western tributary. We flowed downstream into the Andara Rapids, a little-known system of rocky rapids and forested islands that protected the sacred burial sites for the kings of the Hambukushu people.

We switched to inflatable rafts for the 50-kilometer (21 mi) overnight paddle, as our traditional mokoros would capsize in the turbulent rapids. This short section ended with a drop over Popa Falls, a series of five-meter (16 ft) waterfalls into the Okavango River below. It was almost like being in a different, more tropical part of Africa. We saw families of otters and

RIVER GUARDIANS

Children of Africa

An Unbreakable Spirit

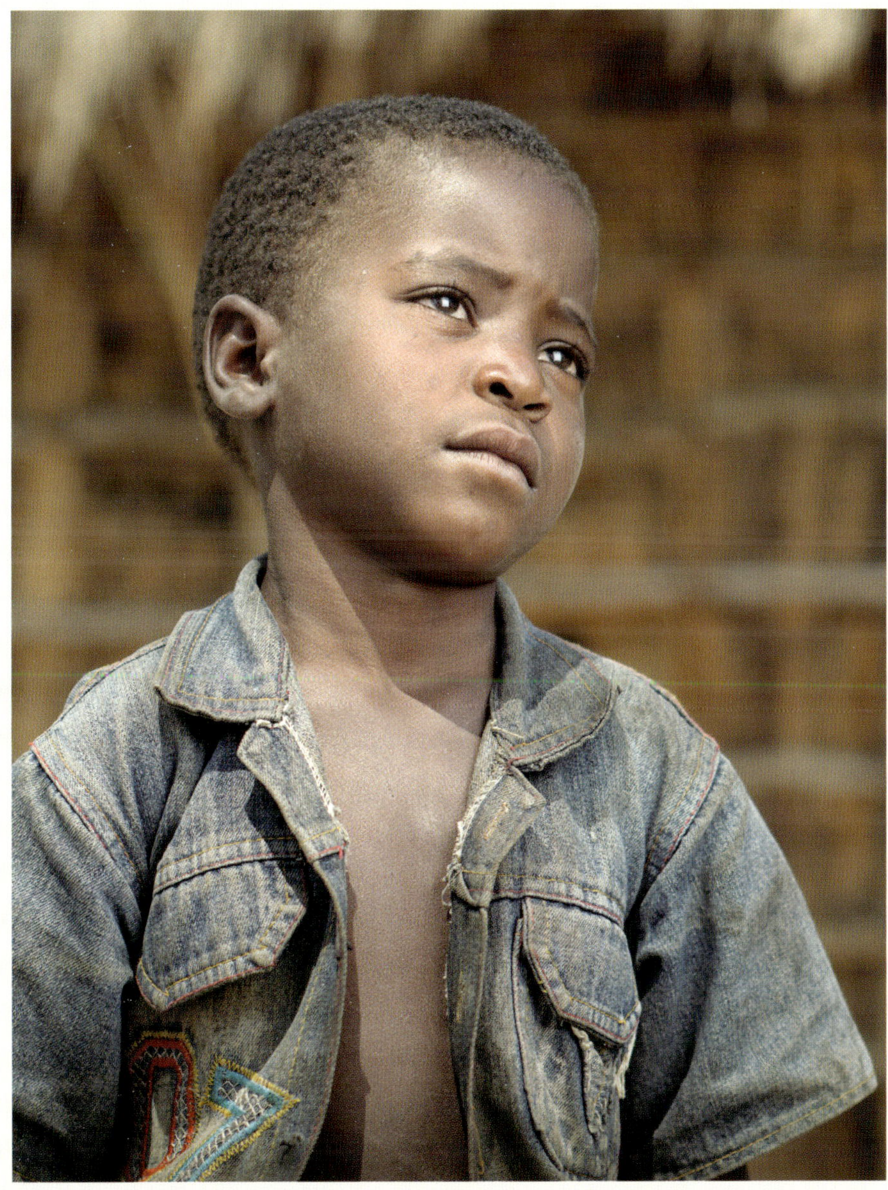

The biggest threat facing young children across Africa, and within the Okavango Basin, is malaria. Eradicating malaria for Africa's children is globally important.

Standing at the threshold of tomorrow, this Luchazi boy is now a young man grown up early, bridging between ancient wisdom and a digital future. His ancestors chose freedom over urban opportunity. His birthright is rooted in language, culture, and tradition. He is the custodian of human experience connected to wilderness, a force of nature, the backbone of an ecosystem. Born in fire and blood, he is the Lisima lya Mwono Landscape quenched by the life-giving waters of the source lakes.

Malaria kills a child every day, a specter that haunts his people day and night. He survived it. Now, at 15, education is a sacrifice far away from home and family. The future lies unwritten, as uncertain as the passage of the rivers that begin their journey in these highlands. Hope blossoms like a wildfire, as bright and defiant as his eyes. Whatever the future holds, he will never yield, and he won't forget.

The mistakes he will make, the triumphs he will relish, the risks he will take, and the love he will enjoy—all reflect a life steeped in profoundly rich tradition and challenged by modernity. I hope my own son, Jack Wild Boyes, gets to meet him one day. I hope they learn to know each other, each one helping the other to be a better version of himself.

Children have an unbreakable spirit that can never be re-created. You see it in the young girl who never liked going to school, the little boy who never fit in. Those whom society considers strange or other are often best nurtured in the wild.

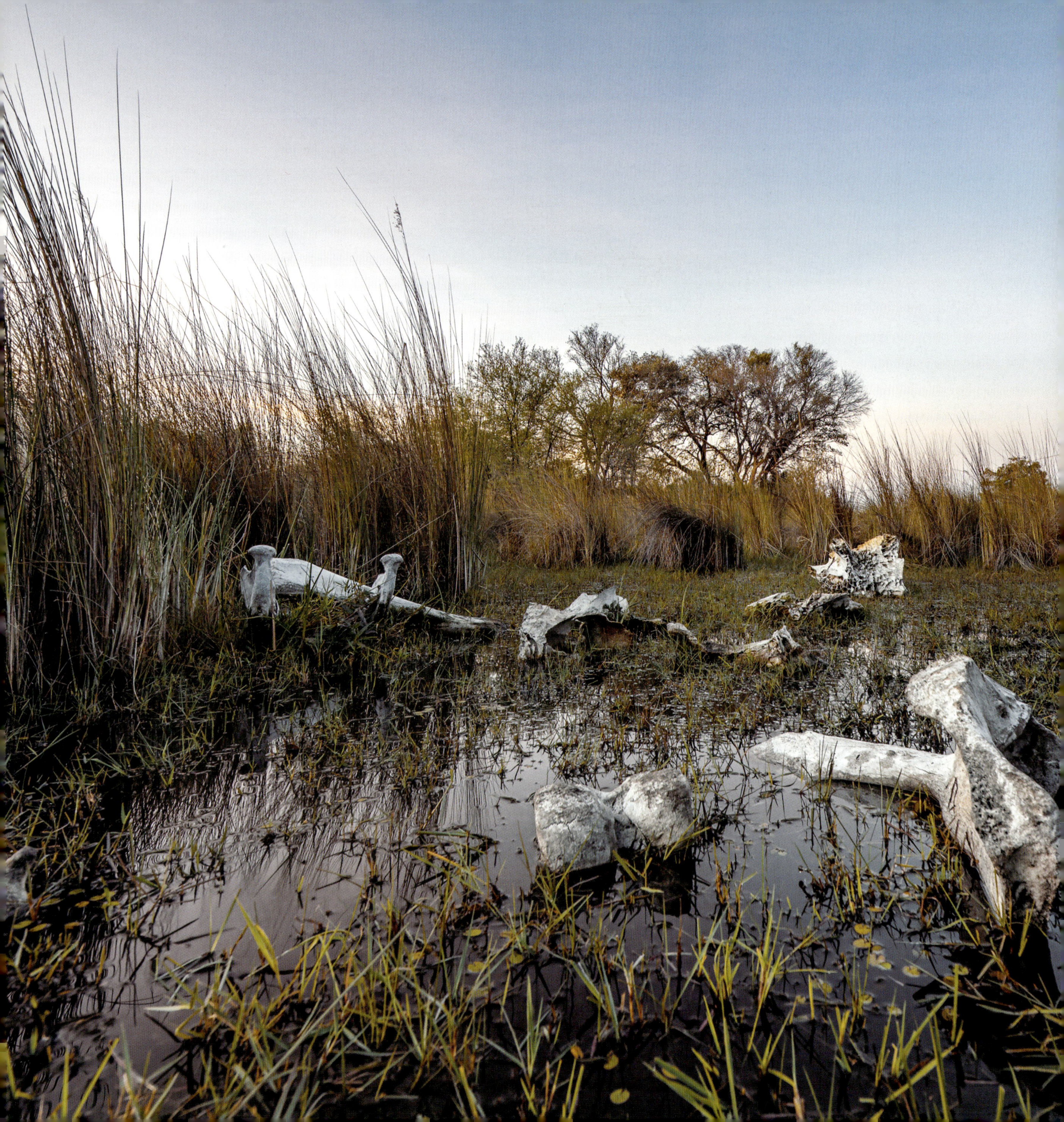

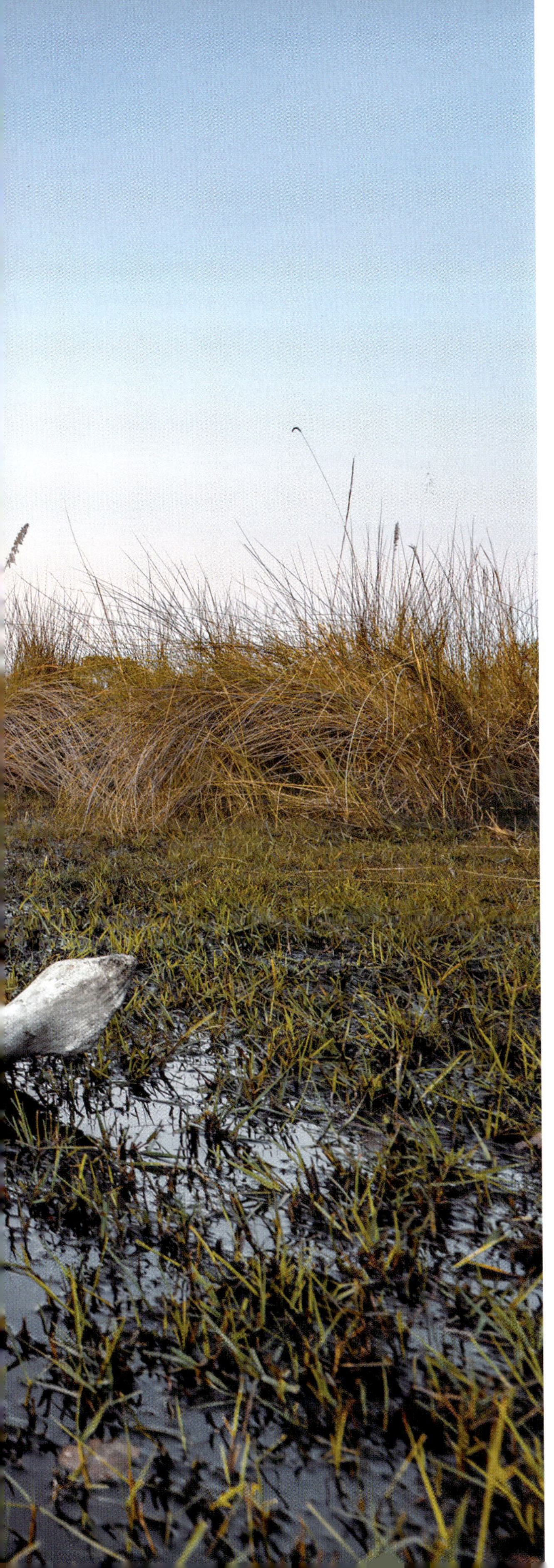

elephants bathing. We caught massive, meter-long (3 ft) tiger fish and explored islands covered in palms and wild ginger. Just before camping the first night, we heard voices raised in song. We paddled around a bend to find an old Catholic monastery, where nuns were singing as they took bedsheets off washing lines.

The number of animals increased as we moved downstream, back in our mokoros. As we passed into the Mahango Core Area of Bwabwata National Park, we saw kudu, lions, warthogs, and bushbuck. It seemed the animals knew they were safe here. Just before entering Botswana, we traveled the wildest section of river we had explored so far. We saw hundreds upon hundreds of hippos and massive crocodiles; we went very slowly, weaving our way between them. This area had the highest concentration of hippos in Africa; many had come here to escape the war in Angola. A local fisheries researcher came up to us in his motorboat and said that we were crazy to be moving among them. By then, we were experts at spotting and avoiding hippos.

That day, we had lunch with a breeding herd of more than 30 elephants swimming and playing around us. It was a beautiful respite from the stress of the previous four hours of dodging hippos and crocodiles the size of our mokoros.

We took a side channel, and Giles ran into the forest to look for a campsite. He returned quickly and whispered that he had seen three men burying a skinned elephant skull. I ran up the riverbank to meet them. They were San Kwe-Kwe skinners in the abattoir of an elephant hunting camp. Eight freshly skinned

You often find elephant graveyards in the Okavango Delta. The elephants often visit them, touching the bones and moving the skulls. Like us, they mourn their dead and remember their ancestors.

elephant skulls, including tusks, were buried nearby. Elephant meat was drying on wires everywhere. The soil was caked with blood. At least 30 elephant skulls lay strewn in the surrounding bush.

I sat down with the skinners, who could speak Afrikaans. The oldest among them, Kobus, told me about working as a tracker during the Angolan Civil War, hunting wildlife to feed the military. He said that during the war, "we were treated like men," but now, "we live like dogs." His people had lost their land, forced to live in villages near the main road. They became skinners for the luxury hunting camp on the ridge above us, overlooking the Okavango River, where hunters pay tens of thousands of dollars to legally shoot an elephant. Kobus tracked and skinned elephants, a job he hated, but he earned about $50 a month to buy maize meal for his family.

When he was a child, Kobus explained, they only hunted one elephant. It was a big event that took months of preparation. They used every little bit of the elephant, and all the clans were there to help.

"This is an elephant graveyard," he said of the clearing where we sat. "The elephants come here at night. You hear the skulls knocking against each other as they touch them. I lie here awake and say to them in my language, which they understand, 'I love you. I'm sorry. I don't want to be here.'"

WE PADDLED into Botswana, through the delta's so-called panhandle. Now we were on the Okavango River, which winds through vast

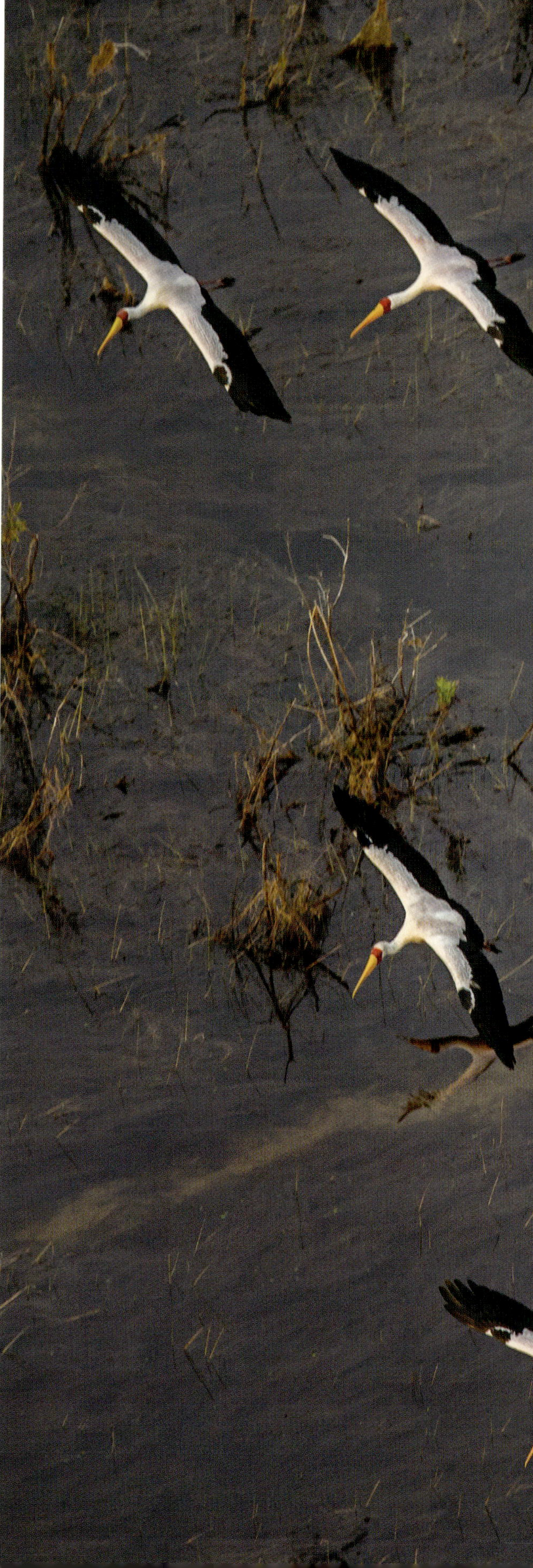

The National Geographic Okavango Wilderness Project started as the Okavango Wetland Bird Survey, using wetland birds, such as these yellow-billed storks, as indicator species for ecosystem health.

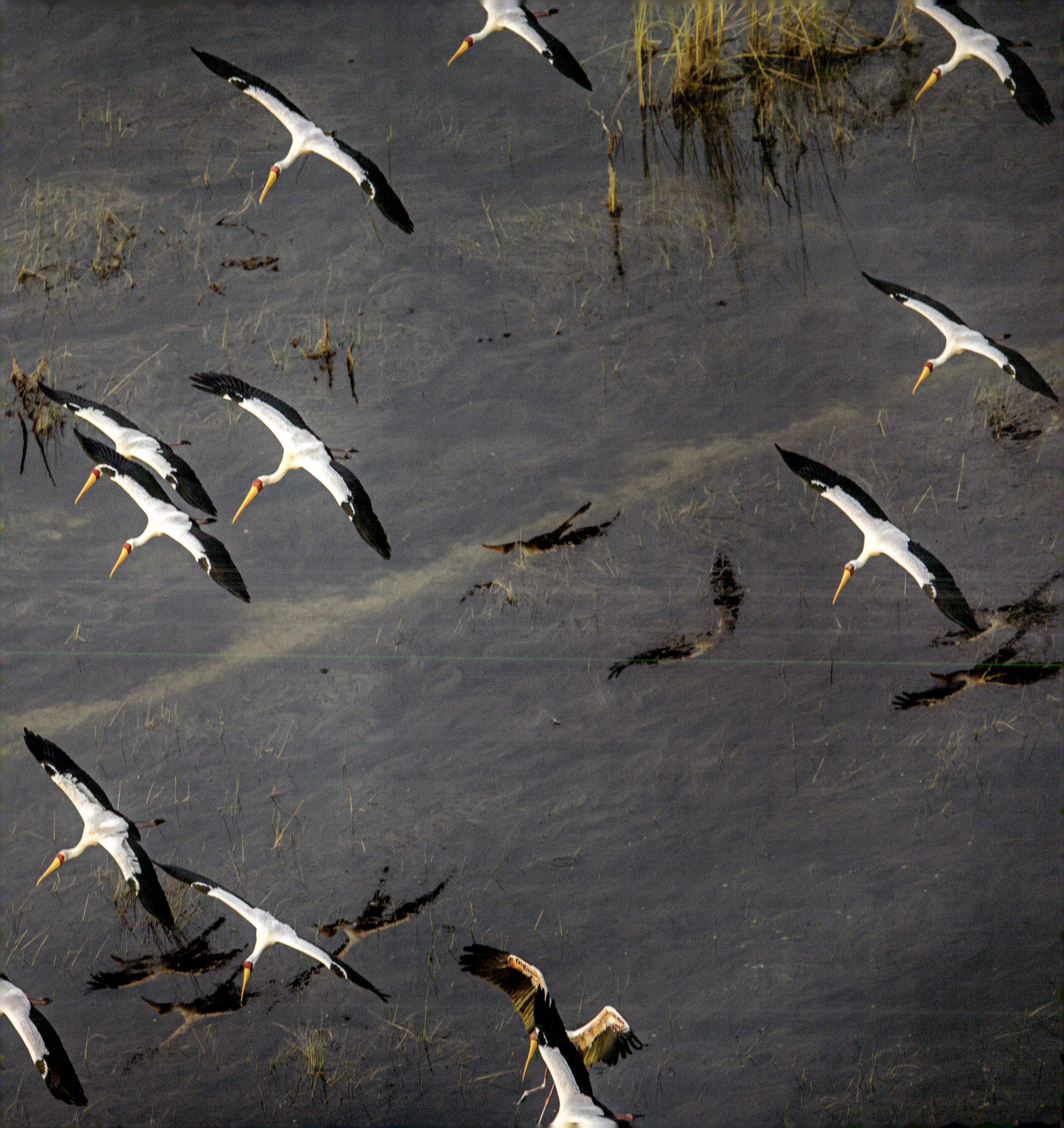

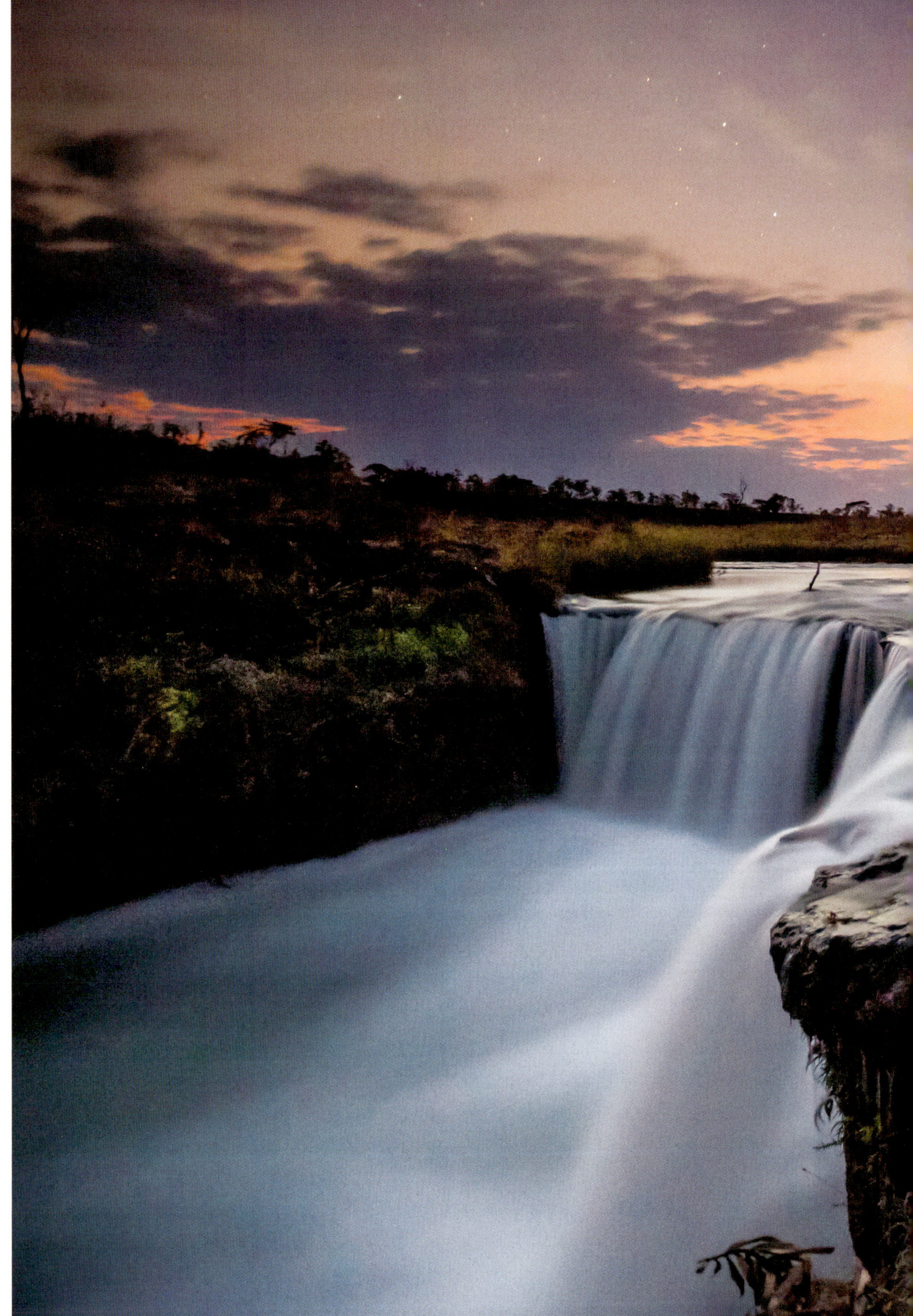

All the experts told us that waterfalls were geologically impossible on the Cuito River; hence, we were astonished to find this previously undocumented waterfall, which we had to portage around. We asked the local Luchazi what they called it, and they said simply, "The Waterfall." We thought that was wonderful.

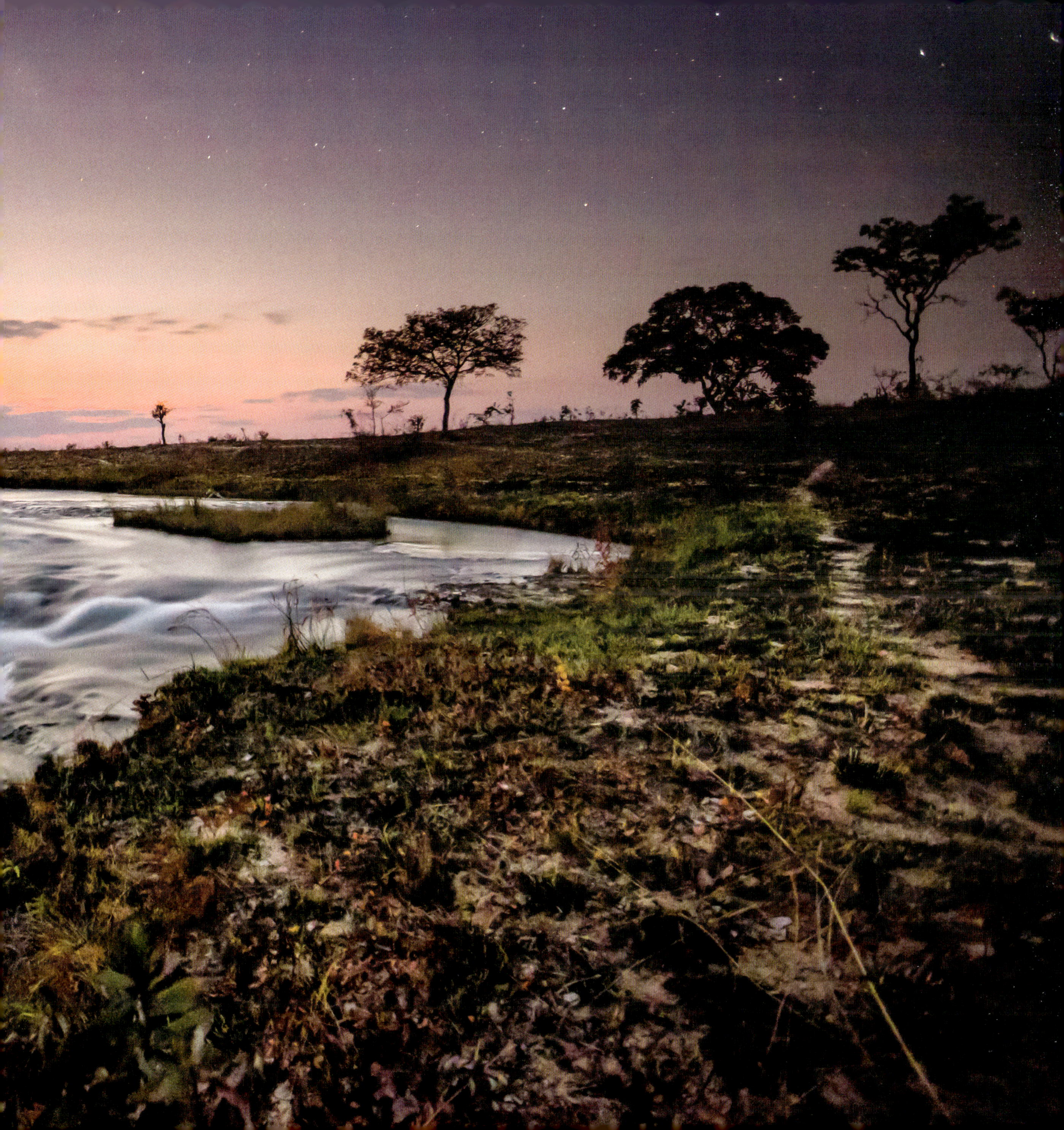

papyrus beds—a massive filter system purifying the floodwaters before they enter the alluvial fan. The location is famous for its "barbel run," an annual migration of sharptooth catfish that draws thousands of fish-eating birds to glut on the small fish chased out by the catfish. We passed houseboats and fishing lodges and met many local fishermen, bathers, and people washing their clothes.

A Tswana lady on the riverbank shouted out, "Where are you going?" We answered, "Maun"—the town on the far side of the delta.

"You are lying," she said, thinking it much too far for mokoros to reach. "Where have you come from?"

"From the source of the Okavango," we answered.

"Muhembo?" she replied, pointing upstream to the town on the border with Namibia. In the minds of most local people, Namibia was the source of the Okavango's waters; or, perhaps, the floodwaters fell out of the sky or bubbled up from the ground. Angola was a place too far to imagine, a place scrubbed from living memory.

We now knew the source lake was a giant sponge holding fresh water that made the delta far more resilient to climate change than we had assumed.

Now we had a new story, carried by our expedition team: GB, Tom, Snaps, and Mr. Water. They had met people along the river; they had seen the source; they had experienced every inch of the winding waters. Their stories would begin to connect the people at the sources of the Okavango with the people in the delta. They were ambassadors of a river.

After traveling 90 kilometers (56 mi) farther, past houseboats, lodges, and farmers, we stopped at the entrance of the Okavango Delta, paradise at the end of the river—the destination we had been dreaming about for weeks.

In a way, we were all home. Seronga, the village here on the edge of the delta, was where I had met GB and launched the first mokoro expedition in 2010. It was also where Tom, KG, and Snaps lived, and where Mr. Water had many family members. They all got to be with their families, and I got to see friends from Vundumtiki. We slaughtered a goat and celebrated.

Everyone in the village was invited. More than 250 people, singing and dancing, shared their joy. Gifts of pumpkins, corn, and fermented millet arrived. GB, Tom, Snaps, Mr. Water, and KG were all famous now. They told our many stories. Fittingly, many were about hippos—about the strange hippos of Angola, the decapitated hippo we found in the Cuito River, thousands of hippos on the Namibian border. The elders gave us advice and reflected on what was ahead.

Upstream may have been lost from living memory, but we had gone on this journey to recover this knowledge for the people of the Okavango Delta. We now knew the source lake, surrounded by previously unknown peatlands, was a giant sponge holding vast quantities of fresh water that made the delta far more resilient to climate change than we had assumed. We had seen vast forests, Africa's largest remaining miombo woodlands, pumping water into the river. We had met the Luchazi, the river's stewards and guardians, who protected the flow of the Okavango's floods using fire and forest. For the people, wildlife, biodiversity, and ecosystems downstream here in the Kalahari Desert, we had found the Source of Life.

RIVER GUARDIANS

João António Ifafe

Freedom in Wilderness

IFAFE TAUGHT ME that humanity's only superpower is being in nature. He is a custodian of this superpower: the wild heart of Africa. His traditional knowledge and skills are the key to his power. When everyone else left Africa's remotest wildernesses, he stayed.

Our expedition team has been exploring the Angolan Highlands since 2015. We arrived in armored trucks and 4x4 vehicles, bringing thousands of liters of fuel, solar power plants, satellite communications systems, and tons of food, equipment, and supplies with us. We bore hundreds of kilograms of gear per person, just to survive, as we considered it.

Ifafe, meanwhile, travels thousands of miles in this land with nothing but a small hand axe. He is proof that we aren't free, thanks to machines and technology. He is free because he doesn't depend on any of that.

In 2018, one of our Luchazi guides had a personal crisis and needed to be replaced on the expedition. We were so far downstream, though, that we couldn't offer our usual modes of help. We didn't know what to do. As it happened, two Luchazi hunters, Ifafe and Armando, walked into our camp at sunrise. Somehow they had sensed our distress and walked through the night to reach us. The same journey would have taken me days on a mountain bike, two days in a vehicle. Ifafe's quiet surrender and wild resolve had transcended time and space between his distant hunting camp and our remote location.

If I have ever witnessed a superhero moment, that was it.

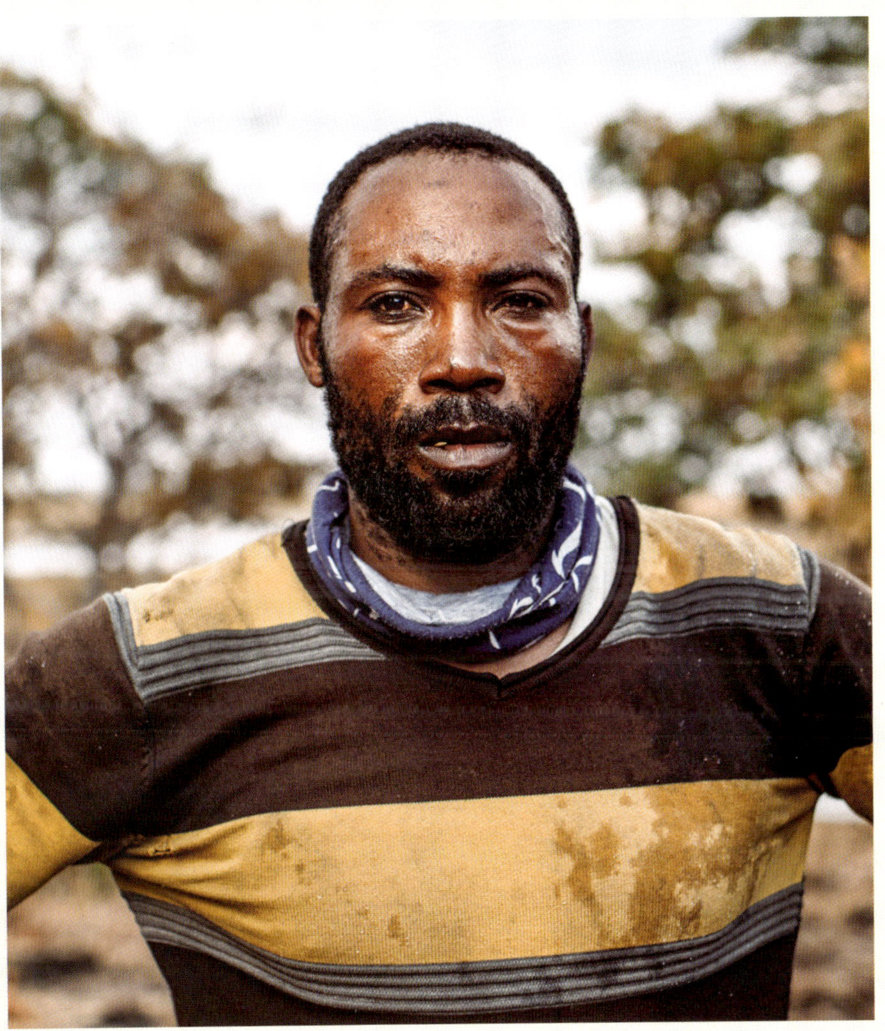

Ifafe is an emerging community leader in the Lisima Landscape and a trusted aide to the traditional leaders, who rely on his advice and judgment.

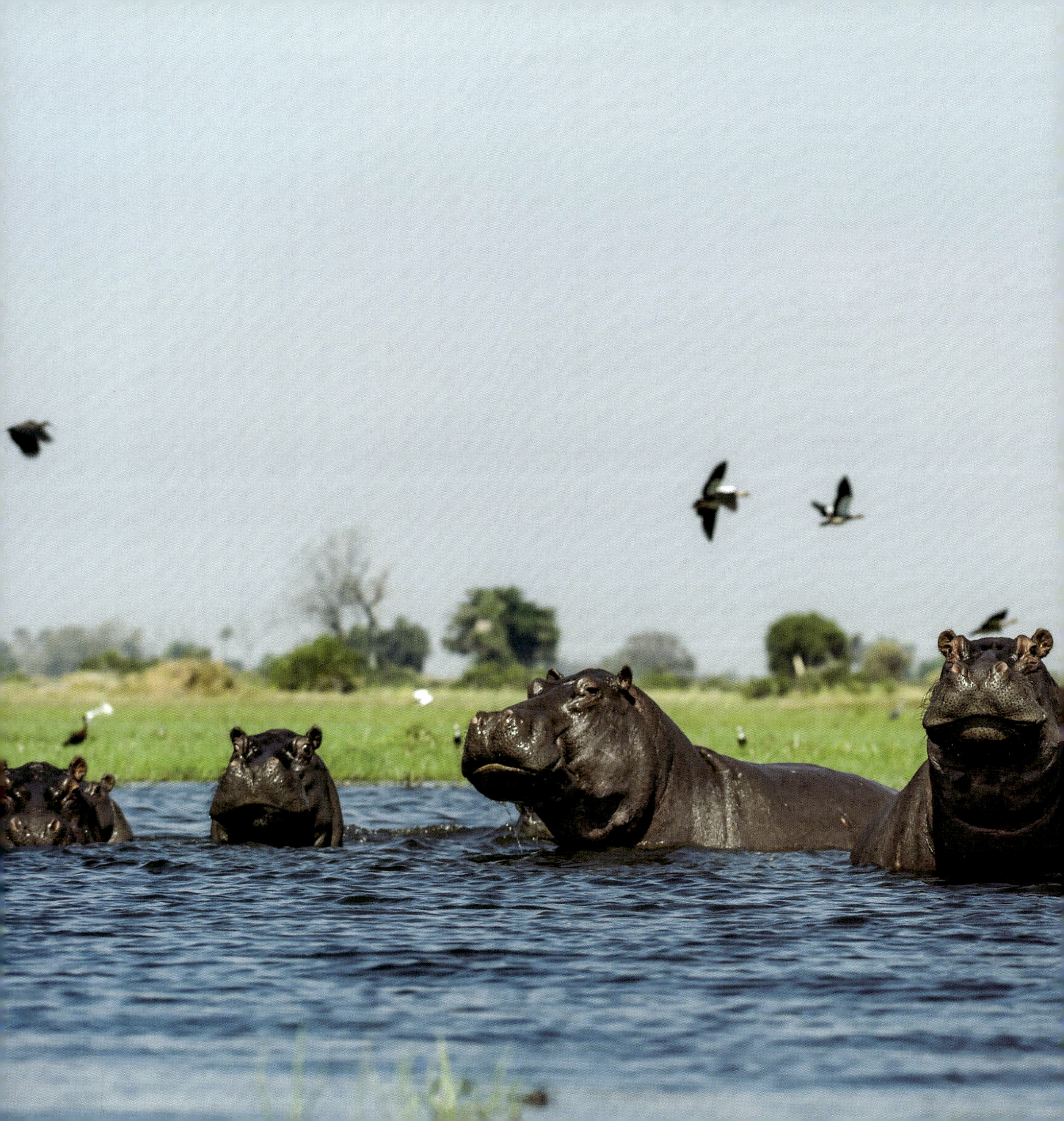

CHAPTER 4

Return to Paradise

CROSSING THE OKAVANGO DELTA, 2015

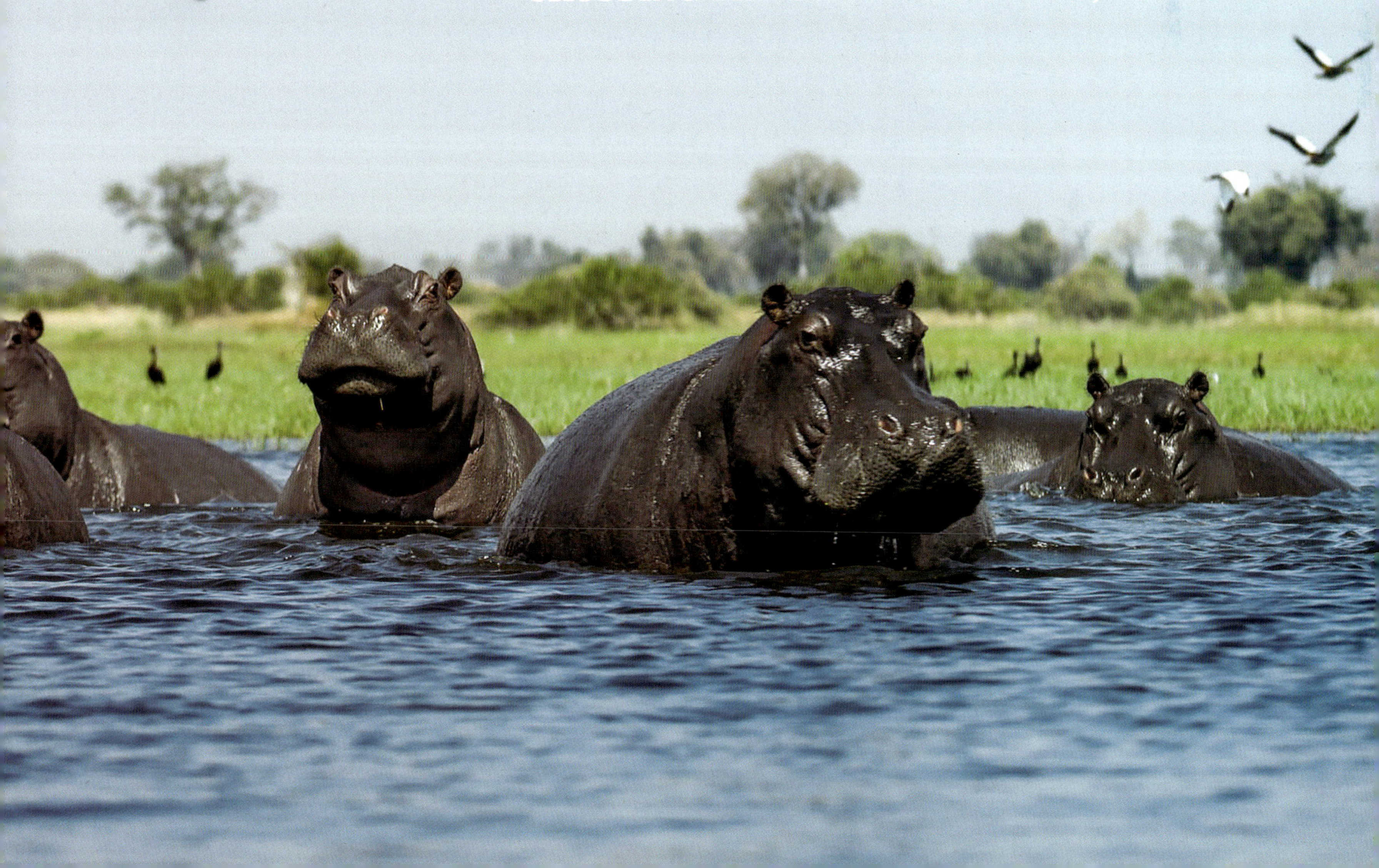

THE FLOODWATERS of the Okavango are estimated to journey 90 days from the source lake of the Cuito River to the Okavango Delta. We had taken nearly that amount of time to travel the same distance: We had been on the water for almost three months (a month over schedule), with a month still to go. The August floodwater entering the delta with us at Seronga was the same water we drank at the source in May.

Before leaving Seronga, we paid rent for the team members who needed it and promised to pay extra if more funds came in. More than anything else, the whole team just wanted to carry on until the end. This was our life now.

IT WAS ENERGIZING to stand up and pole our mokoros. We picked up two Wayeyi elders, Comet Masupa and Judge Dineo, from their homes at Jao Island to join us until we reached Maun. Judge showed us his birth islands, and as we poled away, Comet played the hippo drum—a hollowed log with a leather bellow, which imitates a hippo's grunting—while Tom played the *tswororo*, a mouth organ made from a brandy bush bow with a palm leaf stretched over it.

The flood levels had dropped significantly by the time we got to the delta. We spent days trying to find the main channels and lagoons, dragging the mokoros through ankle-deep water swimming with leeches. The hippos were everywhere, yet they seemed to see us as part of the river now. We didn't look at them, and they ignored us as we passed. Tension only came when we lost sight of them behind reedbeds at the entrance to an

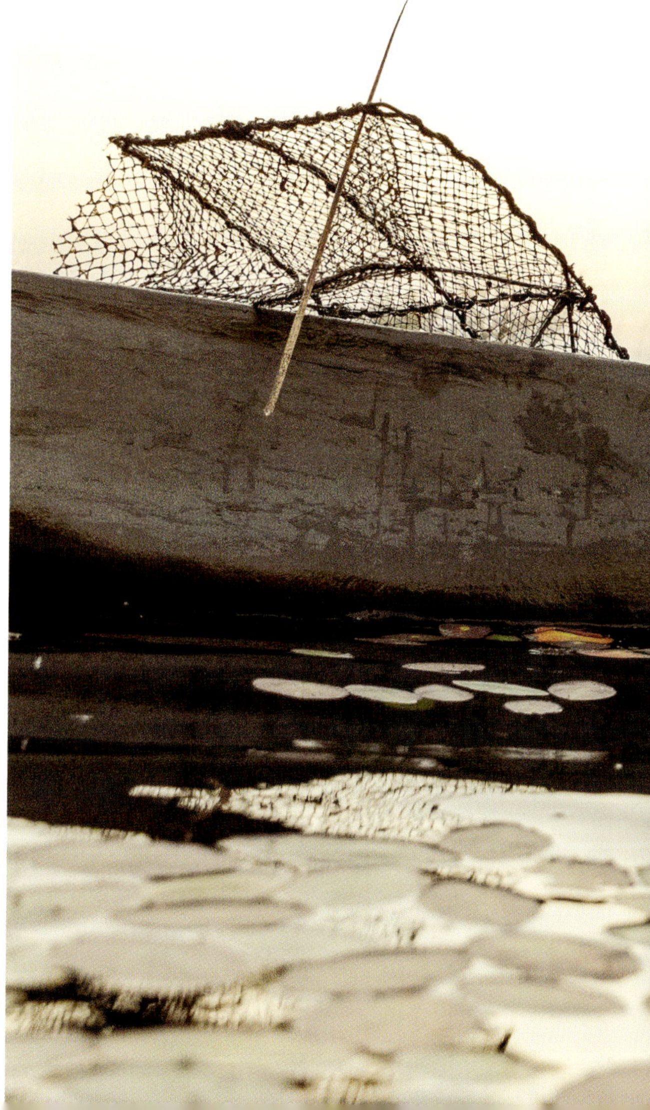

Rainer von Brandis (holding the fishing net) and Götz Neef lead our research team and have been studying the fish populations of the Okavango Delta for a decade, using catch-and-release nets and environmental DNA analyses. **PAGES 98–99:** *Hippos are the guardians of Africa's rivers. They create the channels and lagoons, protect the waters from intruders, and tend the pastures that feed abundant wildlife.*

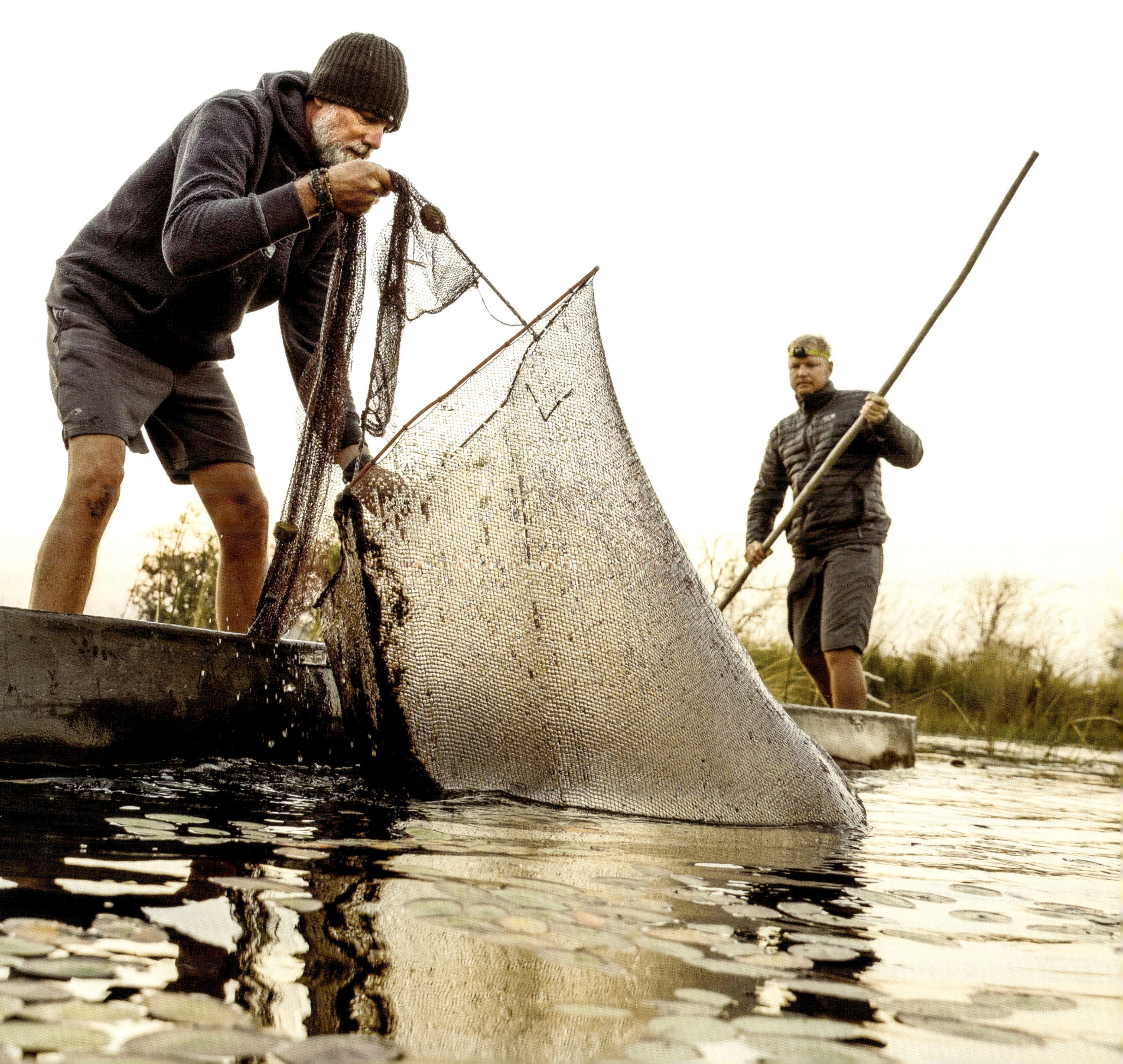

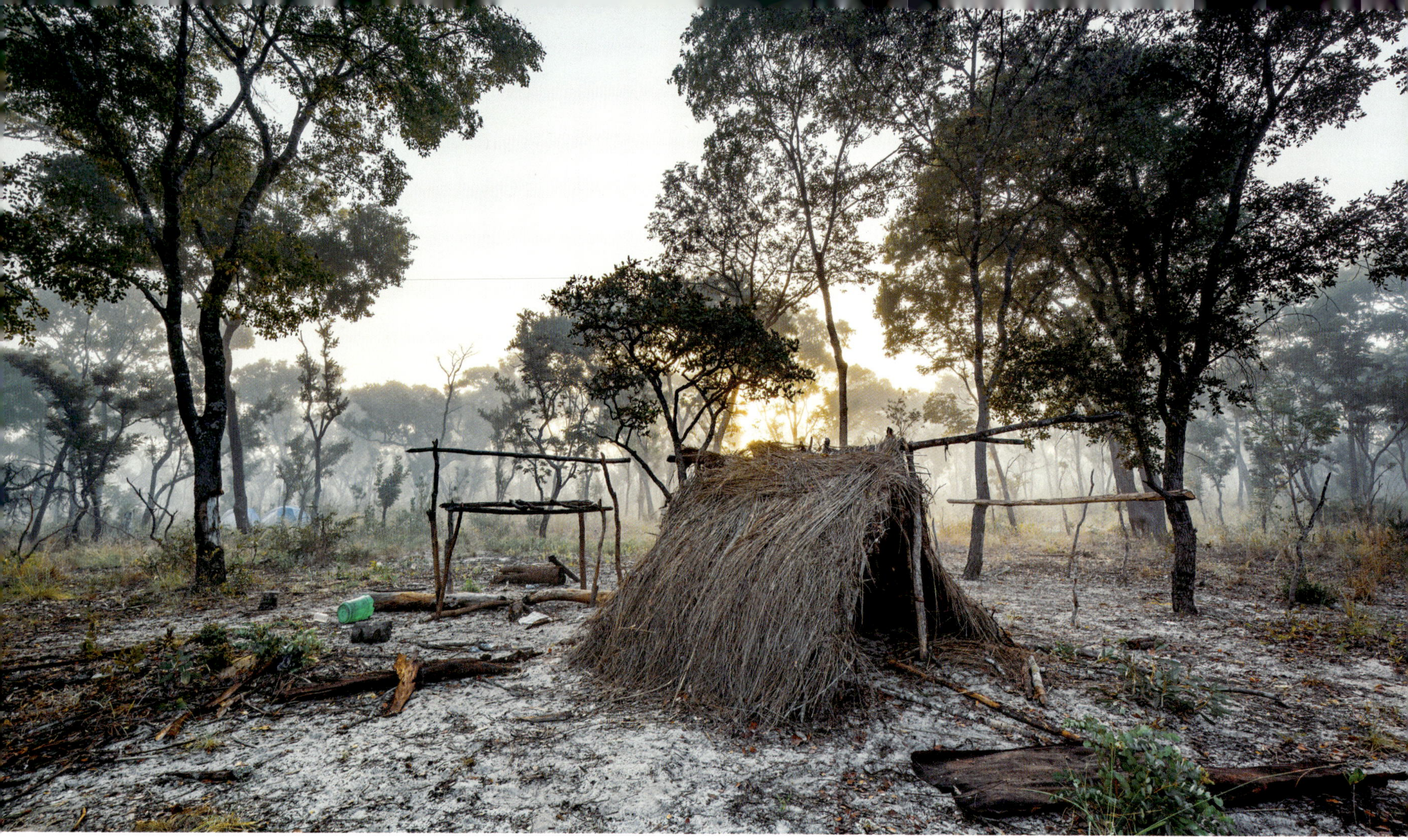

inlet or channel. Lone bulls or mothers with calves were almost always sleeping in these tightly packed reeds, so we often pulled our mokoros around, hoping not to disturb them.

When the reedbeds were simply too big or too dry, and hippos hadn't made paths around them that we could use, we had to go through the middle, ready to evade hippos at the last moment. We had many close calls. We encountered more than 1,000 hippos at close quarters and learned a lot about these majestic animals. They aren't friendly or cuddly as often portrayed. Yet they are powerful, reliable, caring of their own, and sentient. An African river without hippos is a river without voice and character. These animals love the rivers that we explore and will defend them to the death. You could say that hippos have the kind of unreasonable attitude that we must sometimes have to effectively protect these waters.

JUST OVER A WEEK after leaving Seronga, after about 100 days on the water, arriving in the village of Maun was rather anticlimactic. We had become wild, part of the river. Burgers and beer felt too foreign to be enjoyable, and the brash posing of local guides and professional hunters was irritating. Yet we all had a lifetime of friends in Maun, so it was wonderful seeing them after so many days away.

One person was missing though: my wife, Kirsten. Our son, Jack Wild Boyes, was still too young to travel to Botswana. They were all I could think about as I poled closer to home. Kirst also had a Ph.D. in zoology, had crossed the Okavango in a mokoro with me in 2012, and had been foundational to the conservation work of the Wild Bird Trust. She understood the mission, yet caring for a small child while your husband is away for months at a time is incredibly difficult.

I would close my eyes each sunset and think about Kirst and Jack: breathe in and know I'm alive, breath out and smile. I could feel them with me every day. This was all I often had, as our satellite internet was slow and always busy uploading science data. I could only afford to use the satellite phone once a week. On these long expeditions, you might consider whether it is all worth it, being away so long. But you have to believe.

I had been away from home for four months when we poled out of Maun on September 3, 2015. We were ready for the last push, 480 kilometers (300 mi) down the Boteti River to wherever the water ended. A two-decade-long drought had stopped the Boteti's flow between the late 1980s and 2009, so trees had grown in the riverbed and died. The returning water was undrinkable due to decades of accumulated cattle dung and dead animals on the riverbank. Fortunately for us, our support team met us most days and brought drinking water from wells in nearby villages.

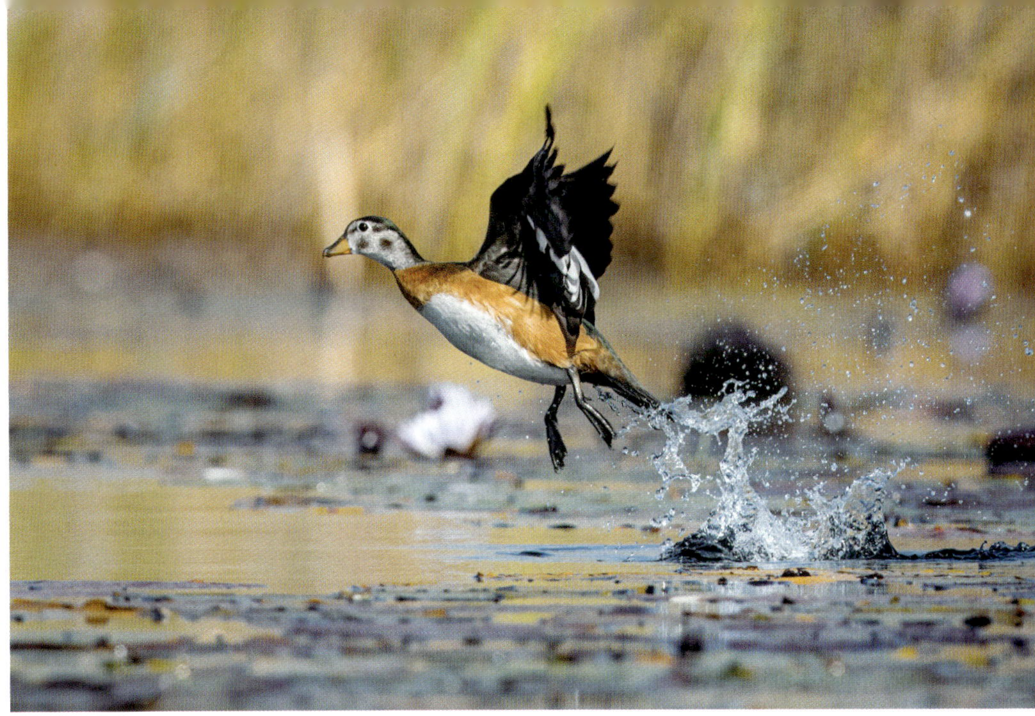

Every day, we encountered long corners with small rocks and fast-moving water too shallow to navigate, so we had to get out and scratch our loaded mokoros over them. To our horror, we found millions of small, multicolored worms under all the rocks. After being out of the mokoro for 10 minutes, we would have to sit at the back of the boat picking these worms out of the dead skin and scar tissue on our feet. You could see little pits where the worms had chewed out skin. This experience, repeated a few times a day, was alarming and saddening. Something was wrong: The river was sick, and it was making people sick. This was the same water that left the source lake 100 days before, crystal clear and pure—a source of life, not death in the desert.

In addition to the drought, another problem came from a seemingly simple addition to the landscape: fences. Millions of wildebeests as well as plains zebras—Botswana's national animal—once migrated up along the Boteti

Pygmy geese feed almost exclusively on water lilies. These beautiful little ducks and the lilies are abundant in the floodplains of the Okavango Delta. **OPPOSITE:** *By the early 2000s, traditional hunting and illegal poaching had been discontinued in the Okavango Delta. Today, all you will find on the remotest islands, accessible only by mokoro, are old abandoned hunting camps like this one.*

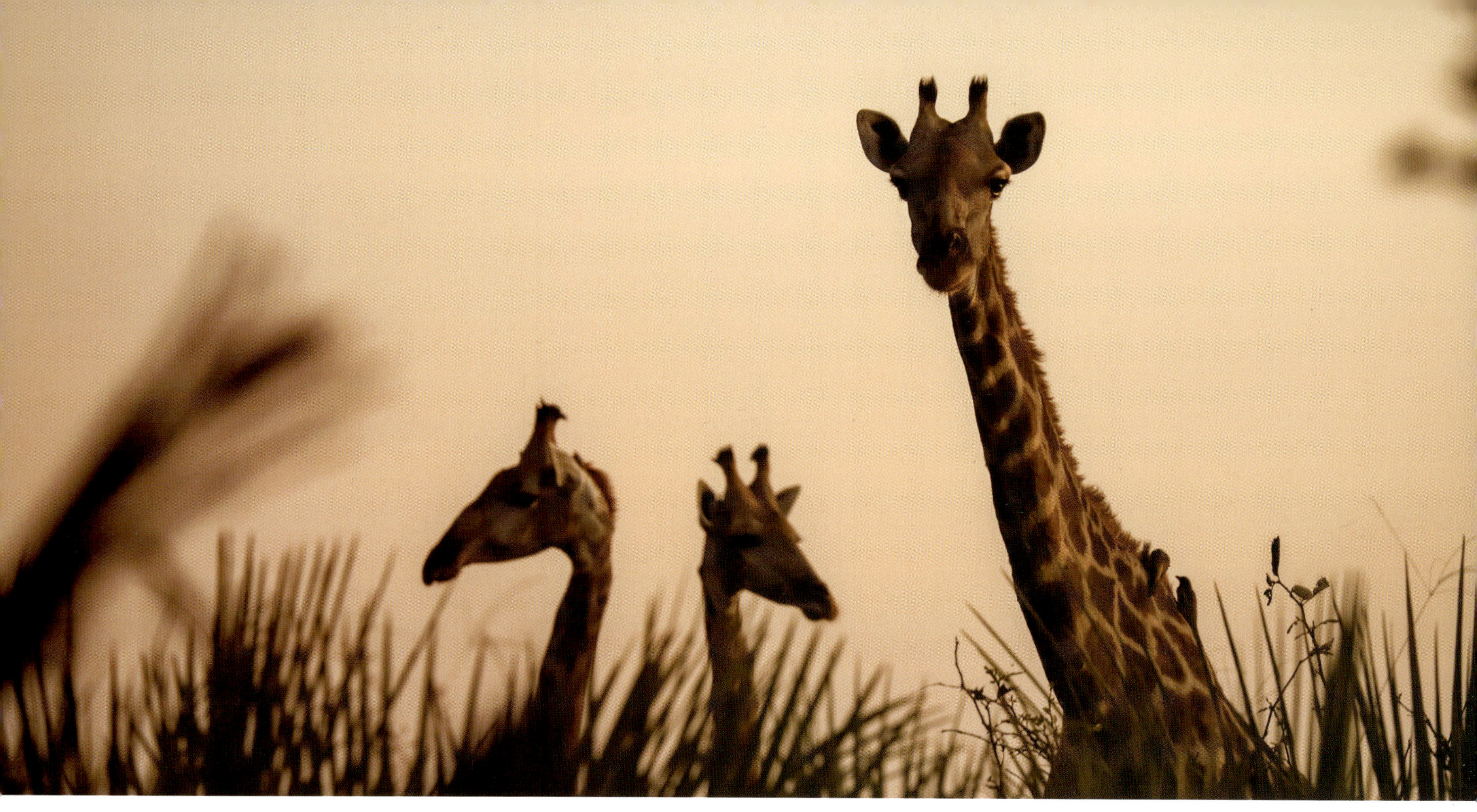

Giraffes thrive on Chief's Island, the largest island in the middle of the Okavango Delta. On our mokoro expeditions across the delta, we saw them on a daily basis, quietly wandering the thousands of islands and floodplains of this wetland paradise.

River into the Makgadikgadi salt pans, Africa's largest land migration. Hundreds of thousands of zebras, resident in the pans below the Boteti River, also migrated between the salt pans and the Okavango Delta and Chobe River. Yet by 1970, cattle fences had cut off these ancient routes. Hundreds of thousands of animals, perhaps millions, died along the Boteti River, trapped behind fence lines.

When fences were removed in the early 2000s, tens of thousands of zebras started using the Boteti River, Nxai Pan, and Okavango-Makgadikgadi migratory routes again after decades. Even so, by the time we reached the Boteti in 2015, it was still a long way from recovery. The cattle industry was still dominant, and the migrations hadn't recovered. As we paddled, we resolved to use the river baselines we were creating to help monitor and better manage the Boteti River for the benefit of people, wildlife, and livestock. The recovery of this delicate ecosystem would take decades of conservation interventions in partnership with local people.

As we paddled with the slow, muddy flow, we had to watch for the constant threat of underwater trees and fence posts, which broke the hulls of our fully loaded mokoros several times. We traveled up to 65 kilometers (40 mi) daily, passing groups of people—sometimes whole families—clearly sick and sitting by the river. At the time, we didn't know what was

IN MEMORIAM

Judge Dineo

The Secret Is Now

A GIANT of a man with a booming voice and many children, Judge Dineo belonged to the wilderness. Humble to a fault and unnaturally powerful, he was a hunter, fisherman, farmer, and community leader. When he died in 2022, we lost a regal African who chose a traditional life, poling his mokoro in freedom.

Judge's defining trait was his lack of self-consciousness. As a young man, he worked for a safari company. He had never learned English, so his job was off-loading baggage. One day the famous Richard Branson came on safari and gave Judge a Virgin T-shirt as a thank-you for his services. Judge asked what was written on it, and Sir Richard told him: "Virgin." Judge thrust the T-shirt back at him, saying, "I am not a virgin!"

Our modern society is stuck in a constant cycle of self-criticism and doubt, but Judge taught me that the point of being alive doesn't lie in some future destination—some abstract ideal yet to be attained. The secret is always in the right now, wherever we are, together. In towering silence or a quiet chuckle and a subtle smile, he showed us we were living fulfillment every day. Nothing else was of any real importance. He was completely integrated; there was no gap between his thoughts and feelings. All of us who worked alongside Judge recognized how important it was to reclaim the naturalness and assuredness that he lived every day.

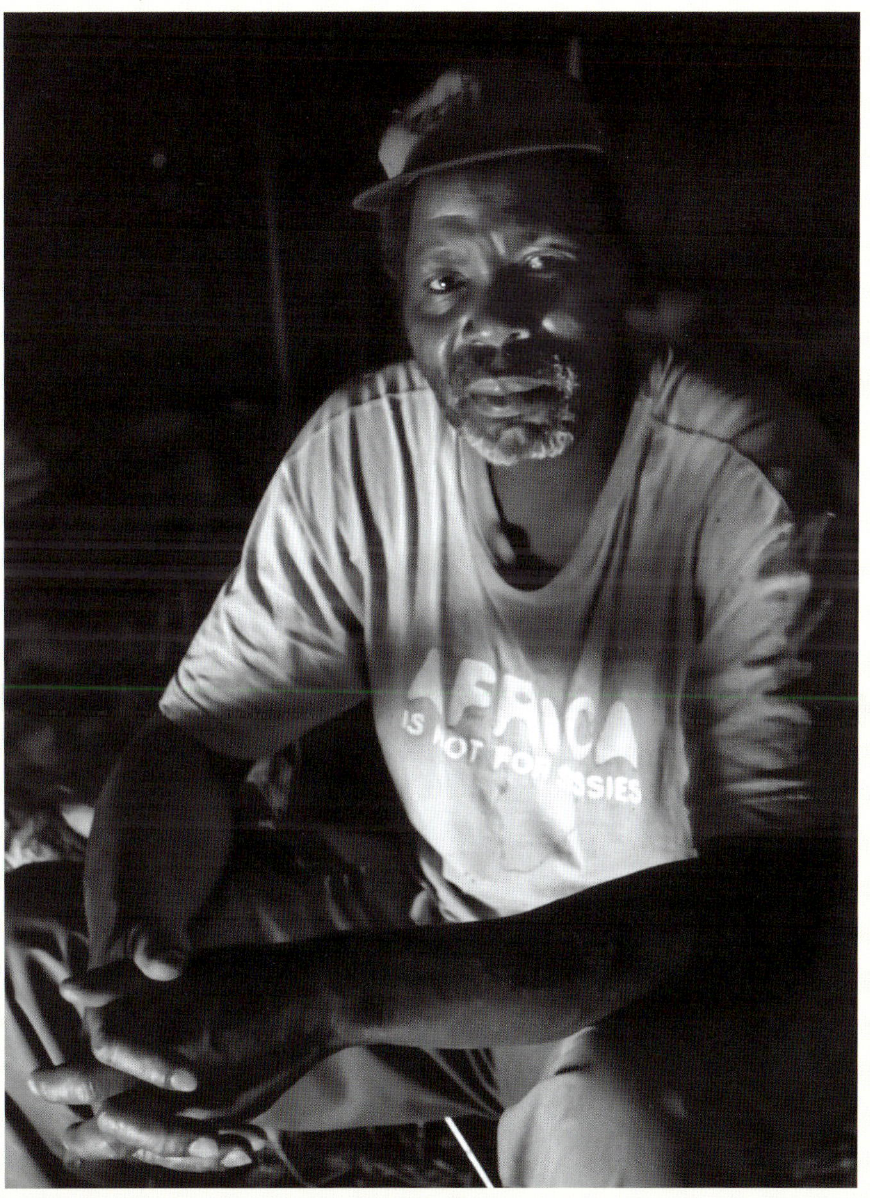

A central figure in our annual mokoro expeditions across the Okavango Delta, Judge grew up and had lived his whole life traditionally.

wrong with them. Later we learned the condition was bilharzia, or schistosomiasis, a disease caused by parasitic worms, though not the worms on our feet. A study in the 1980s, when the river last flowed properly, had predicted a bilharzia outbreak within 30 years, correctly foreseeing that the river would be completely polluted with animal dung when the waters fully returned. And, in fact, we all had bilharzia when we got home and needed multiple rounds of medication to clear the worms that had given us rashes, fevers, joint pain, and muscle aches. Finishing our expedition this way was very difficult.

On our final day, we woke at 5 a.m. and paddled until sunset, portaging over two bridges and finding a way through impenetrable reedbeds at the inlet to Lake Xau, the end of the water. Nothing could have stopped us that day. We saw boisterous hippos and massive crocodiles just before the lake, there to congratulate us and let us go. It was sunset as our group of 13 explorers dragged the seven mokoros onto the muddy southern edge of the lake, covered in cattle hoofprints and dung. Many of us cried in each other's arms.

By September 21, 2015, after 121 days of nonstop paddling and poling, we were done. After traveling 2,476 kilometers (1,243 mi) on the water, Water Setlabosha had the words: "This is the place we were looking for." From source to sand, we had completed an unprecedented megatransect across one of the world's largest undeveloped river basins.

Returning to so-called ordinary life was no easy task. I had been living on rice and beans,

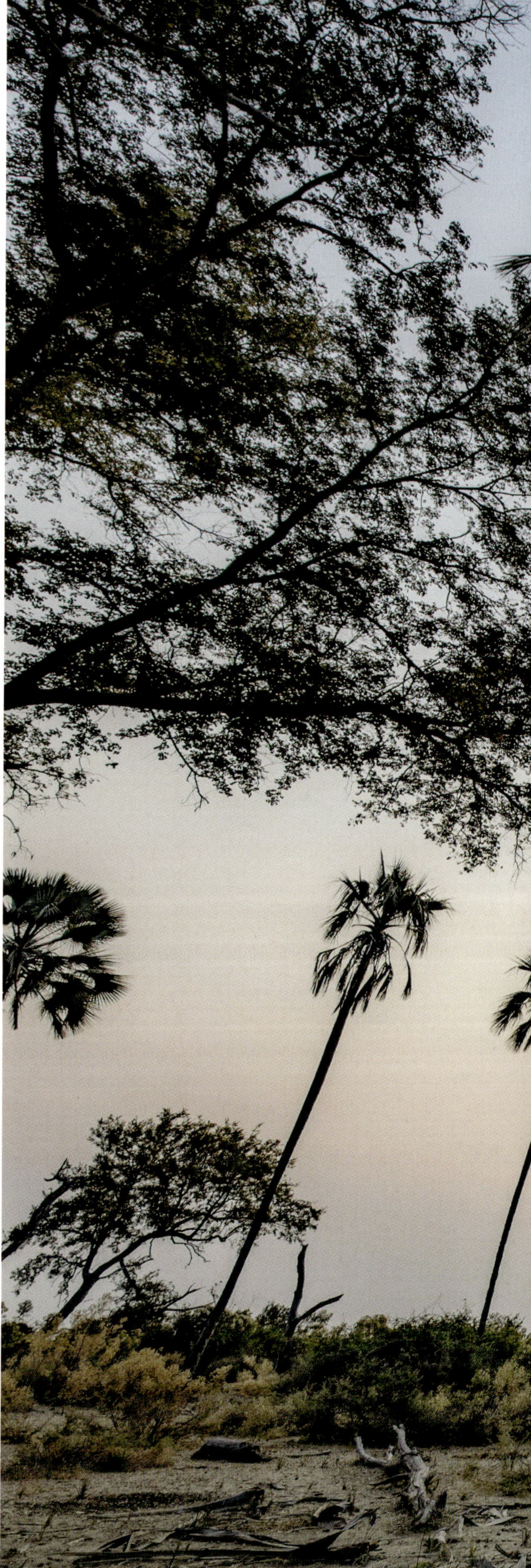

The stresses of climate change have killed most of the oldest baobab trees in the Okavango Delta, but a few hundred ancient ones remain—often the primary features on an island, surrounded by real fan palms. Some are more than 2,000 years old, and most have been named.

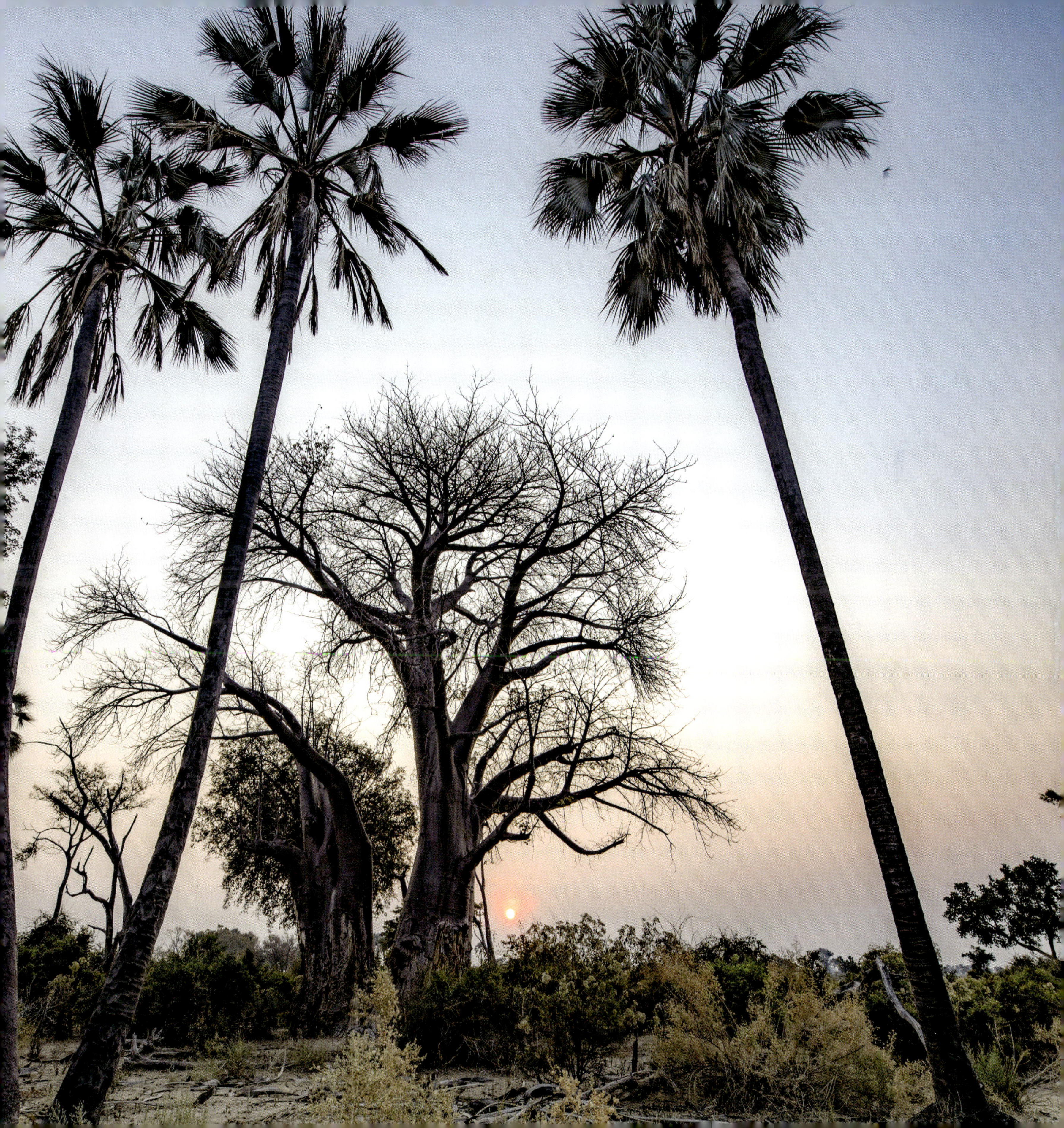

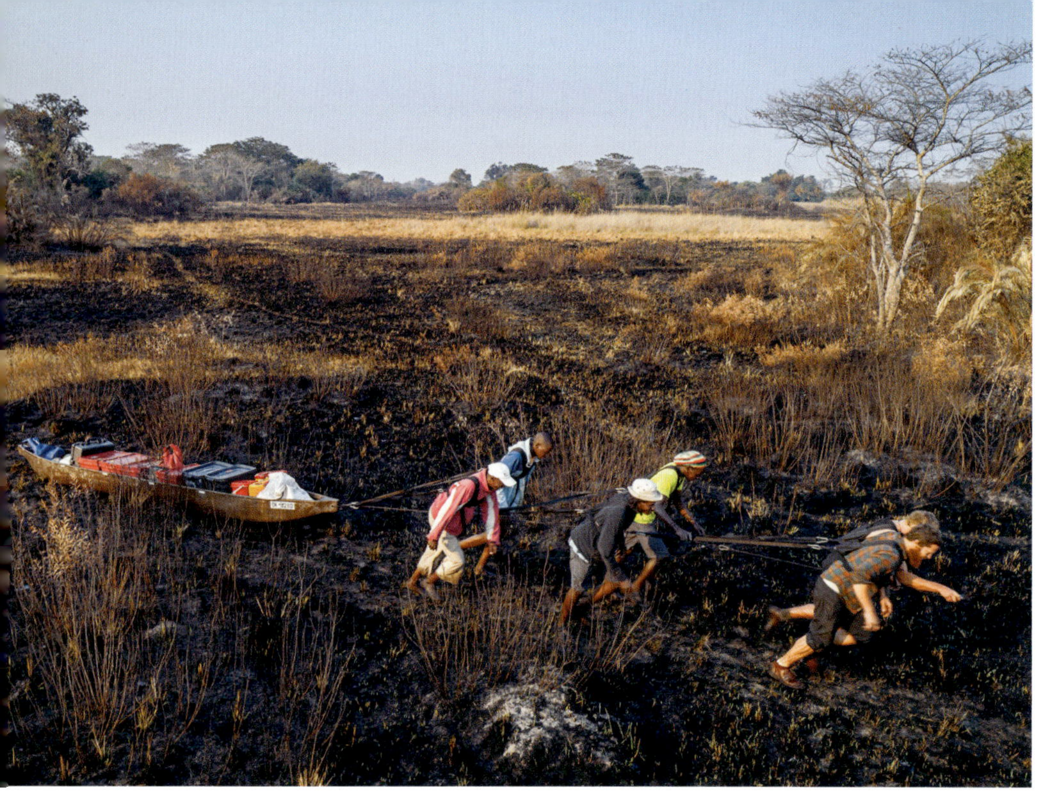

In the Okavango Delta, fires purposely lit play an important role in the ecological balance, as they help maintain fertile pastures, both for cattle and for wildlife. We have often had to portage over burned floodplains during the dry winter months.

bathing in a bucket of cold water, and paddling a marathon six to eight hours every day. When we got to Maun, I literally couldn't pay my hotel bill because after four months, I had forgotten my PINs or logins for any bank cards or online accounts. I had no presentable clothes other than the "pajamas" I slept in every night and hadn't washed. I didn't want to hang around. I wanted to get home.

At the last minute, the support team was able to get me a business-class ticket for the flight back to Cape Town. A nice American couple next to me on the four-hour trip seemed a little concerned about having a feral creature beside them, but they enjoyed hearing my stories.

Waiting for my expedition bags in Cape Town, I was shaking with nerves. I hadn't seen Kirst and Jack for more than five months; even just landing in the busy Cape Town airport was overwhelming. By the time I made it through customs, I was crying and must have looked like a different man—hence the bewildered looks on their faces when I arrived. I had dreamed of this moment for months. We kissed and cried. Almost two years old, Jack was scared; he didn't recognize me. My hair and beard had been bleached by the sun, and Kirst had to scrub me for two days to get me clean. A few days later, Jack mistook my brother for me when he visited looking similarly frayed.

Yet even as I delighted in being home, I knew this wasn't the end. After six years of taking detailed research transects of the Okavango, the 2015 expedition was the first hydrological, ecological, and socioeconomic river baseline captured across an entire river basin. We had done what was considered impossible. Even so, the Okavango Megatransect wasn't finished. We were only halfway, and we still needed to survey the Cuanavale and Cubango Rivers, the two other unstudied tributaries of the Okavango Delta in the Angolan Highlands. Within weeks, we were planning our return.

Despite the four difficult months we had just spent nonstop paddling and poling, within six months, we set out again—driving back up to the source lakes of the Okavango Basin in the remote eastern Angolan Highlands. This time we went in search of the source of a mythical river, one that Luchazi traditional leaders we'd meet later that year would tell us was steeped in magic and mysticism. This was the Cuanavale, and soon we were plunged into its magic as we scouted a route to discover where this river began.

Hankuzi the Hippo Hunter

Living the Legend

CENTURIES AGO, the great Barotse Kingdom was a powerful state with a culture of trade, river exploration, war, and conquest. From there, legend has it that Hankuzi the Hippo Hunter led the Wayeyi people downstream, seeking a peaceful home far from what early Portuguese explorers called the "land at the end of the earth." Oral histories say that the Wayeyi people arrived in the Okavango Delta in the 1770s, departing from present-day Angola and Zambia by navigating the Cuando River and nearby tributaries. They found heaven on Earth in this wetland wilderness, a wild paradise where they say God lived among humans before joining the ancestors.

The Wayeyi became river Bushmen, forever connected to the waterways. They arrived with Iron Age technologies—axes to carve their mokoros and nkashi poles to propel them. They united with the local Basarwa people and never farmed the central wilderness. They traded tobacco for iron with the Lozi people in the north.

When the dominant farmer and pastoralist Tswana people arrived, they subjugated the Wayeyi and Basarwa, imposing their language and culture on them. By the early 20th century, European ivory hunters further threatened the Wayeyi. The transcendent song of the tswororo mouth bow retreated into the wilderness, and by the late 1980s, the last Wayeyi families were moved out of the delta.

But the way of Hankuzi the Hippo Hunter carries on. The elders we met on our expeditions still cling to ancient traditions—and to hope. In the words of our team member GB, a proud Wayeyi man, "We don't know where we are going, but we know that we will get there."

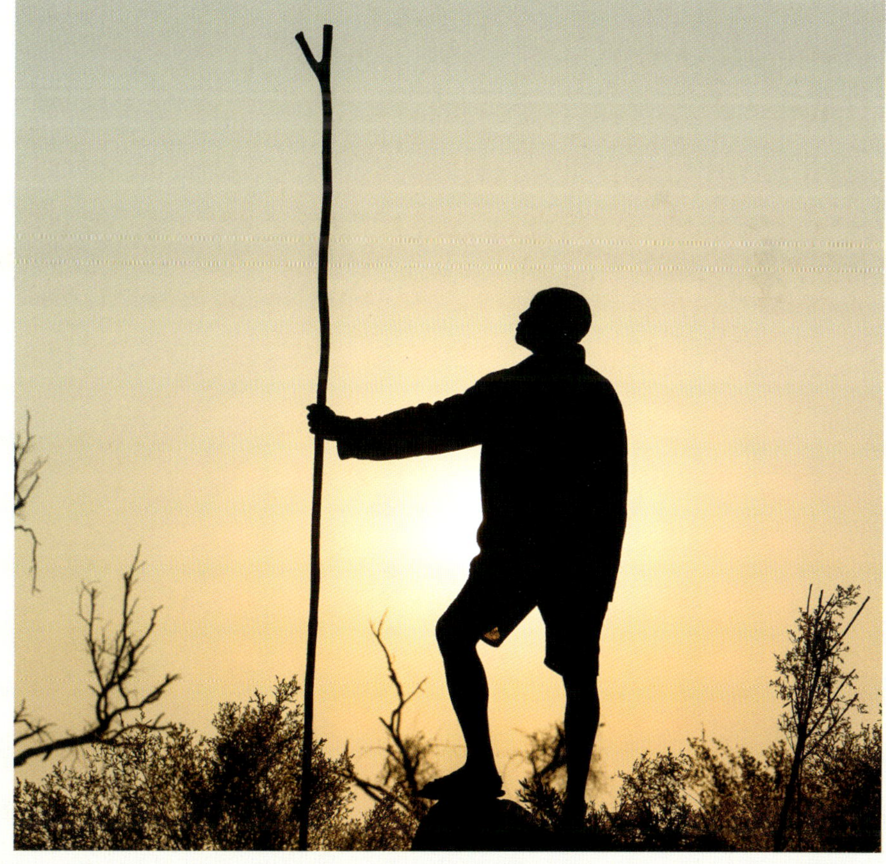

Hankuzi the Hippo Hunter was the legendary founding leader of the Wayeyi people who led them down from Angola and Zambia into the Okavango Delta.

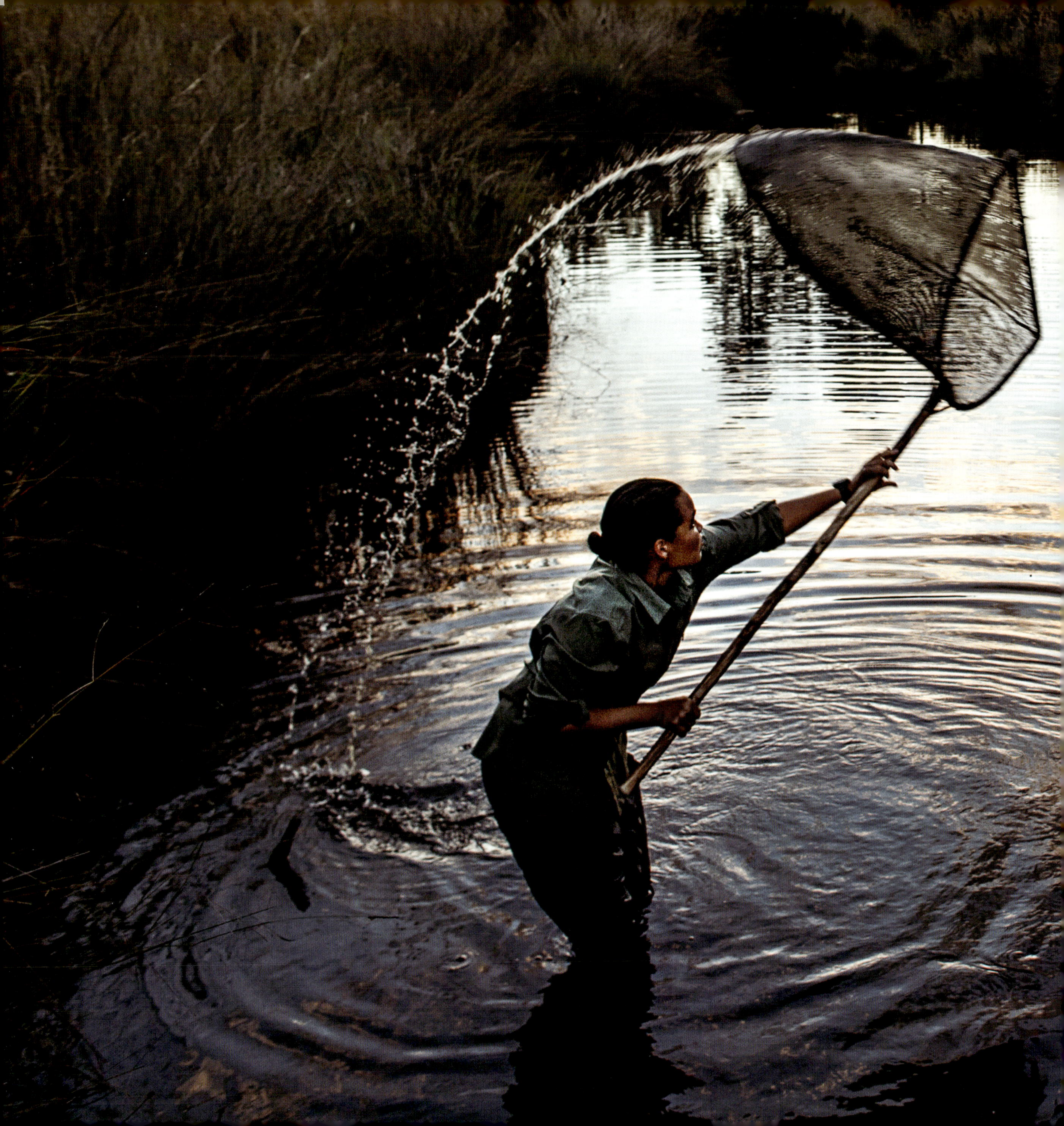

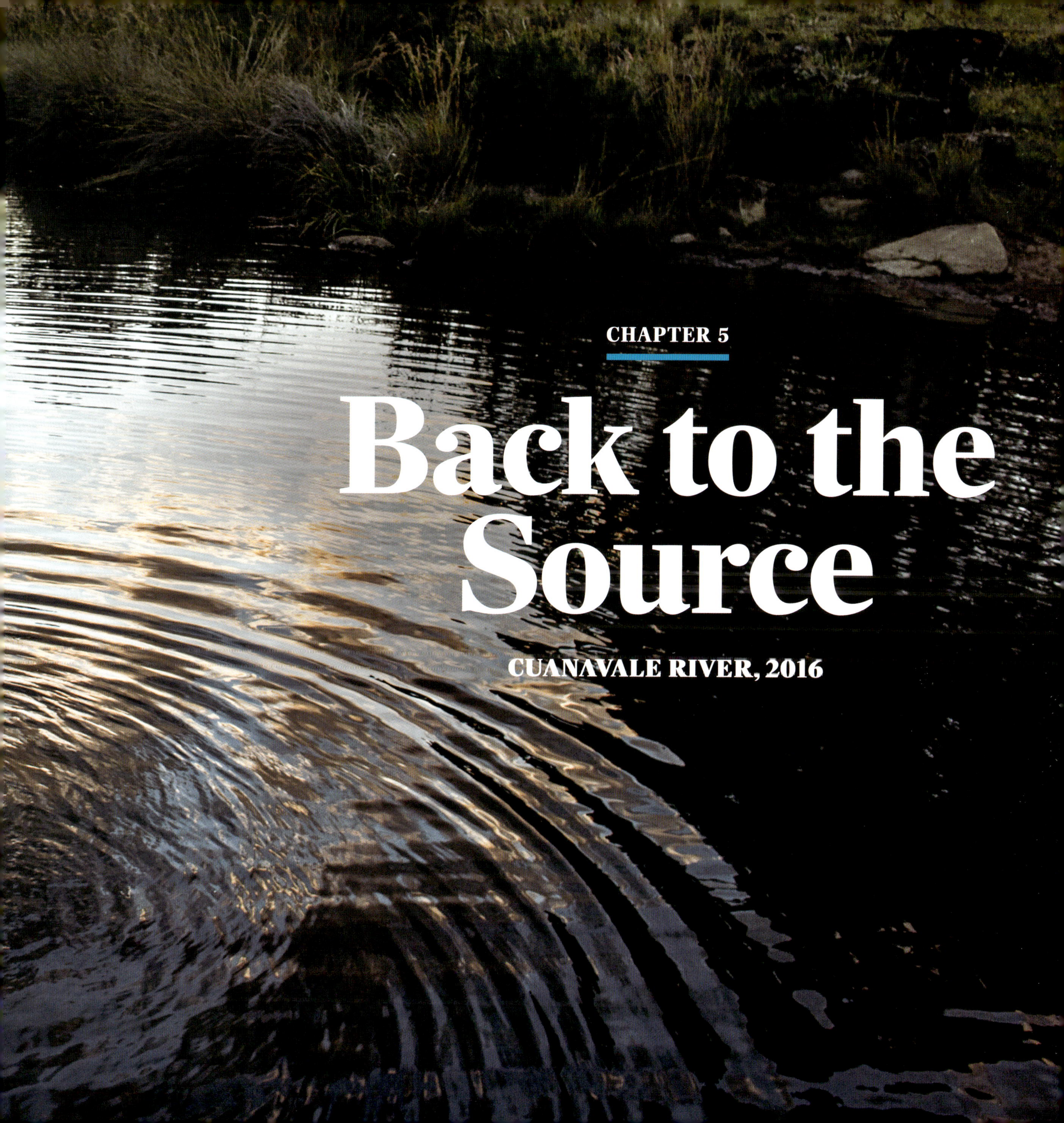

CHAPTER 5

Back to the Source

CUANAVALE RIVER, 2016

Every few kilometers along the Cuanavale River, we saw families of pied kingfishers perched on dead trees looking for small fish. **OPPOSITE:** *GB is a master poler, able to expertly move a fully loaded mokoro backward down the narrow, fast-flowing channel of the upper Cuanavale River.* **PAGES 110–111:** *Adjany Costa casts her net, collecting small fish and aquatic invertebrates as part of the Okavango Wilderness Project biodiversity studies.*

IN 2016, JUST BEFORE we departed to survey the Cuanavale River, the National Geographic Society signed a five-year commitment with significant financial support. The mission was to continue exploring, monitoring, and better protecting the little-known Angolan watersheds and rivers that sustain the Okavango Delta, all for the first time in history. This partnership with the Wild Bird Trust represented the National Geographic Society's largest single financial investment in any conservation project in its 128-year history. It was extraordinary, and our team felt the simultaneous pull of excitement and stress: We now had five years to dream big, get the megatransects done, do the biodiversity surveys, and work with the Angolan government and local people to protect the sources of the Okavango and Cuando Rivers.

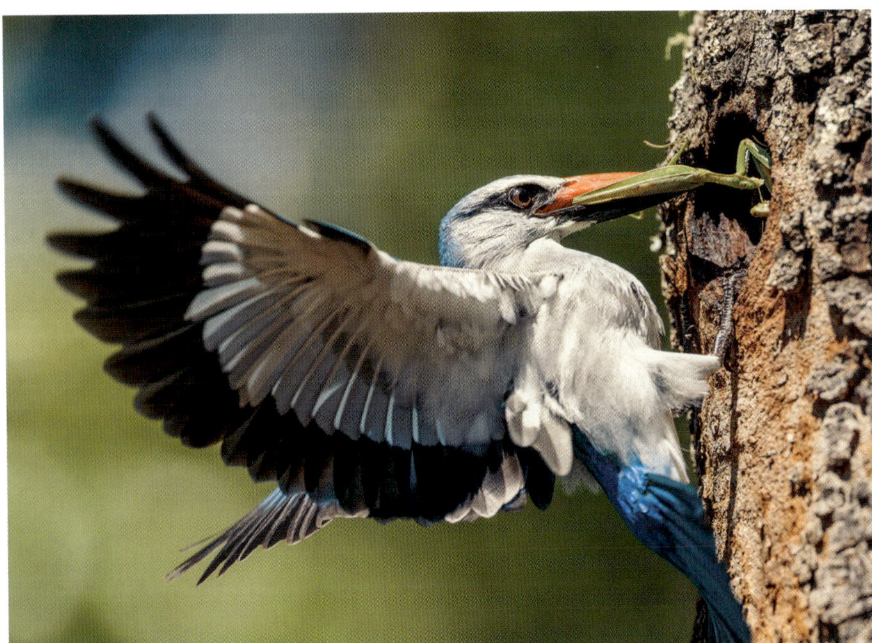

We had completed an unprecedented scientific transect across the Okavango Basin, but now we had to complete baseline river surveys spanning the entire river basin, gathering hydrological, biodiversity, and socioeconomic data we would use to monitor the waterways' and ecosystems' health into the future. The Cuanavale, a major tributary of the Cuito River, became our next goal. This river joins the Cuito at the bridge at Cuíto Cuanavale, fought over heavily during the war. But the origins of the water upstream were, like all of this area's waterways, still unknown.

In early April 2016, we launched seven mokoros at the source lake of the Cuanavale River, back in the Angolan Highlands. We'd had a much easier time getting there than on our previous trip to the Cuito source lake: We had partnered with the people of Samanunga village, located down a much easier track from Munhango, and learned from them how to get to the Cuanavale source lake, driving off-road through the bush. What we discovered was the most beautiful of all the source lakes, crystal clear all the way down to the bottom. This lake is sacred to the locals, who believe it is protected by Mukhisi, a serpentine water spirit said to be a python that never stops growing and eventually sprouts legs and crawls into the lake.

This area, much like the source lake of the Cuito, was surrounded by dense, carbon-storing peatlands, which we measured as deep as five meters (16 feet). Once again, we were finding in these hills a significant carbon sink of global importance that had been previously overlooked.

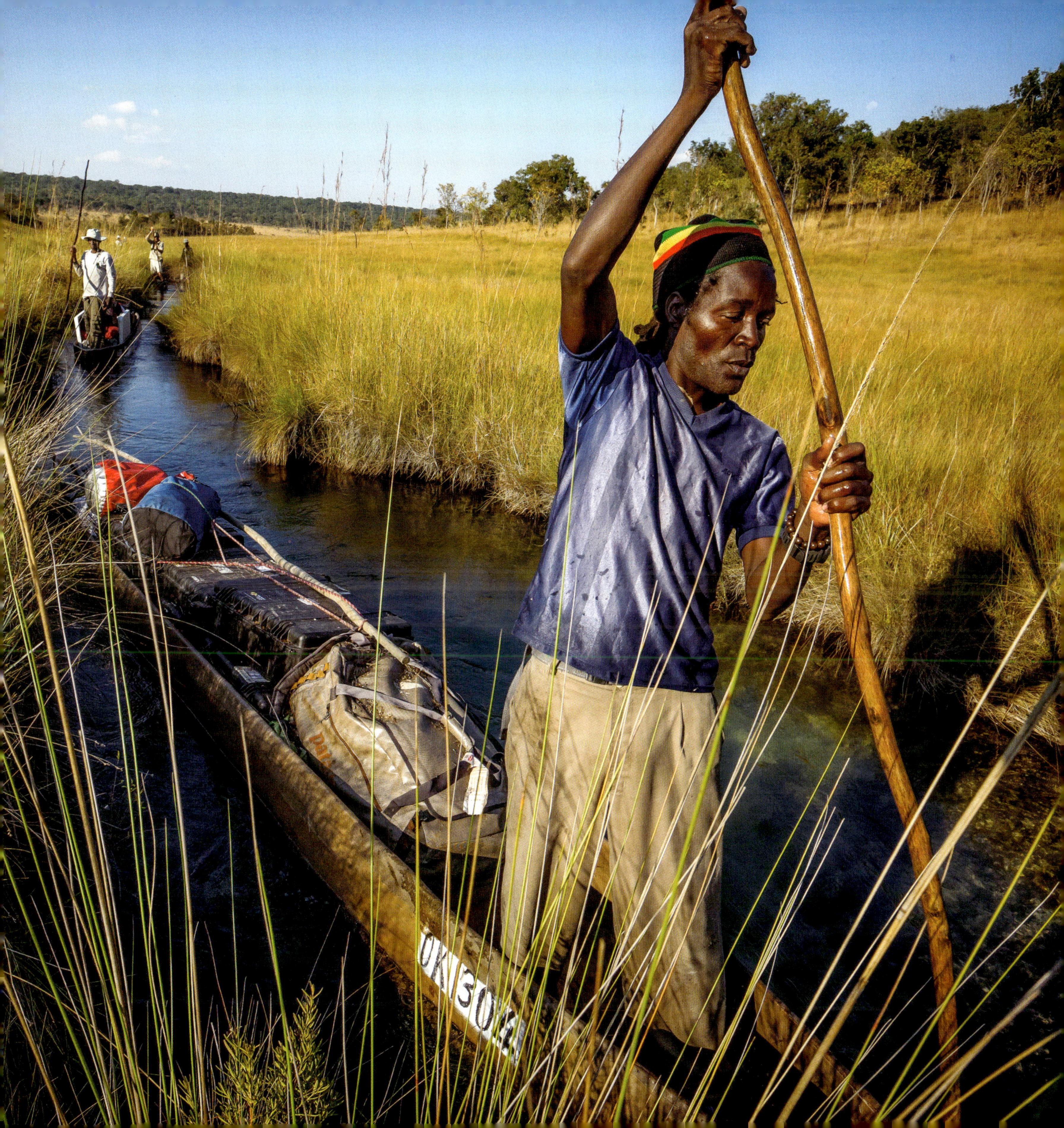

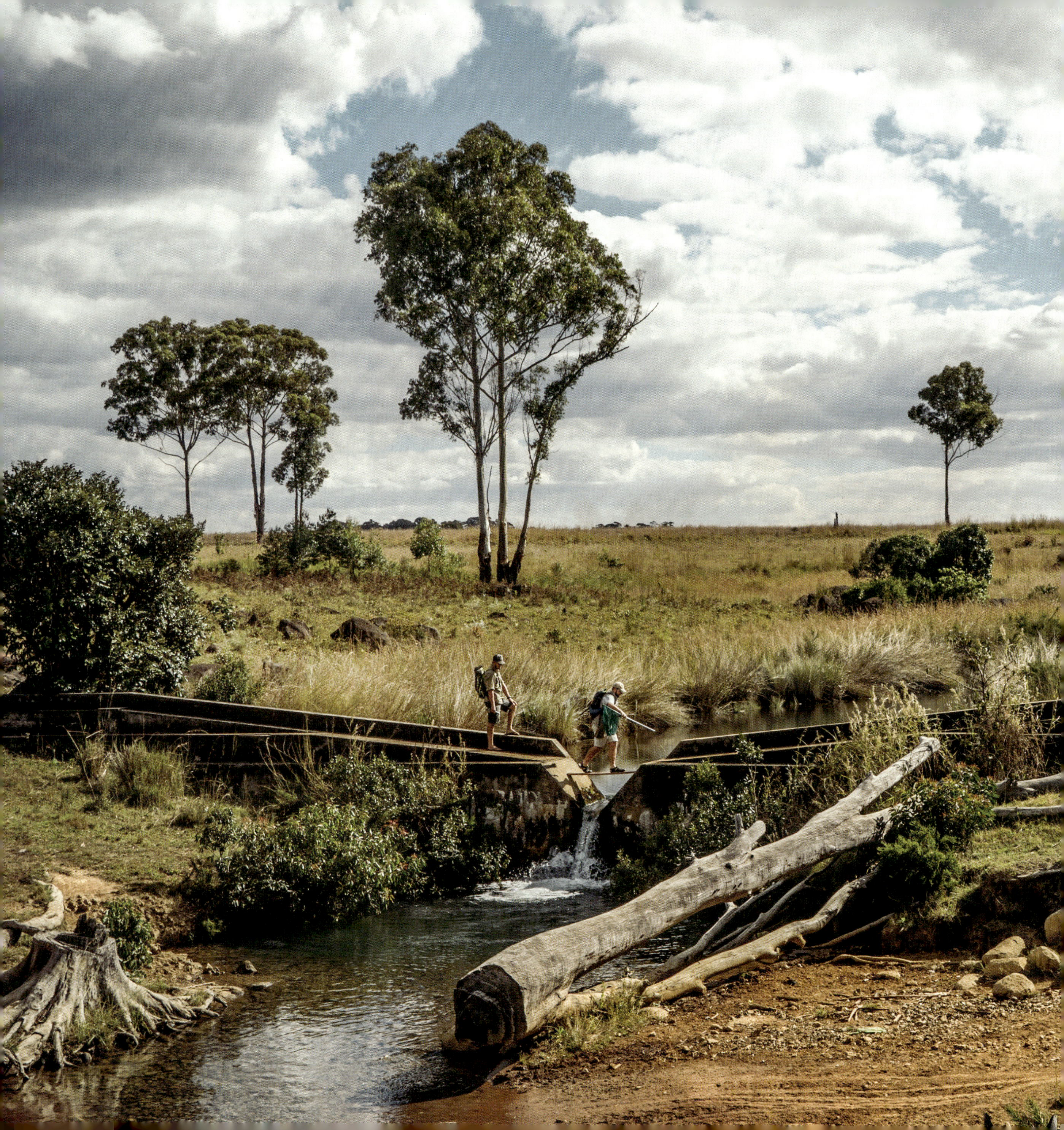

It was the end of the rainy season, and the river was still flowing very strongly. It felt like someone was trying to climb from the water into the mokoro on every corner; capsizing was a constant threat. Yet the main issues we faced on the Cuanavale River were sweat bees and honeybees. We had been filming a short, 25-minute film for the Angolan government. The producer, Zach Vincent, said that although the pestering bees had made filming hard, editing the film was even harder: In every scene, people were waving wildly at the air, slapping themselves, and cursing furiously and repeatedly. The film turned out well in the end, with lots of voice-over and music to cover the incessant buzzing.

I had never experienced bee densities like we saw that year. Every time we stopped near the forest, sweat bees would choke the air around us. Stingless, sweat bees defend their hives by hounding and frustrating their target. At the end of the summer, bee populations were at their peak, and the flowers were gone. The high-altitude watersheds had been highly leeched due to heavy seasonal rainfall on sandy soils. With no available salts on the surface, trees were holding on to their minerals very carefully, surrendering them only for pollination and seed distribution. Both bee species need these mineral salts to prepare their nests for winter, so our sweat became their primary focus.

We woke to bees swarming our tents. We would get up before sunrise, and as soon as the first bead of sweat dripped down our faces, they were on us: inside our clothes, in our hair. If you killed them, they swarmed more. If you ignored them, they went up your shorts into your underpants or into your ears or got caught in your eyelashes or under your eyelids. I began to feel dread and anxiety as soon as I heard them buzzing; it was like living with an alarm clock going off all the time.

We were getting stung by bees at least 15 times a day—not because the bees were aggressive, but because they got stuck in our clothes and panicked. Unrelenting and seemingly infinite in number, the sweat bees were infuriating. It was also unbearably hot and humid. Afternoon thunderstorms were an alternate misery, even though they kept the bees away. I have never been so uncomfortable surrounded by such incredible beauty.

Despite all this, our time on the Cuanavale was one of the best expeditions I have ever led.

We often saw baby crocodiles swimming below us in the Cuanavale River, especially near the source lake in the heart of the Lisima lya Mwono Landscape. **OPPOSITE:** *While the river expedition crew is busy, teams of scientists and biodiversity experts hike nearby rivers and tributaries to collect samples of plants and small animals.*

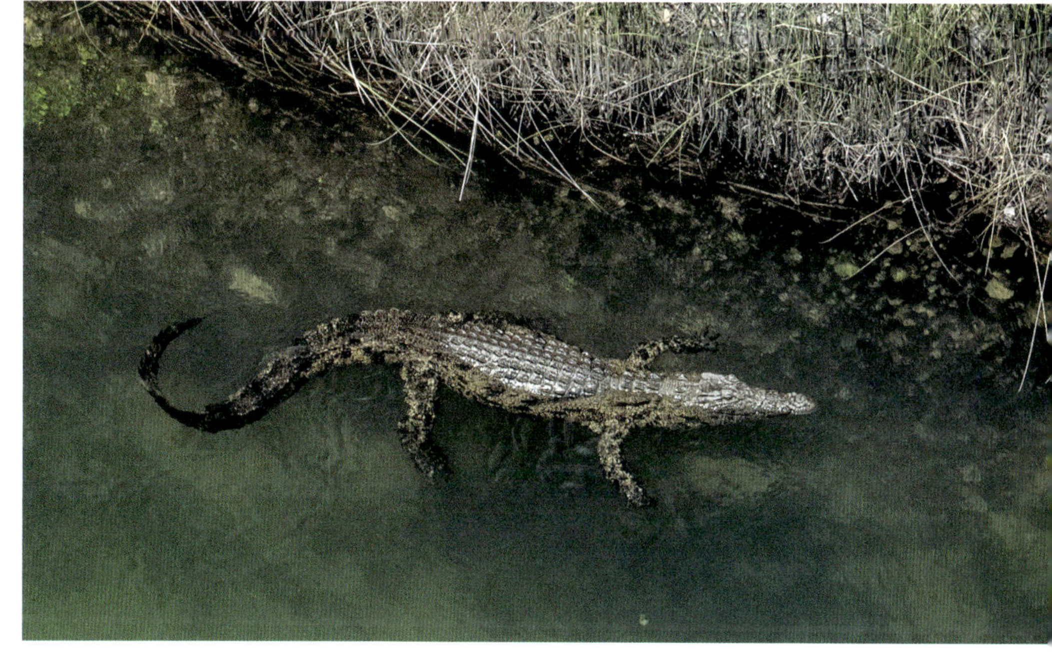

The landscape we were traveling was absolutely stunning. Something about the Cuanavale River made us not want to get off the water—not just because it was the only place where the bees couldn't find us, but also because the river valley was so wild and beautiful. We could imagine tourists coming here to experience nature and celebrate these amazing, forested watersheds.

We learned a lot about ourselves during that trip on one of the most magical rivers on the planet. On one of our most deeply uncomfortable days, I looked back and saw Abhi Mandela Ravivarma, an Indian conservation technologist acting as one of our research technicians, sitting, eyes closed, with butterflies perched on his fingertips as he meditated through the heat of midday. The feeling of having insects inside your clothes and touching you softly all the time is infuriating, but Abhi had broken through to the other side. He looked at peace.

We woke to bees swarming our tents. As soon as the first bead of sweat dripped down our faces, they were on us.

Wildness began to beckon and call. Just looking at the surrounding forests, broken by grassland plateaus created by frequent fires, you could see that these little river valleys were filled with unseen wildlife. Every time I went for a walk, I would find animal tracks and droppings along the narrow grassland corridors between the floodplain or peatlands and the forest. The forests had been opened up by the activity of elephants: Large trees had been pushed over, and lion, leopard, and hyena prints became more common in the dirt. These were the changing seasons at the Source of Life.

Just below the confluence with the Tempué River, we found a rubbing on a tree by an elephant. More than two meters (6.5 ft) off the ground, it must have come from a big elephant. We heard lions and hyenas calling that night and met hunters along this remote stretch of river who told us about leopards raiding their camp. These Luchazi hunters said that young people weren't interested in the hardships of hunting bushmeat in these incredibly remote forests. They acknowledged that the Lisima Landscape held many secrets, and that these rivers and forests must be approached with deep reverence. Second time around, the upper reaches of the Okavango Basin were starting to share their secrets. Instead of spying on us, the people and wildlife were beginning to show themselves.

For the next two days, we saw signs of elephants and animal tracks along all the trails. We stopped to hide from a passing thunderstorm under a tarpaulin and, while hunching out of the rain, found lechwe droppings. I got out my binoculars and scanned the distance. There they were: red lechwe, aquatic antelope common in the Okavango Delta, but—along with several other large game species, like zebras and giraffes—thought to have gone locally extinct during the civil war. A small herd with a beautiful male, some females, and calves watched us. This was the nucleus of a population's recovery, if allowed to live safely. A few days later, we saw lechwe again and filmed them as they grazed. We also saw several sitatunga, another wetland-dwelling antelope, jumping into the river and swimming to the safety of nearby reedbeds at the sight of us.

A few weeks earlier at Samanunga village,

IN MEMORIAM

Soba Andre Samanunga

The End of an Era

Soba Andre was the chief of Samanunga, a village hidden deep in the remotest part of the eastern Angolan Highlands. I met him in 2016, when I was visiting his village to ask permission to explore the Cuanavale River from its source lake. He was the guardian of the river, responsible for protecting its ancient crystal clear source lake. He passed away in 2024.

In life, Soba Andre was a living ancestor, a custodian of Luchazi language, history, and tradition. His people have lived in these high-altitude miombo woodlands for at least 700 years. Clean air, carbon storage, fresh water, and abundant life have been theirs for a millennium.

Originally, the Luchazi had a well-structured democratic system and no paramount leader. They established a close relationship with the Chokwe people, who had also migrated to the Lisima Landscape. During the war, most Luchazi people fled Angola. After 2002, some returned to their traditional way of life as hunters, farmers, and beekeepers. Soba Andre worked every day in his fields. He was quick to trust, smiled generously, spoke humbly, and helped where he could.

When Soba Andre sat down with you, you felt his powerful radiating presence as an overwhelming calmness. He lived his life in one of Africa's last wild places in limited contact with the outside world. Our arrival heralded the beginning of a new era—a new way of life that his sons will have to navigate without losing themselves and their traditions.

I still carry the traditional knife he gave me.

Soba Andre was the visionary leader of Samanunga village, near the source lake of the Cuanavale River, and one of our earliest supporters.

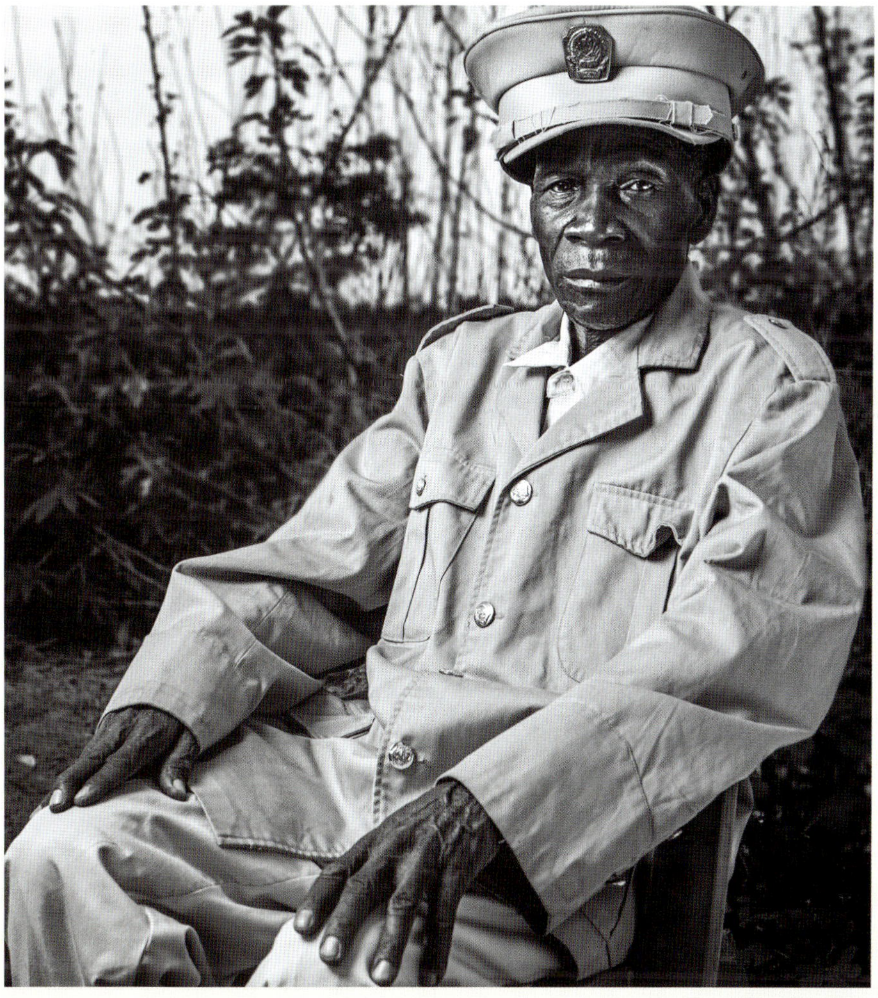

near the Cuanavale source lake, we had been told that African wild dogs, also known as painted hunting dogs, were common in these remote forests during the summer months. This was hard to believe, more than 805 kilometers (500 mi) from any other known populations of this endangered species. We had seen dog tracks, but we thought these were from hunting dogs used by Luchazi bushmeat hunters. Then, while on the Cuanavale River expedition, a biodiversity survey team at the Cuito source lake sent us an extraordinary photograph: four wild dogs that had just killed a common duiker in front of them. It was evidence of a new population of an endangered species. This was truly a sanctuary for biodiversity. There was hope here, a flicker of hope, deep in the Cuanavale wilderness.

In 2015, we had budgeted 19 days to complete the Cuito River from its source to the confluence. That trip had ended up taking us six weeks. We got luckier the second time around: 19 days is exactly how long it took us in 2016 to navigate the Cuanavale River to the confluence, a trip of about 600 kilometers (370 mi). We had come prepared and found a crystal clear, free-flowing river in the remotest part of the eastern Angolan Highlands. There were hardly any foot or motorbike bridges; the adjacent valleys were steep and lined up one after the other, making the region impossible to traverse. Like many places in these highlands, the Cuanavale's remoteness had preserved it from the outside world.

We didn't linger. We planned to return in winter, when the bees were hibernating and

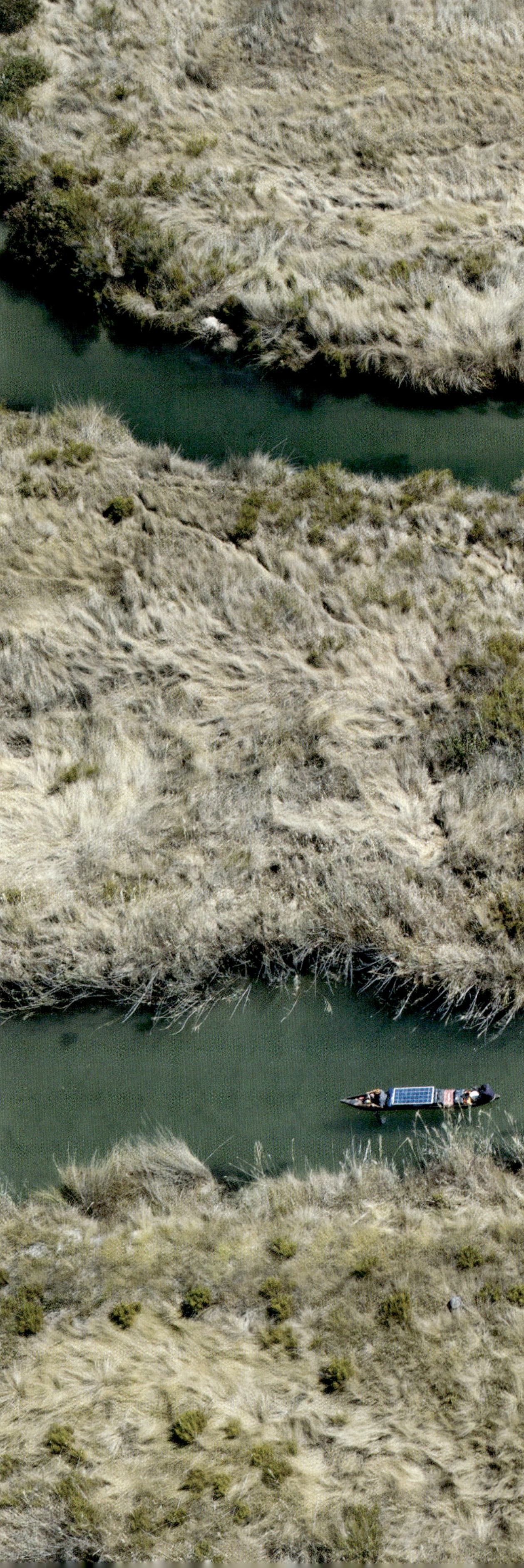

The Cuanavale River makes a gradual descent through the peatlands of the Angolan Highlands, causing the river to wind upon itself. Navigating this river, you feel like you are going in circles.

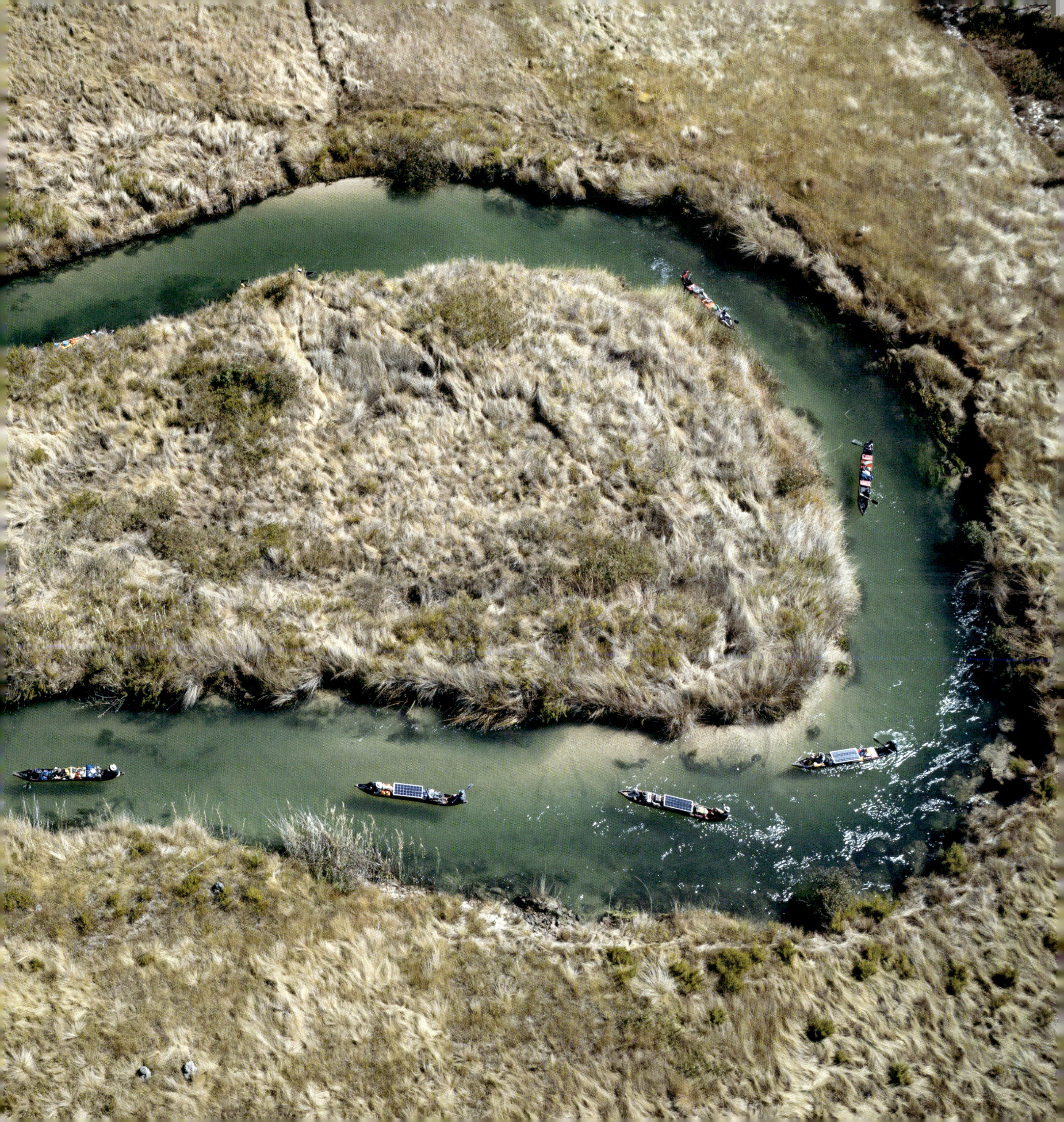

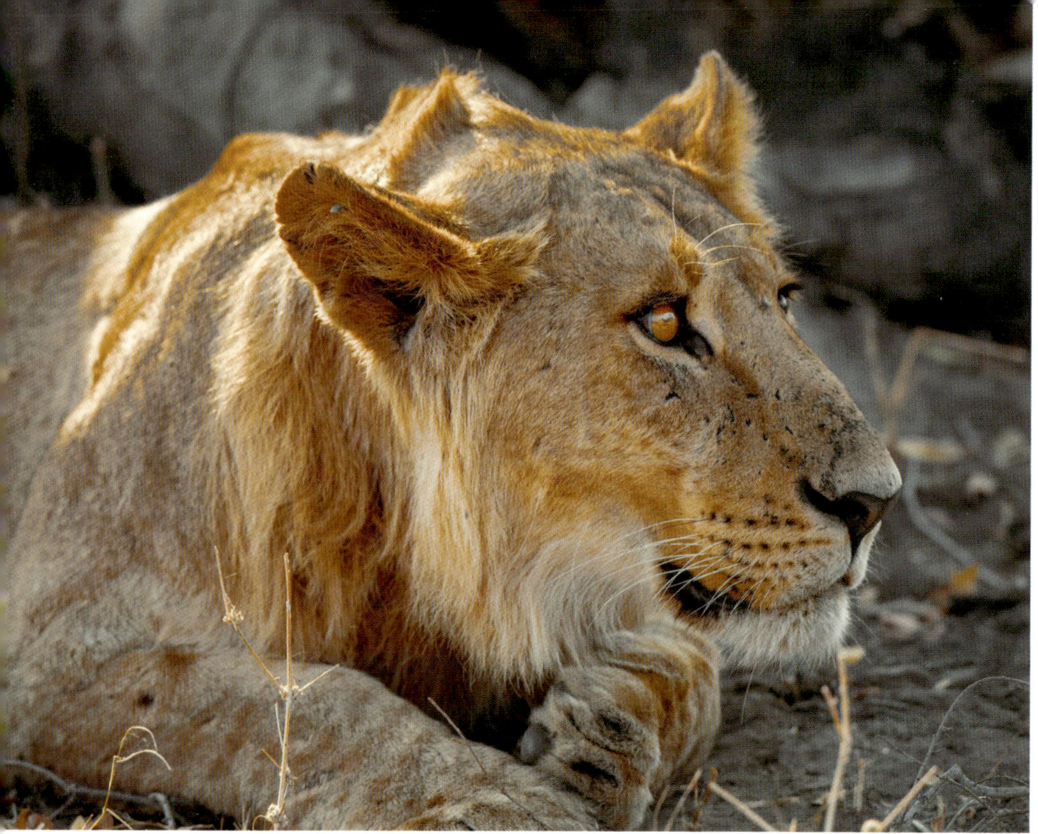

Below the confluence of the Cuanavale and Tempué Rivers, we heard lions calling at night, found hunting camps, and saw many signs of elephants.

when much of the wildlife had retreated, no longer aggregating around the river as they sensed the coming rainy season. In wet weather, migratory and nomadic species like elephants and wild dogs followed the floodwaters downstream to escape the bitter cold at higher altitudes.

Later that year, we cycled on fat-tire mountain bikes back into these remote valleys and set up a large network of motion- and heat-sensing camera traps, which began revealing an extraordinary profusion of wildlife taking refuge in this remote wilderness.

IN 2016, the government of Angola asked us what to call these lands of vast forested watersheds, with their previously undocumented source lakes and peatlands, hundreds of rivers and streams, and Africa's largest remaining intact miombo woodland. The local hunters had many names for the different parts of the eastern Angolan Highlands. During our travels on the Cuanavale, we began holding lengthy consultations with local communities—the Chokwe, Luvale, Mbunda, Luchazi, Lunda, and Ovimbundu—consulting their hereditary chiefs, the *sobas* and *regidora*. By the end of that year, the traditional leaders had decided on a Luchazi name for the sources and watersheds of all of these rivers: Lisima lya Mwono—the Source of Life. Most of the villages between the source lakes had been abandoned during the civil war, but since 2012 the government had encouraged people to return to their ancestral lands. "Lisima lya Mwono!" soon became a social movement, founded in pride and ownership of a landscape now coming back to life after having been traumatized and isolated by decades of warfare.

WE HAD NOW COMPLETED baselines of the Cuito and Cuanavale Rivers, as well as the major channels of the Okavango Delta. To complete the picture of the waterways feeding the Okavango, we had one major tributary left to document: the Cubango River, the most distant source and accessible only by a tarred road. This expedition would take us through the most heavily fought-over part of Angola during the civil war, but it would also reveal a landscape desperately in need of the science—and conservation—that the National Geographic Okavango Wilderness Project had been entrusted to carry out—the last piece of the puzzle.

Soba Augusto Tchinjanga

Peaceful Warrior

SOBA AUGUSTO, a community leader from the village of Chinjanga, knows his ancestors. He knows where he comes from. His age-old wisdom springs from wildness. He has lived in the city; he has witnessed war; he is enough a part of the modern world that he has a smartphone. Every time we meet, he chuckles. Perhaps I look frantic to him. We sit together, and I can feel us being present.

Soba Augusto was a commander in the UNITA rebel forces during the civil war in Angola, and he worked closely with the charismatic leader Jonas Savimbi. No signs of this war-torn past remain in him today. The atrocities he witnessed seem to have made him more sensitive to the value of life. Spending time with him, you very quickly understand that he is a conservationist. His village is next to a bridge over the Cuanavale River, and the valleys below his village are among the last with elephants and lions. He told me that he doesn't allow just any bushmeat hunters into these valleys, as the wrong people will upset the lions and elephants, causing them to come up to his village to destroy his crops and kill his livestock.

An unyielding character, he has watched his village empty. Now, only his old mother and a few young boys remain. Soba Augusto has traveled widely in Angola and learned many things about traditional medicine and magic. As a result, he is feared, and he leverages that fear to protect the wildlife he so values. He knows the next generation will be grateful when they venture down into his valleys and find grand elephants and great lions not only in myth and legend but still living in the nearby wilderness, commanding respect and bringing the forests to life.

Soba Augusto was a prominent figure in the UNITA rebel forces during the Angolan Civil War and, as the leader of Chinjanga village on the Cuanavale, has a deep knowledge of this important river.

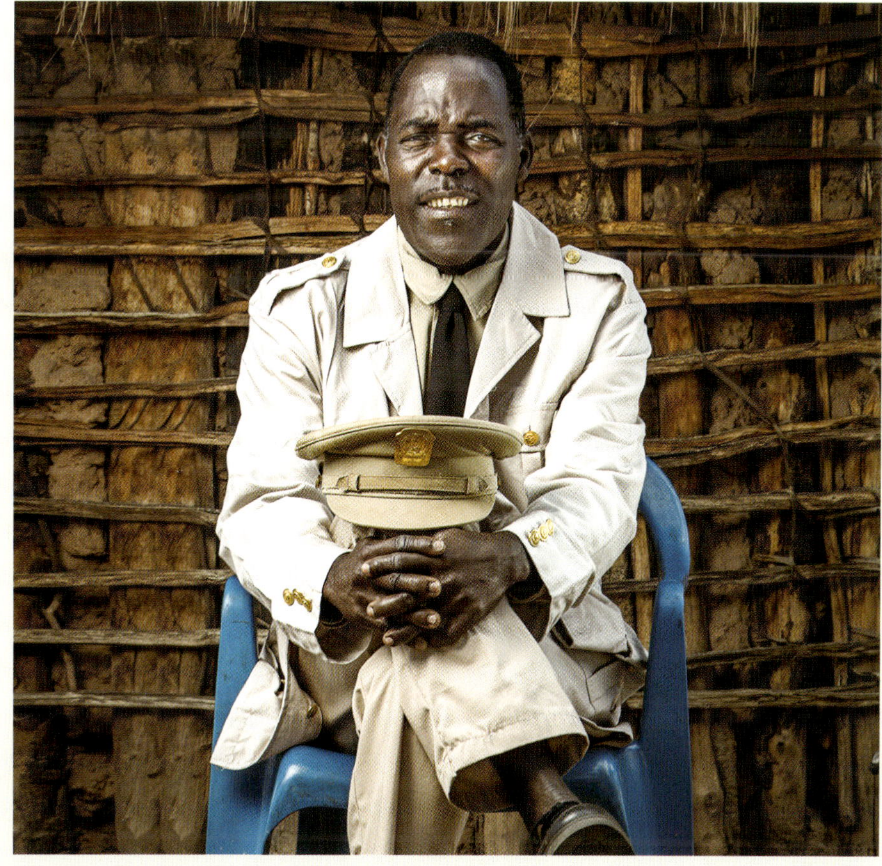

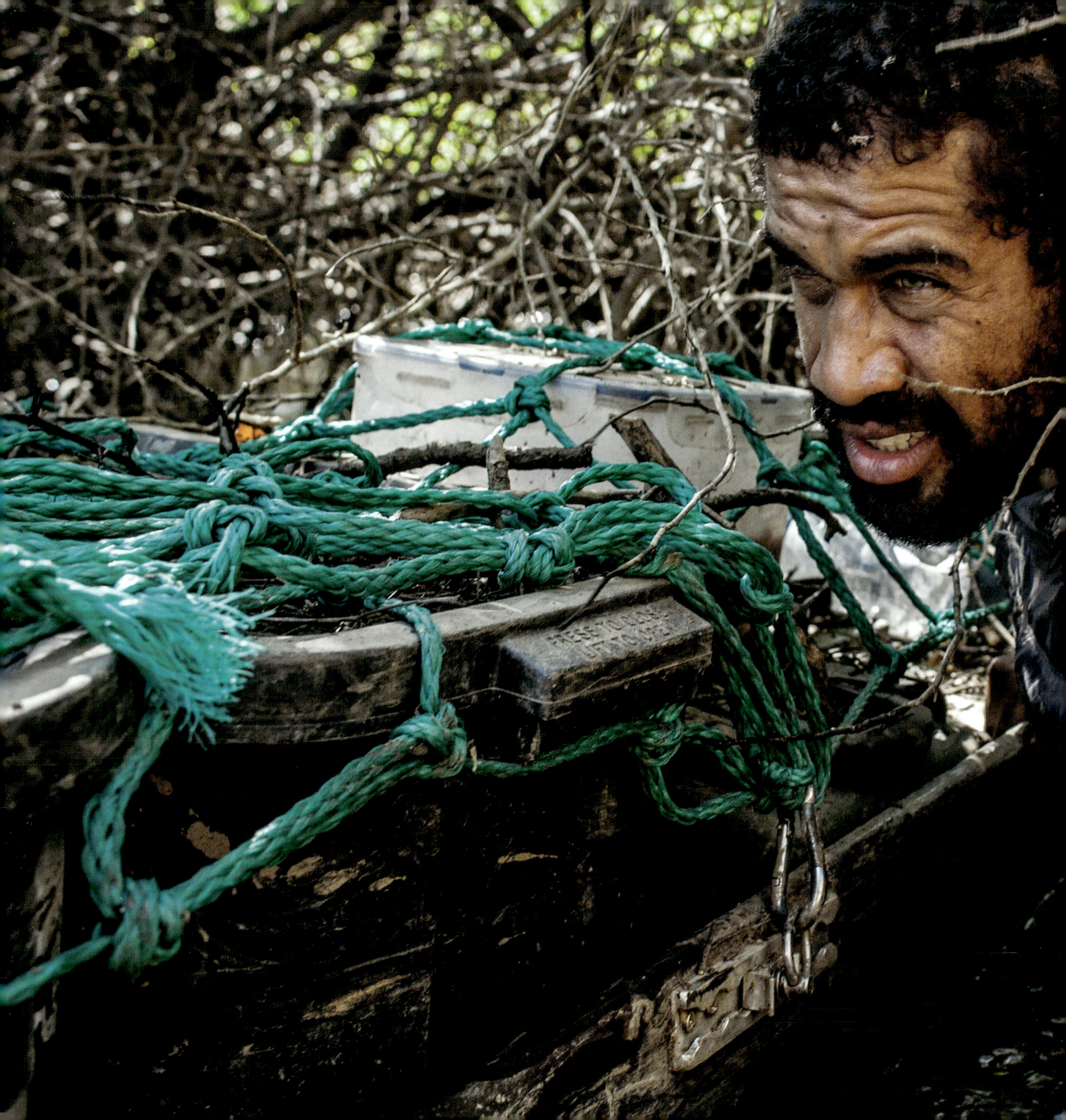

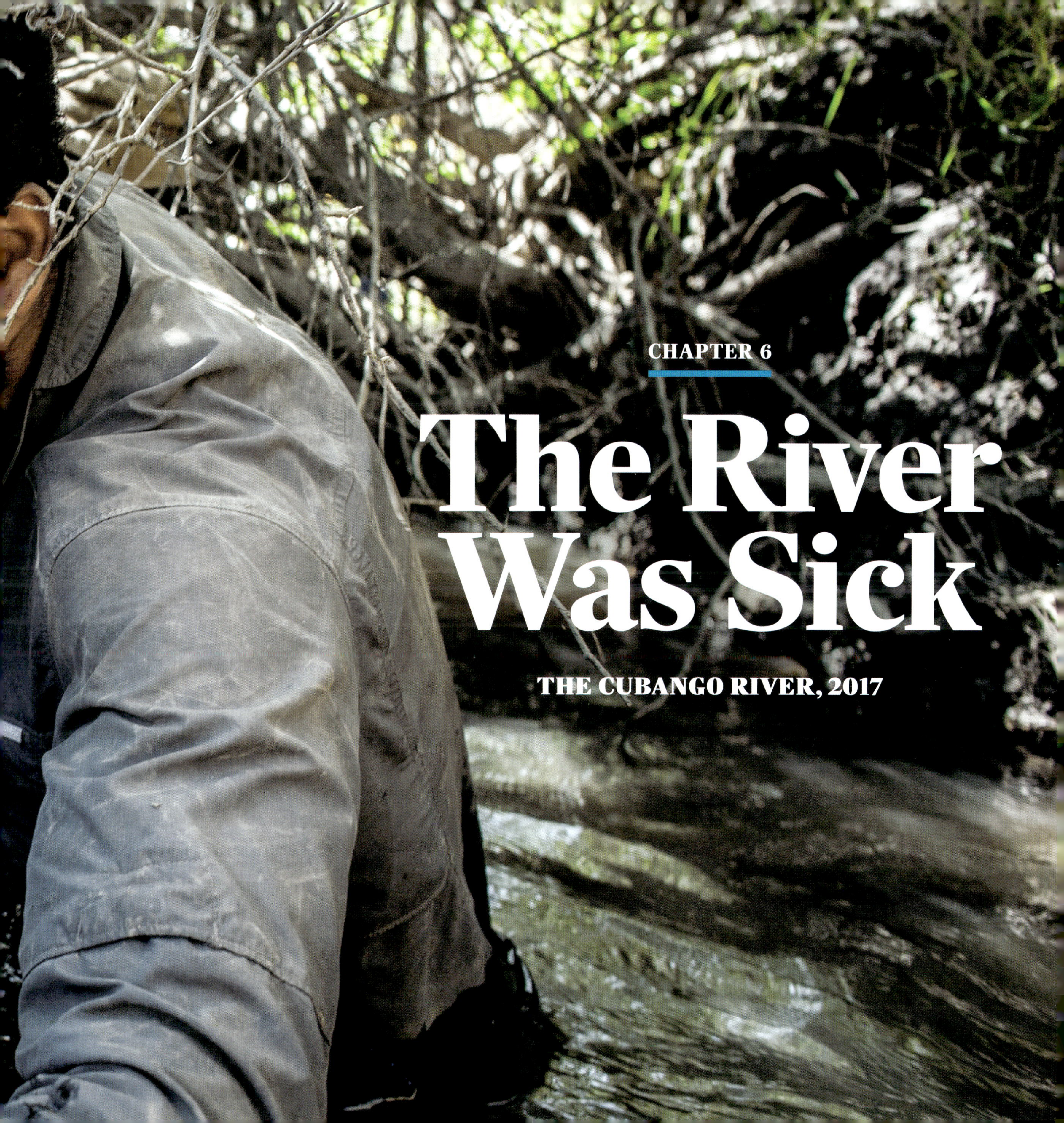

CHAPTER 6

The River Was Sick

THE CUBANGO RIVER, 2017

As in the upper reaches of the Cuito River, mornings near the source of the Cubango River, a place of wonder, were cold and misty. **PAGES 122–123:** *Kerllen Costa, Adjany's brother, pushes a fully loaded mokoro through a tree blockage. We had to push through this sort of densely woven undergrowth for weeks on the Cubango.*

THE CUBANGO RIVER was not officially part of our project area: The region had already been well documented and explored by government departments and resource planners. Following the civil war, most small towns along the river were surrounded by land mines, and all bridges were suspected to be too dangerous to portage around. Yet we also knew that the Cubango provided an estimated 55 percent of the water in the Okavango Delta and, as the most populated river, was much more impacted by human activity than the other remote source waters we had explored. Without information on the health of this river, our knowledge of the Okavango Delta was incomplete.

From the very beginning, we found ourselves facing a completely different type of river. In the eastern Angolan Highlands, the water that flows into the Cuito and Cuanavale Rivers moves slowly. The land slopes more gradually, creating its characteristic deep peatlands and lakes that continually supply the delta with water through the year. The waters to the west that form the Cubango, on the other hand, are in a rush. Although the Cuito and Cuanavale flow over sand, the Cubango flows over rock and compacted soil. Seasonal rains rush down the hard, steep landscape and leave the area in a matter of weeks. Between December and March, the Cubango River almost flash floods, with large volumes of water carrying high sediment loads crucial to the functioning of the Okavango Delta's alluvial fan, setting up the flood pulse dynamic that sustains the delta's abundant floodplains.

We discovered that for the first few hundred kilometers of the upper Cubango, starting from its source near Huambo in the Angolan Highlands, dense tree blockages between 3.7-meter-high (12 ft) sandbanks cut into the landscape, followed by large-scale rocky rapids that, due to their size, backed up the river to form flooded reedbeds. These areas provided some degree of water storage that could be used in the future for sustainable agriculture or urban development. All the way down the river were small-scale subsistence farms, with much larger fields cleared near towns, villages, and settlements. These local people had built amazing wooden water mills to grind cassava and grains in the largest rapids, with systems of fishing baskets and traps made from wild sage bushes on the calmer waters around them. Like so many other places in Angola, traversing this place was like going back in time.

We had to portage around the impassable rapids using the HALO Trust's Kamaz trucks, while our land-based support teams stayed busy conducting biodiversity surveys all the way down. From our mokoros, we studied biodiversity and ecosystems health, took extensive measurements of water quality, and documented river flow dynamics. Everyone on the team wanted to evaluate how humans were impacting this river, so that we could compare it to the other source waterways and monitor the Cubango over time.

As we traveled, human impacts to the river became more apparent. The biggest threat we found was a dam development at Mucundi, which, in 2017, the engineers on-site said was

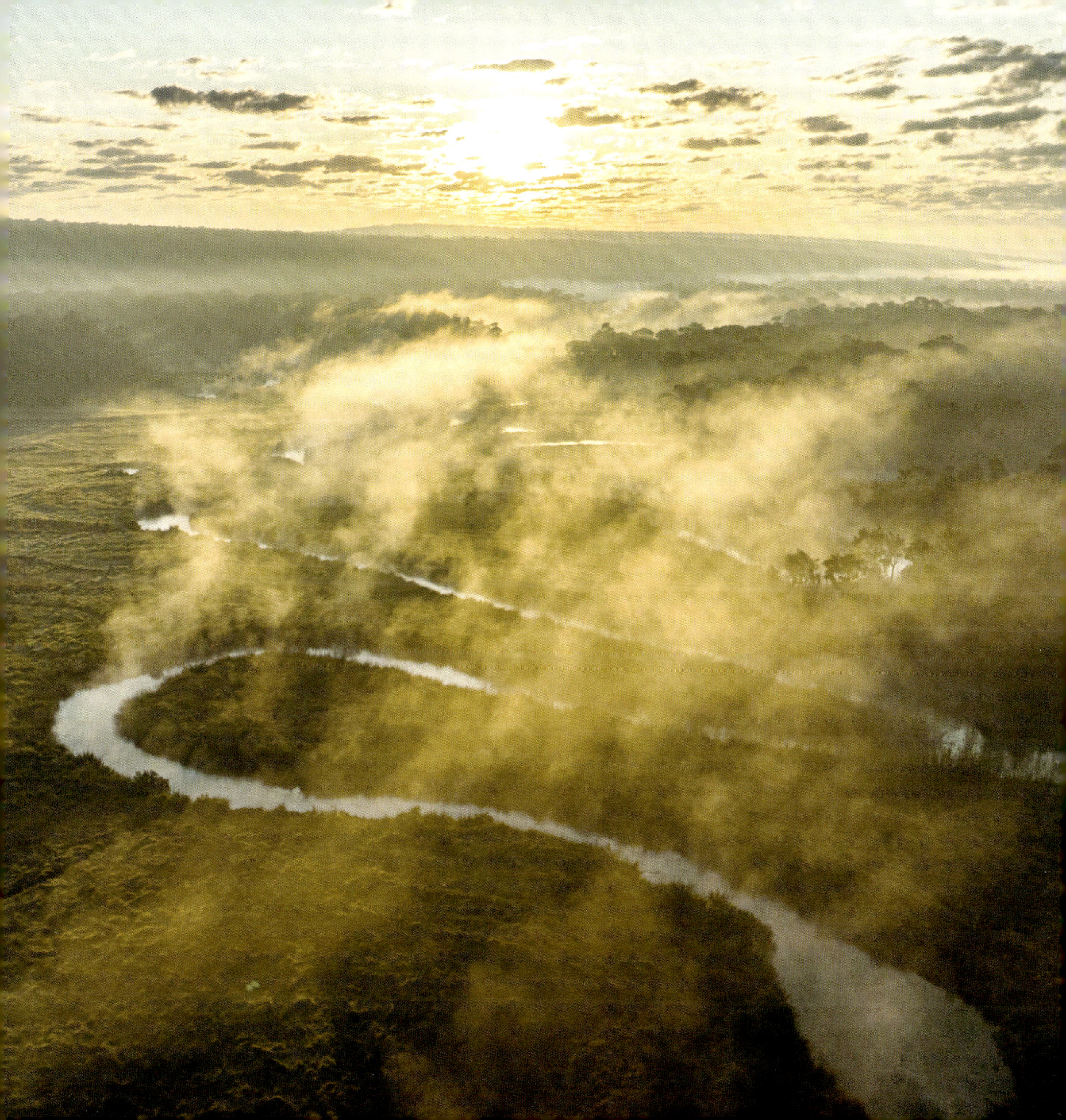

in the "pre-feasibility" stages of planning. Even a small hydroelectric dam would halt the passage of crucial sand and sediment into the Okavango Delta. We have since installed permanent hydrological and meteorological monitoring stations above and below this potential dam development to better inform policymakers and potential investors.

In Angola, the catchments and tributaries of the Cubango River were threatened by large-scale logging of rosewood and teak, commercial rice farming, and forest clearing for charcoal production. Rice farming introduced fertilizers and pesticides into the Cubango River and diverted vast amounts of water, while logging and forest clearing caused erosion and siltation into the river. Forested landscapes create rivers: Without forests, rivers would likely dry up and the landscape would turn into dry grasslands vulnerable to annual fires.

After 39 days, we crossed into the Namibian stretch of the river, where it is known as the Kavango. Although still in near-pristine condition in 2017, the Kavango was clearly the most threatened major section of river in the Okavango Basin. Most of Namibia's rural population lives along the Kavango River, on thousands of small farms with hundreds of unregulated water pumps as well as some large-scale farms with massive irrigation systems, all drawing from the water that feeds the Okavango Delta.

Hippos were almost missing from the river. Local subsistence farmers had dug extensive trench systems along the river to stop them from raiding their crops at night. We found and disarmed hippo snares set by hunters and

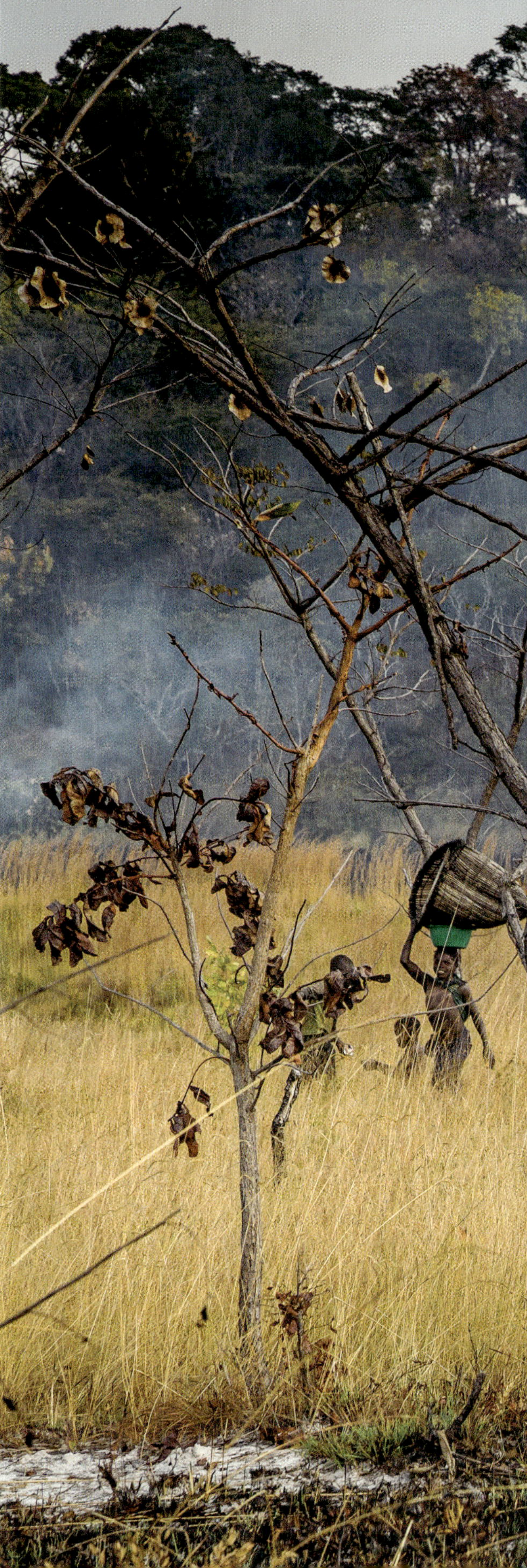

In Angola's remote miombo woodlands, women take their young children with them to live far from their villages and use slash-and-burn techniques to harvest cassava fields and traditional basket traps to catch fish. They sing as they walk, at one with the forests and rivers. Culture and tradition protect these forests for future generations.

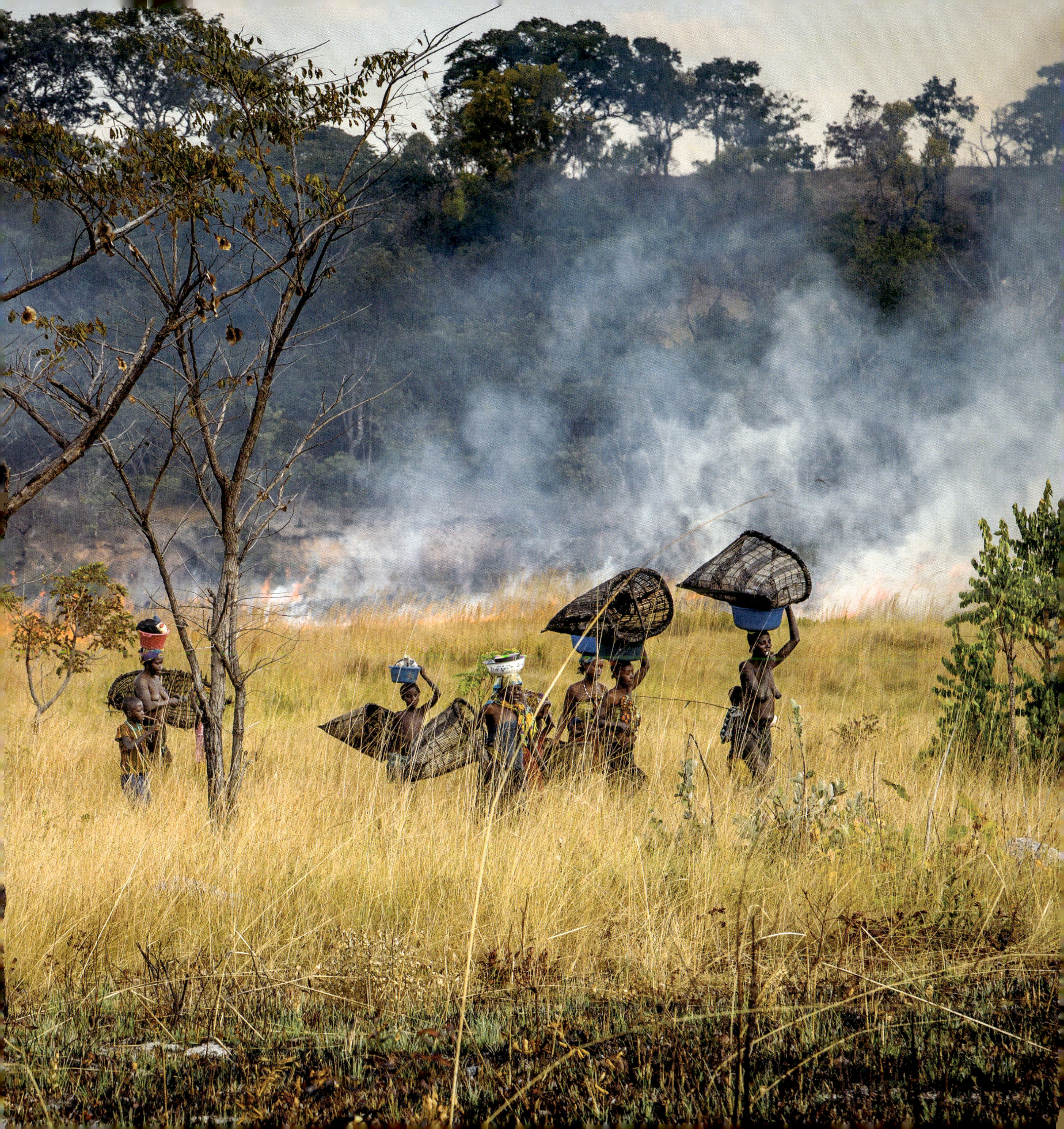

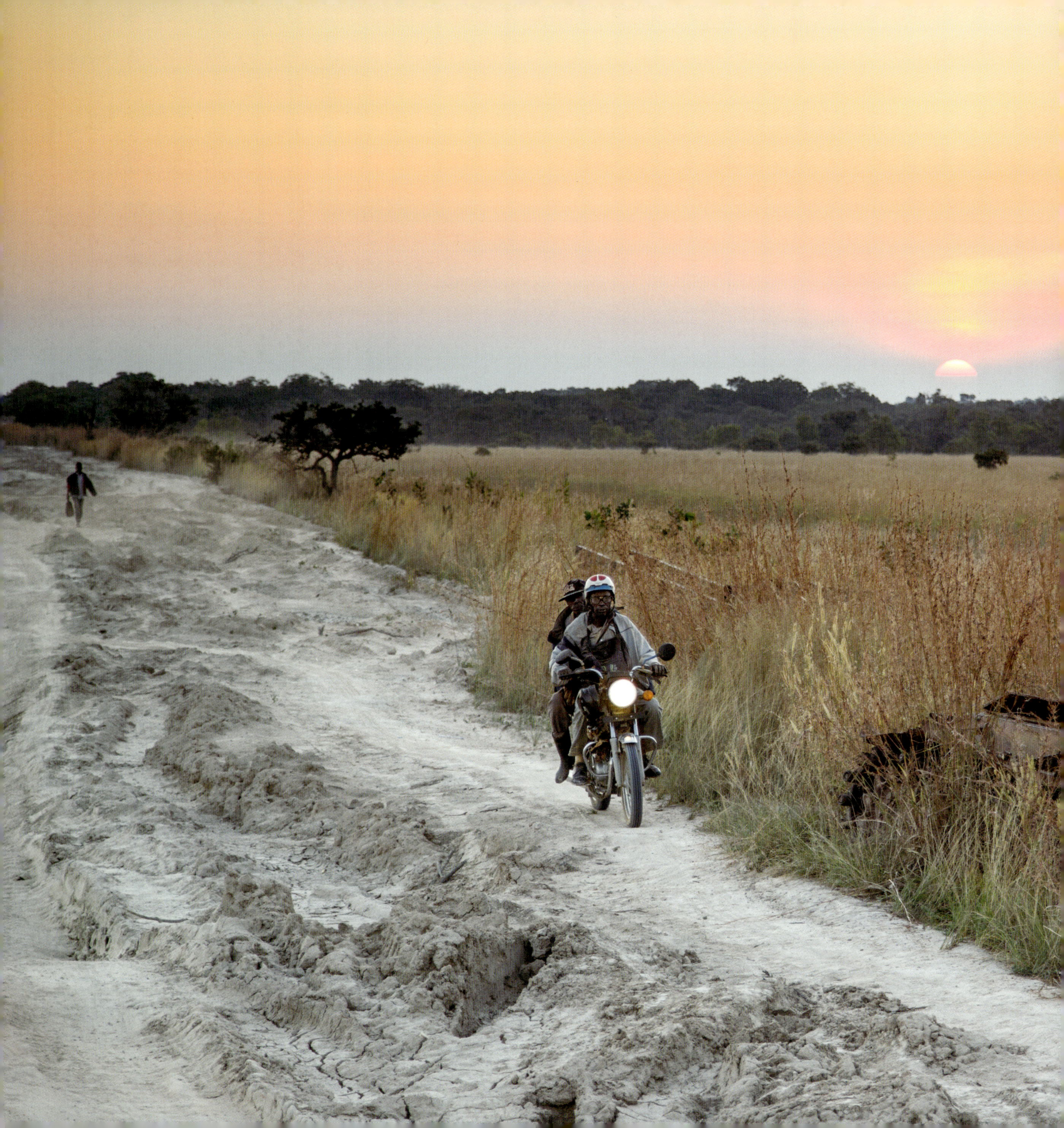

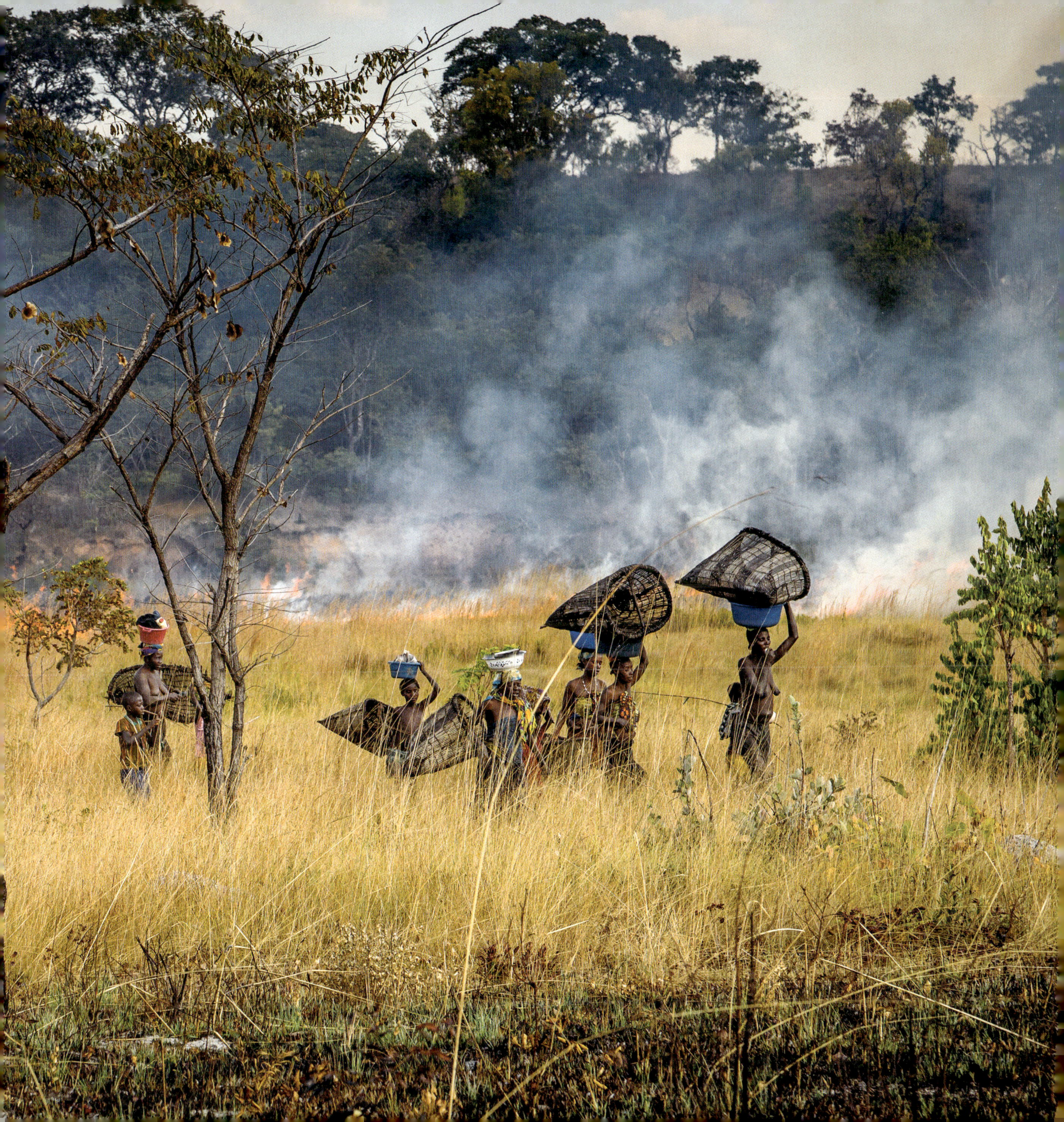

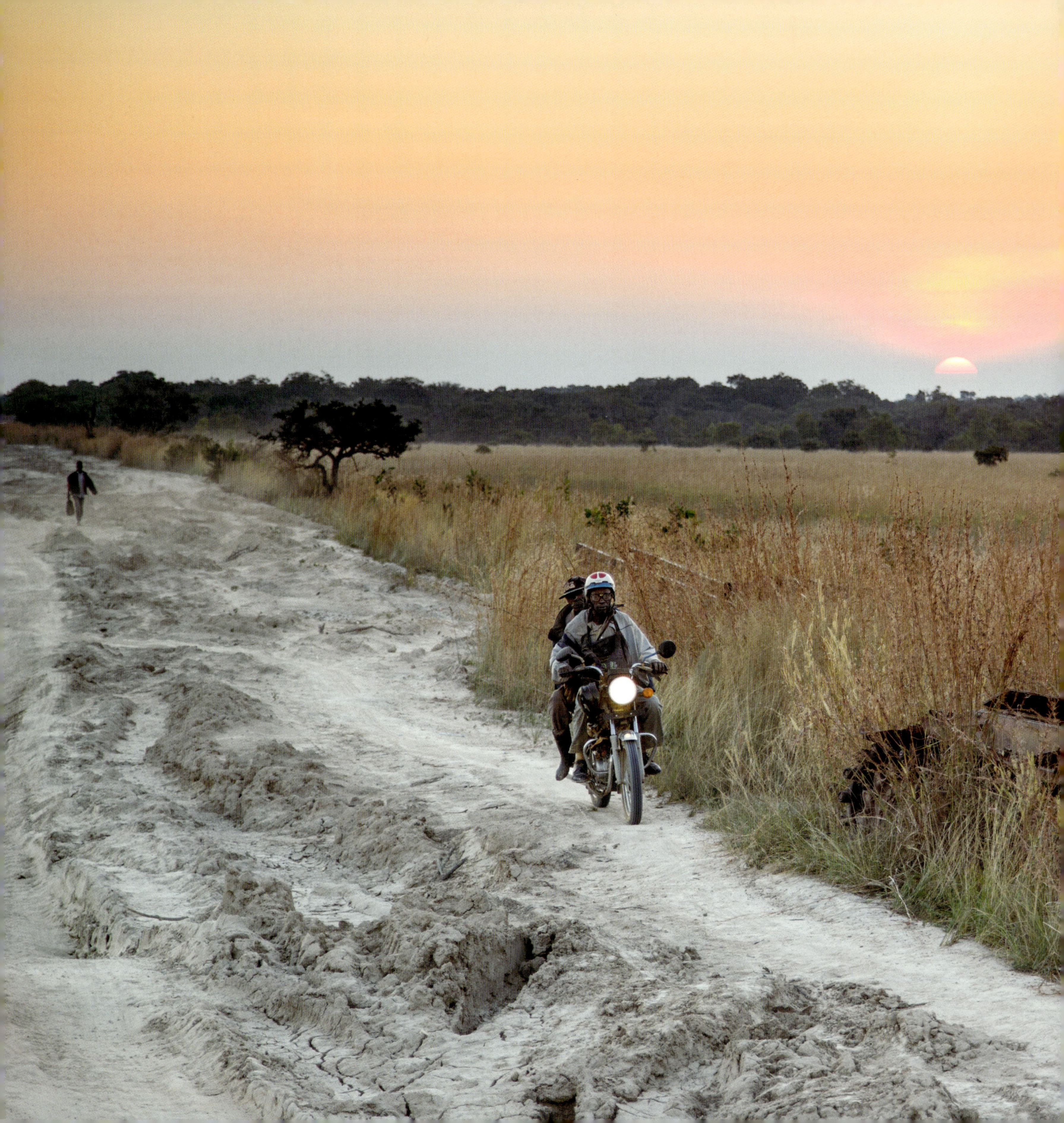

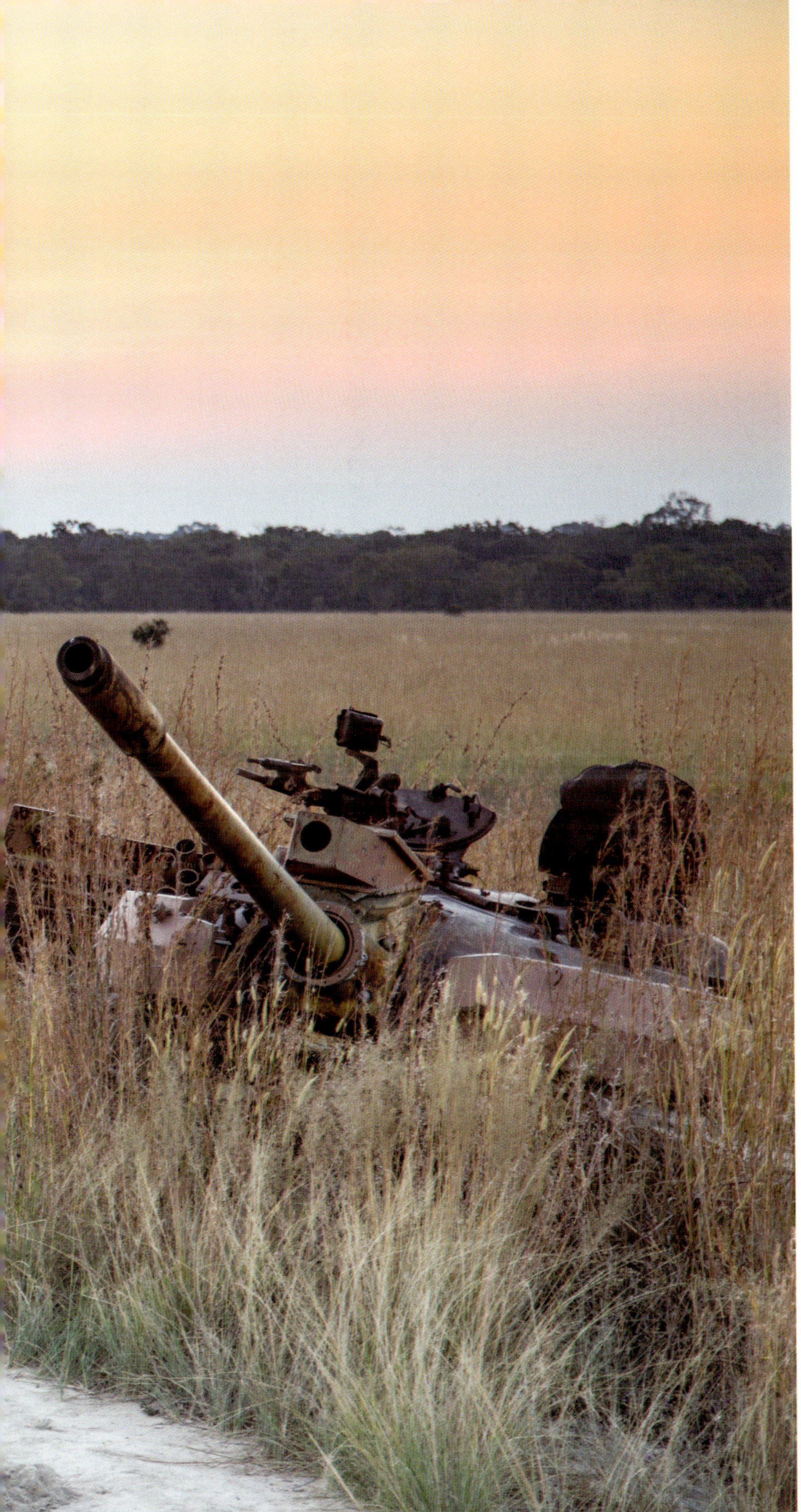

farmers on small islands in the middle of the river. This, in part, explained why their population appeared so low. We found hippos only in front of the police stations in small settlements—the only places where they were safe from hunters and farmers.

Along the way we also documented large-scale rosewood smuggling operations using pontoons, as well as significant declines in fish populations as we progressed downstream. Debris—diapers, aluminum pieces, chip packets, and washing powder bags—littered the river. People fished with washing soap at night to catch big catfish, which had apparently developed a taste for the stuff—at least someone was enjoying the pollution we were documenting. Seeing the world's largest undeveloped river basin succumbing to the pressures of modern society was difficult.

On July 30, 2017, after a difficult 59-day river expedition down the full length of the Cubango River, we reached its confluence with the Cuito River. We ended the trip simultaneously dispirited and inspired: After spending so much of our previous trips in seemingly pristine ecosystems, it was hard not to feel discouraged by the damage that humanity had done. Yet here was a place where we knew protections could make a real difference for the health of people and wildlife. The Cubango expedition was a call to action, one that took us to Capitol Hill in Washington, D.C., where we would advocate for a federal bill to help protect the Okavango by investing in local community projects. We had meetings with leadership of the major global conservation NGOs and

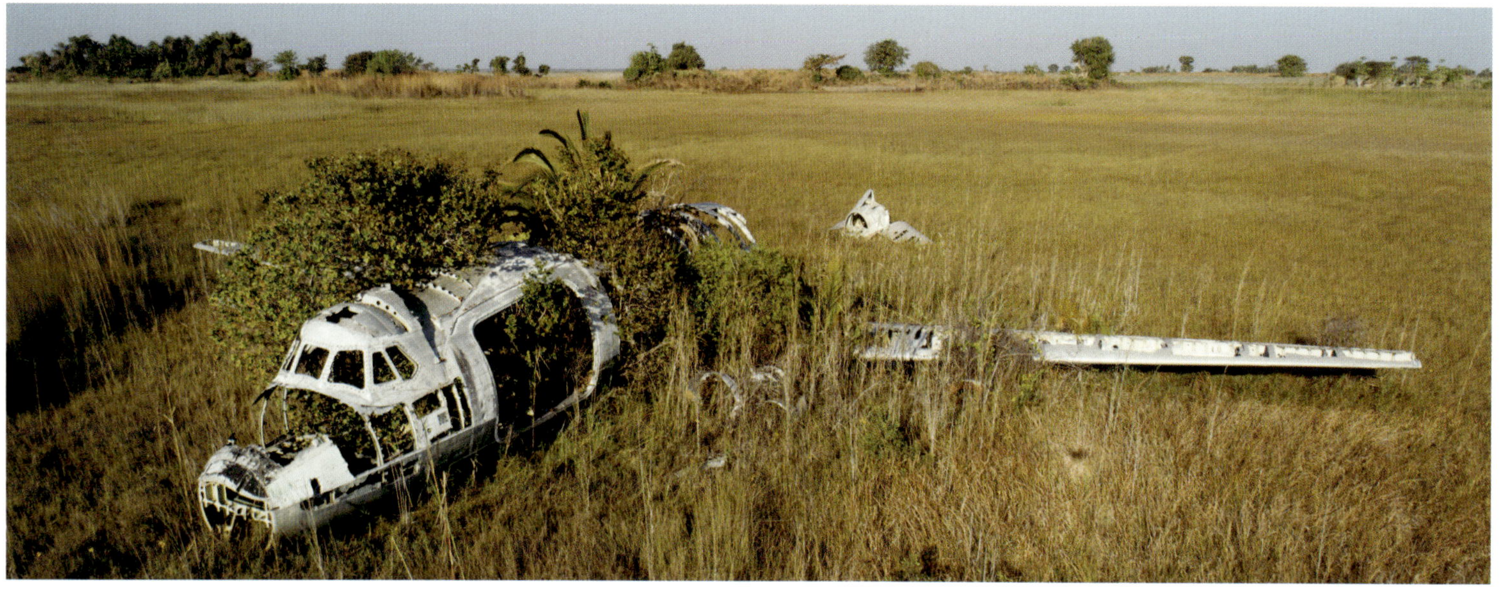

The stark reality of four decades of armed conflict still lies strewn across the Okavango's watersheds in the Angolan Highlands. We found this aircraft in a remote floodplain. **PAGES 128–129:** *You often see blown-up tanks and armored vehicles in the highlands, especially where the bridges were heavily fought over. Land mines are a constant threat.*

shared our findings with regional secretariats: that the threats facing the Cubango River are the biggest threat to the Okavango Delta, while the Cuito and Cuanavale Rivers are the lifeline. Combined, these three represent the biggest conservation opportunity in Africa in decades.

In 2021, we returned to conduct detailed repeat hydrological and ecological surveys down the Kavango River in Namibia. The results were catastrophic: Nitrates from commercial farms had risen to exceed the quantity safe for human consumption, and water levels had plummeted to lows that may dry up the river in coming years. The COVID-19 pandemic had pushed more people onto the river after losing their jobs, and the number of cattle polluting the river had increased exponentially.

River baselines are snapshots until they are repeated, and another repeat transect in 2023 showed us that ecosystems health has gone from bad to worse. Plastic pollution had risen, and escalating *E. coli* outbreaks threatened human populations.

The only good news was that our hydrological modeling had revealed that the Okavango Basin is still far from its breaking point in regard to long-term water flows, which suggests that the farmers and urban developers can use more water during the wet season. However, to keep these waters safe to use, people and farms need to move farther away from this sensitive wetland ecosystem and stop burning adjacent reedbeds. This stretch of river will need a delicate balance of protection and sustainability to ensure that people can subsist on their farms without drawing too heavily from these life-giving waters. Additionally, the El Niño–Southern Oscillation is becoming more unpredictable and severe, making it all the more important to closely monitor the hydrological impacts on the Okavango system.

RIVER GUARDIANS

Three Africans

Brothers in Wildness

KERLLEN COSTA, Chris Boyes, and Götz Neef are brothers in wildness—three humble African heroes who have saved each other's lives, and mine, many times. We have paddled thousands of kilometers together, down wild rivers in Angola, Namibia, and Botswana. Together on expedition for months on end, we would complete a river marathon every day and recharge around the campfire every night for months on end.

Kerllen is a child of Angola, an environmental anthropologist, and a student of the ancestral ways of the Luchazi, the people of the Lisima lya Mwono Landscape. A servant-leader and philosopher, he thinks deeply and acts boldly. Kerllen has taught me that because we are free, we are also inherently responsible.

Chris, my brother, grew up in South Africa but has become a child of the Okavango Delta's waterways. An unbreakable character who chuckles at his own jokes because he thinks no one else will laugh, he delights to see that we all do. Chris is humble and open and knows more about the sources of Africa's last wild rivers—what we call the Great Spine of Africa—than anyone I know.

Götz is a child of Namibia. At 34 years old, he is one of Africa's greatest expeditionary scientists ever. By paddling across river basins, he has recorded the most comprehensive ecological and hydrological baselines ever captured. His name will go down in the history of African river exploration.

Together, we celebrate the glory of Mother Okavango with those who have guided us down the most pristine rivers in Africa. We are all family.

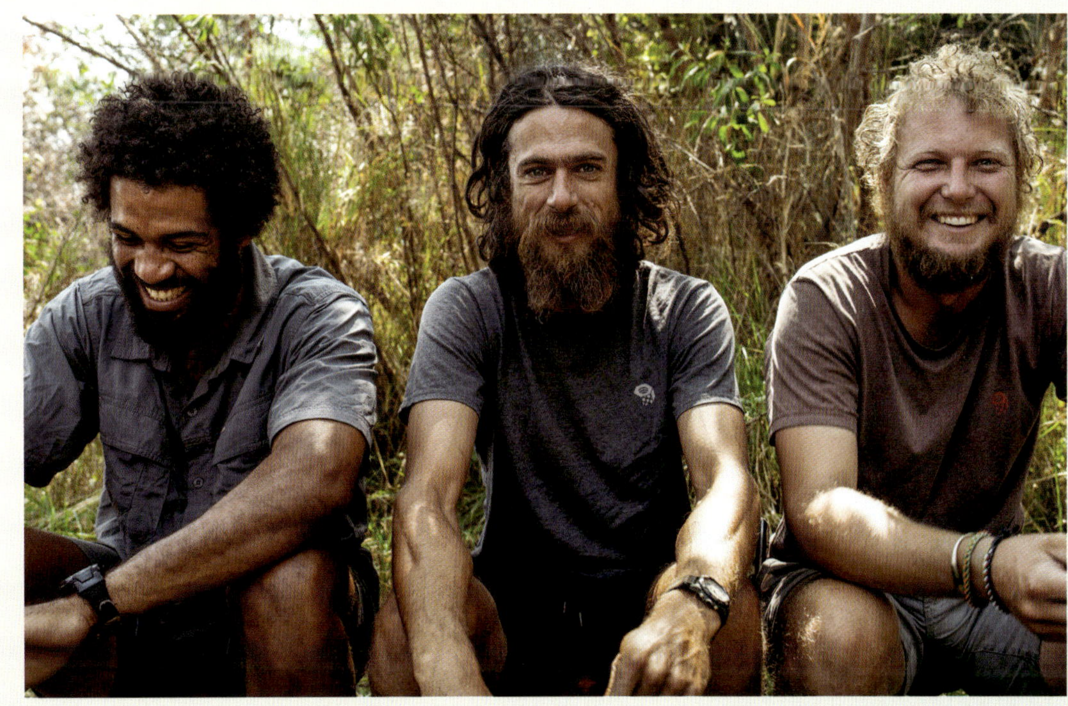

Kerllen Costa (left), my brother, Chris Boyes (center), and Götz Neef (right) led the Cuanavale River expedition together and became close friends in the process.

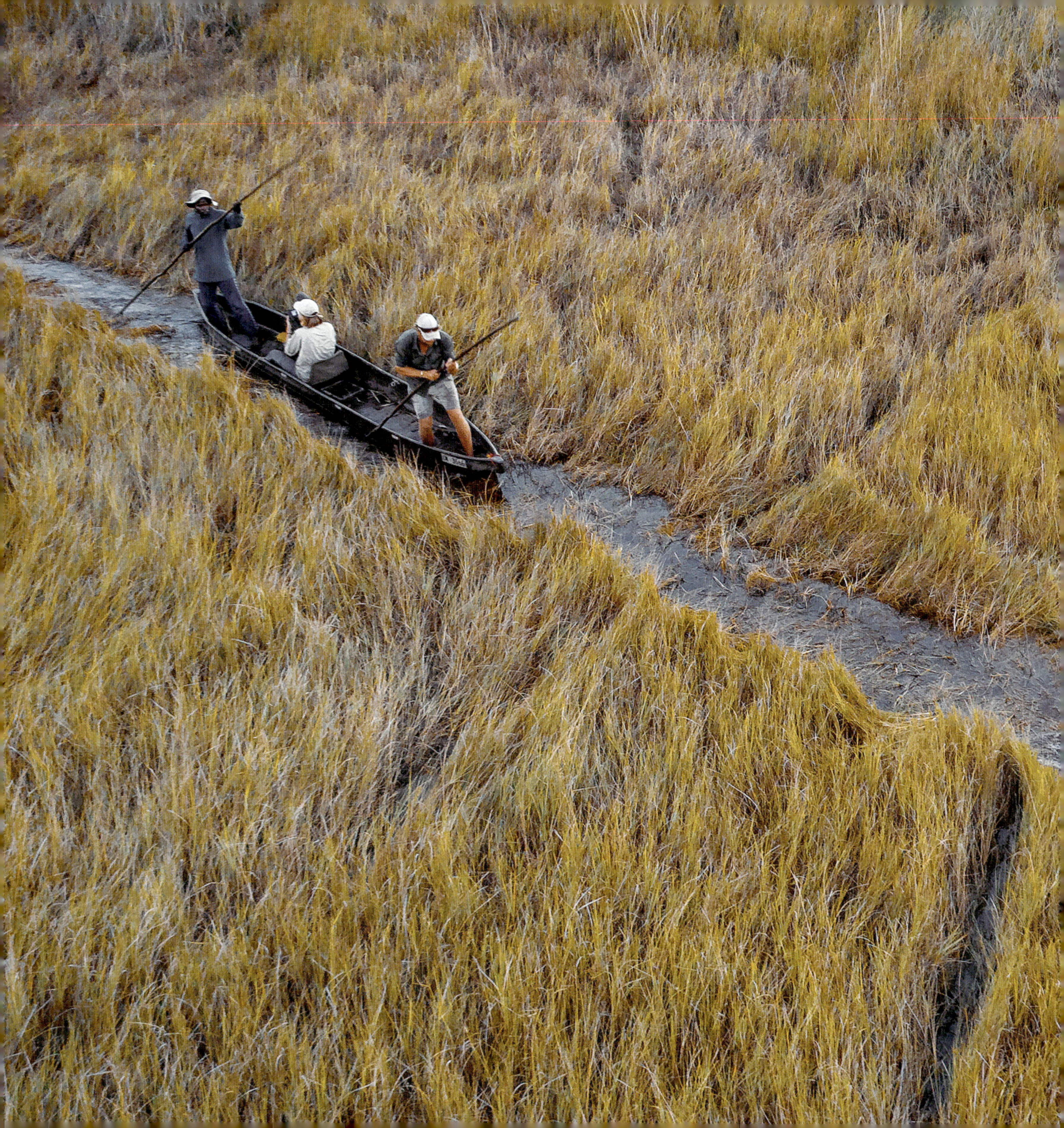

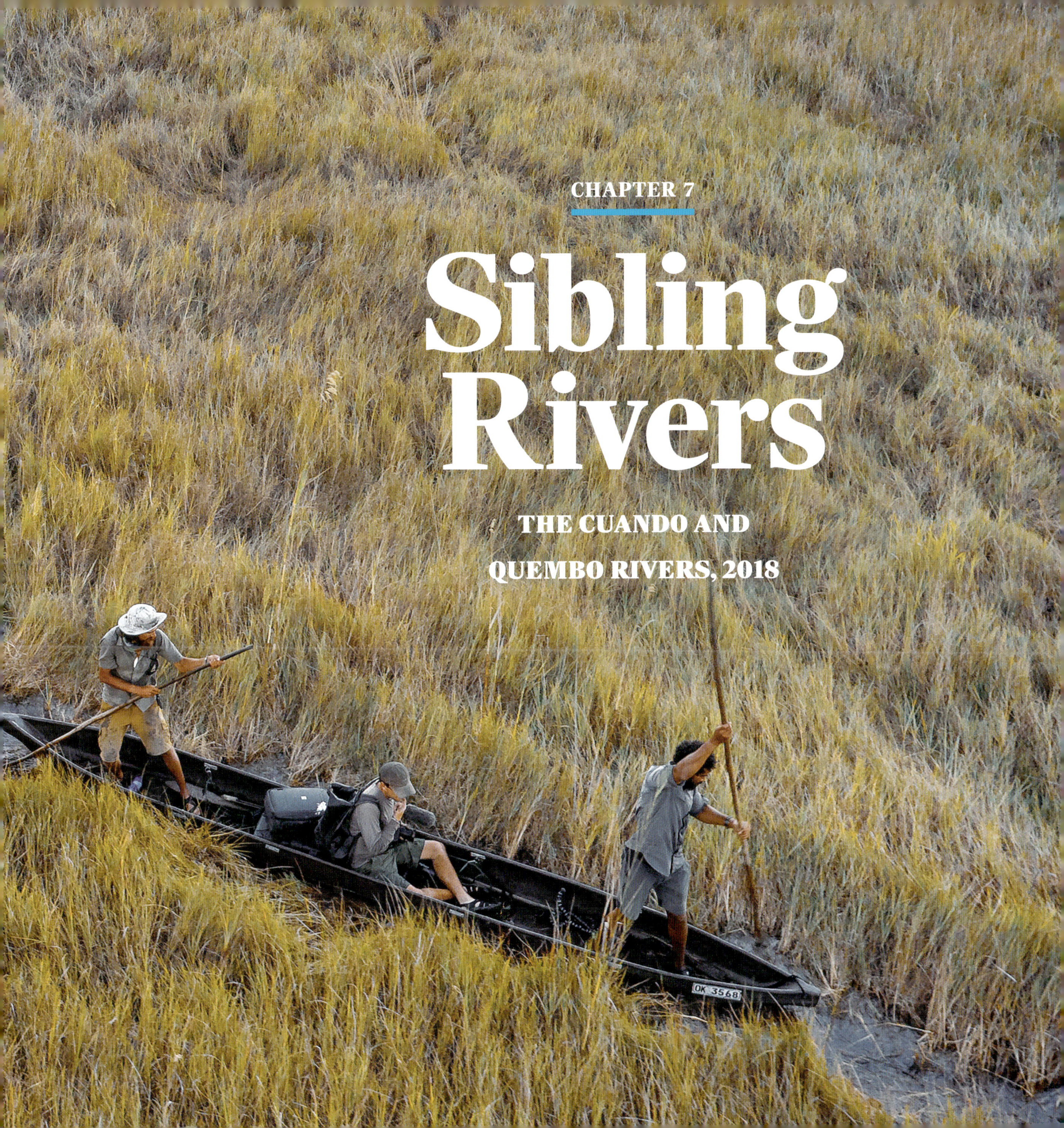

CHAPTER 7

Sibling Rivers

THE CUANDO AND QUEMBO RIVERS, 2018

SINCE 2006, a little-known spillway from Namibia has been flowing intermittently between May and October, connecting the Okavango Delta and the Linyanti Swamp. Known as the Selinda Spillway, this link between the Okavango and Kwando Basins can flow in either direction, with inflow from the Okavango Delta to the south or from the Cuando and Quembo Rivers, originating in the Lisima Landscape, to the northeast. After being guided to their sources by Luchazi hunters, we decided to explore both rivers in 2018. The Selinda Spillway and Linyanti Swamp have abundant wildlife and are crucial to migratory elephants.

I led the Cuando expedition while my brother, Chris, led the Quembo. The Cuando River was bigger and free-flowing, but it had hippos—lots of hippos—and they didn't like people. Two months in, we abandoned the trip after four hippo attacks in two days.

The Quembo was famous for crocodiles, known to be big man-eaters. Crocodiles have stomachs as big as the things they eat, so a crocodile longer than four meters (13 ft) must eat man-size creatures. As Chris described it to me, one five-meter (16 ft) croc picked up Chris's mokoro in its mouth, latching on just behind Carmen Ferreira, our data recorder, who froze, just inches from its jaws. After a few agonizing seconds, the croc released the mokoro, and the team escaped the encounter, their canoe the only casualty.

We returned to the Cuando and Quembo in 2022 and 2024, respectively, filling in all the pieces. To save the sources of the Okavango

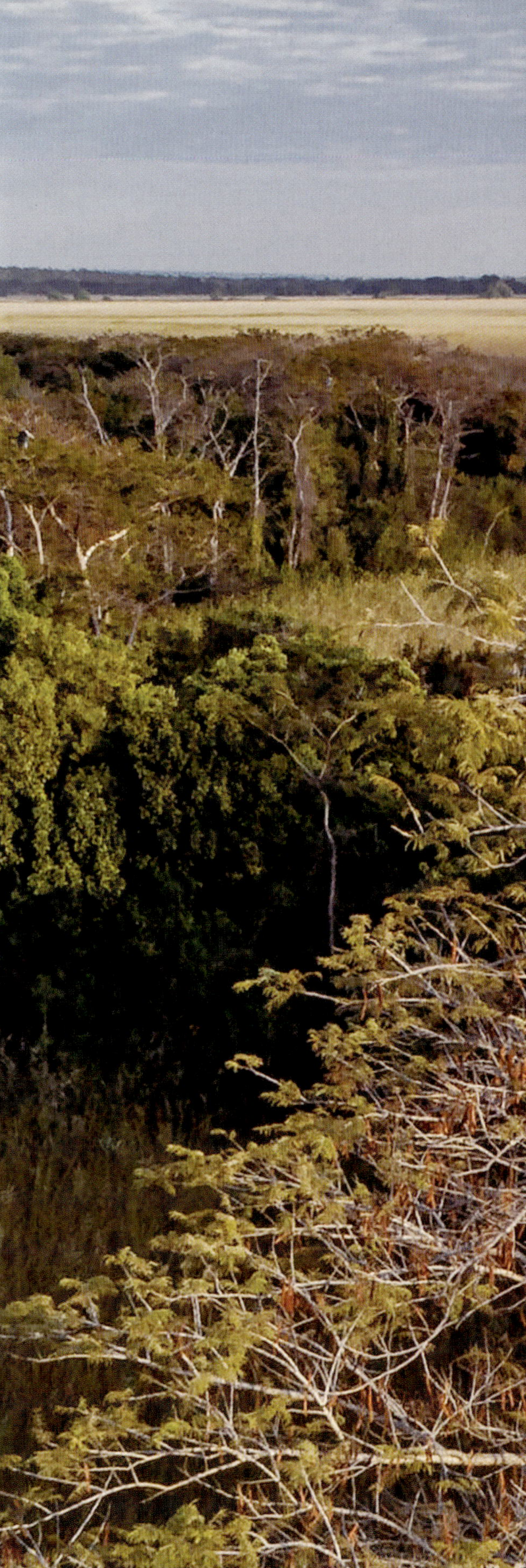

We found marabou storks nesting in large trees along the remotest stretch of the Cuando River. Considered Africa's ugliest bird, they looked like angels to us in this lost wilderness.
PAGES 132–133: *We had to pole our mokoros across the vast floodplains of the Cuando River to find access to land where support vehicles could reach us. Four hippo attacks in two days meant we couldn't continue the expedition.*

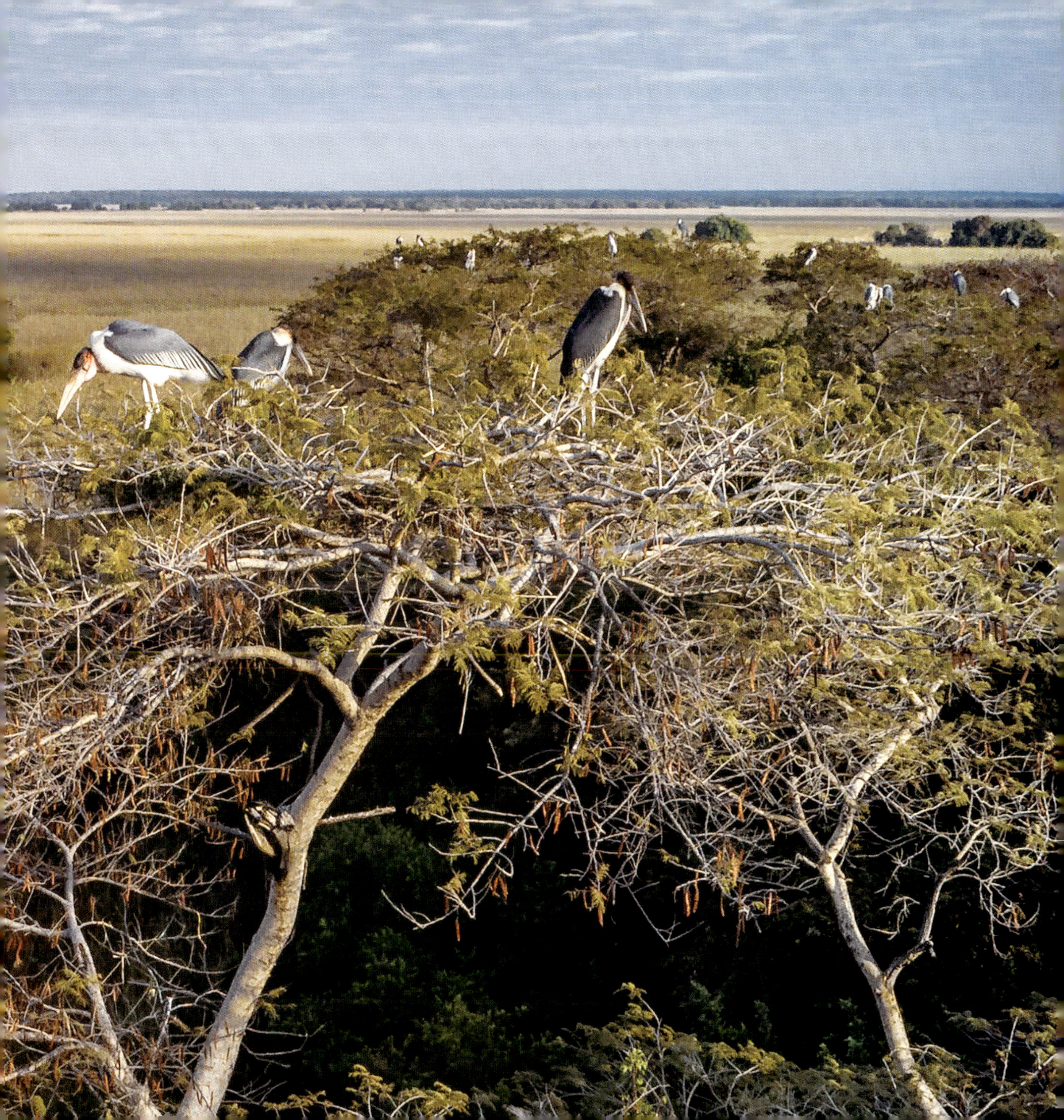

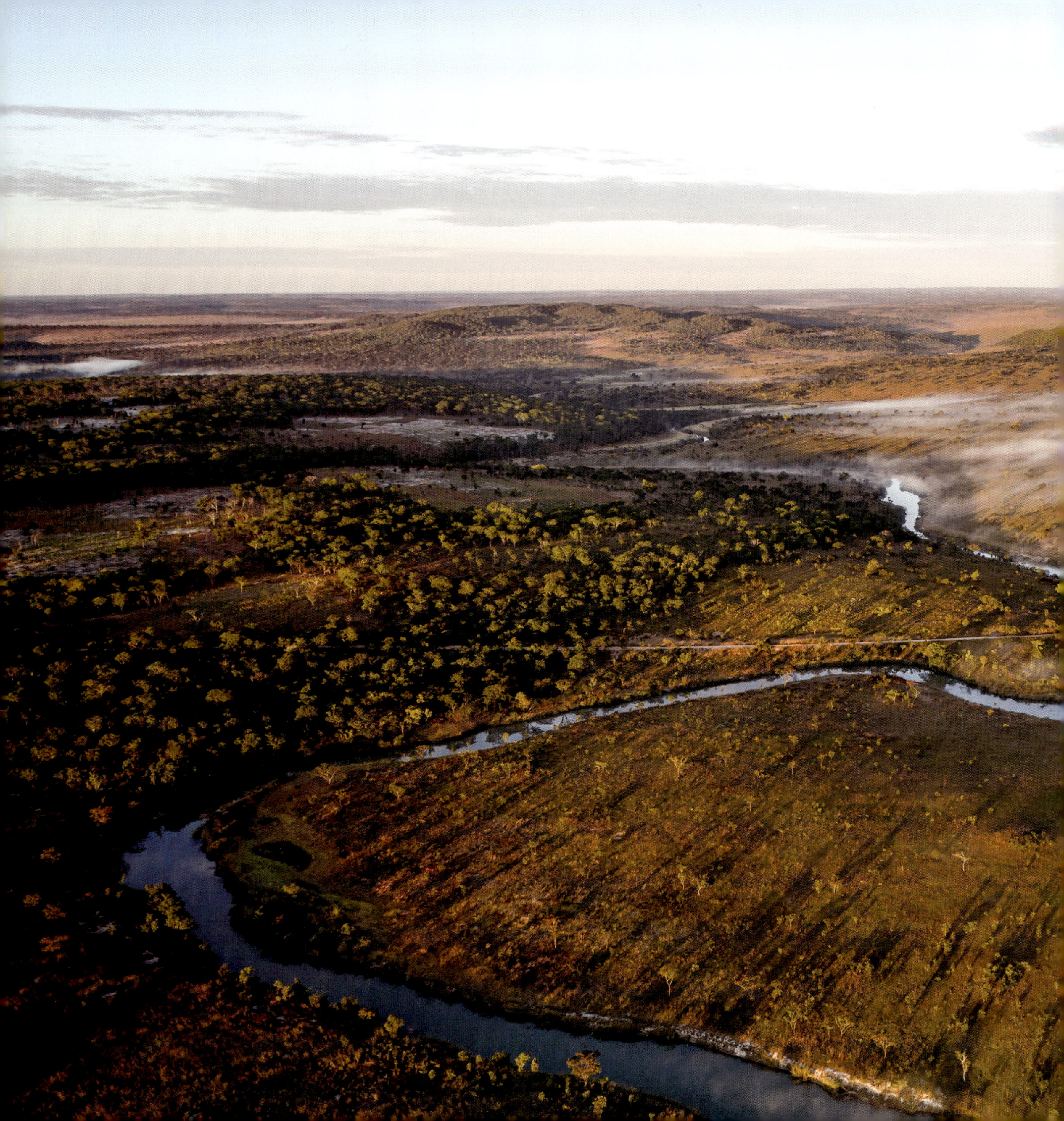

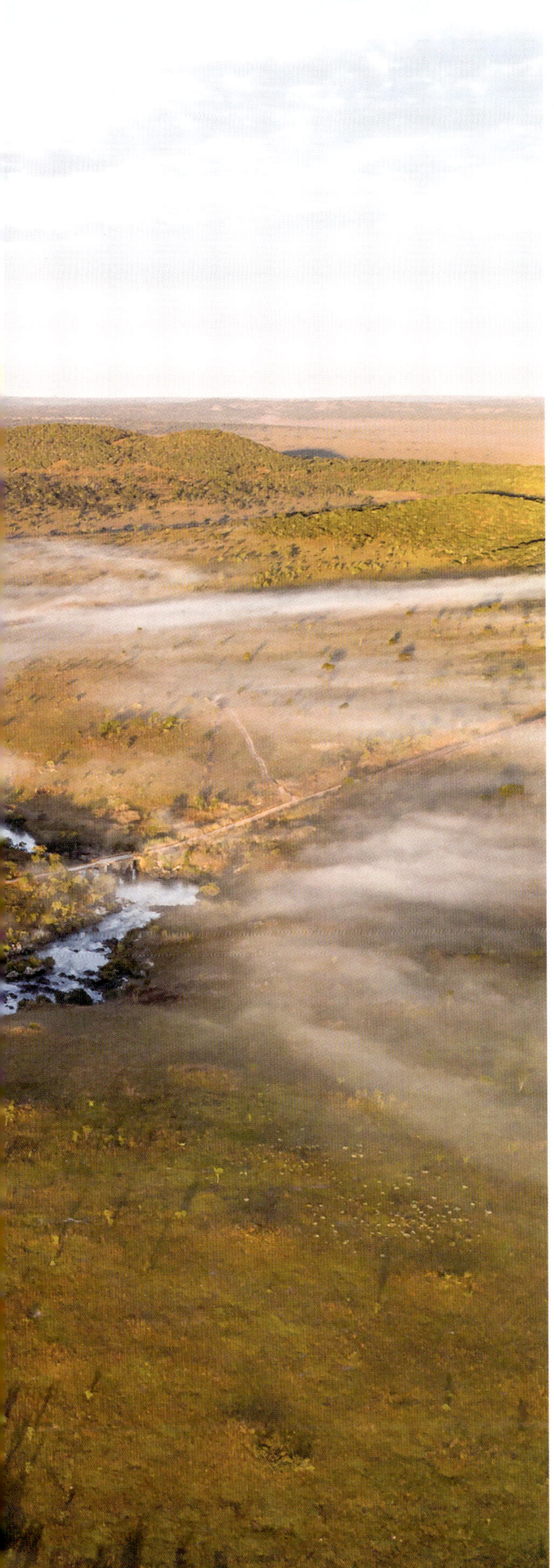

Delta, we needed to save the river sources, all in the Lisima Landscape. The Cuando and Quembo Rivers taught us that the past and future of these river systems are intertwined. The people living between them are intertwined as well.

BETWEEN 2016 AND 2018, we had explored the remotest valleys of the Lisima Landscape on fat-tire mountain bikes. At a slower speed, we really got to know the forests well.

Our starting point was somewhere southeast of the Cuanavale source lake, in the heart of the Lisima Landscape. The climb was brutal: first, a four-hour climb to the top of the ridge, followed by a two-hour descent to a stream for water that was both exhilarating and terrifying. Whenever we stopped, sweat bees would swarm us. I inhaled at least a dozen every day.

These contiguous miombo woodlands are Africa's largest remaining, spanning an area larger than England. Biking through, you can hear sounds of the forest and see habitats change. I still far prefer paddling down the rivers of Lisima and hiking its trails, but there is something about traveling 50 kilometers (30 mi) a day, with hundreds of kilometers of forest wrapped around you in every direction. In Africa, everything is about scale: Everything is far away, every destination takes longer than expected to reach, and the path is always far more difficult than anticipated. It feels like a place for lost things—like us and all the unique species we have discovered there.

IN 2017, we returned to the Cuito, Cuanavale, and Quembo source lakes to conduct the first

The upper reaches of the Cuando and Quembo Rivers are among Africa's remotest lands. These rivers flow into the lower Kwando River, which connects to the Okavango Delta by the Linyanti Swamp and Selinda (or Makwegana) Spillway in northern Botswana. To protect the delta, preserving the near-pristine forests that sustain these rivers is critical.

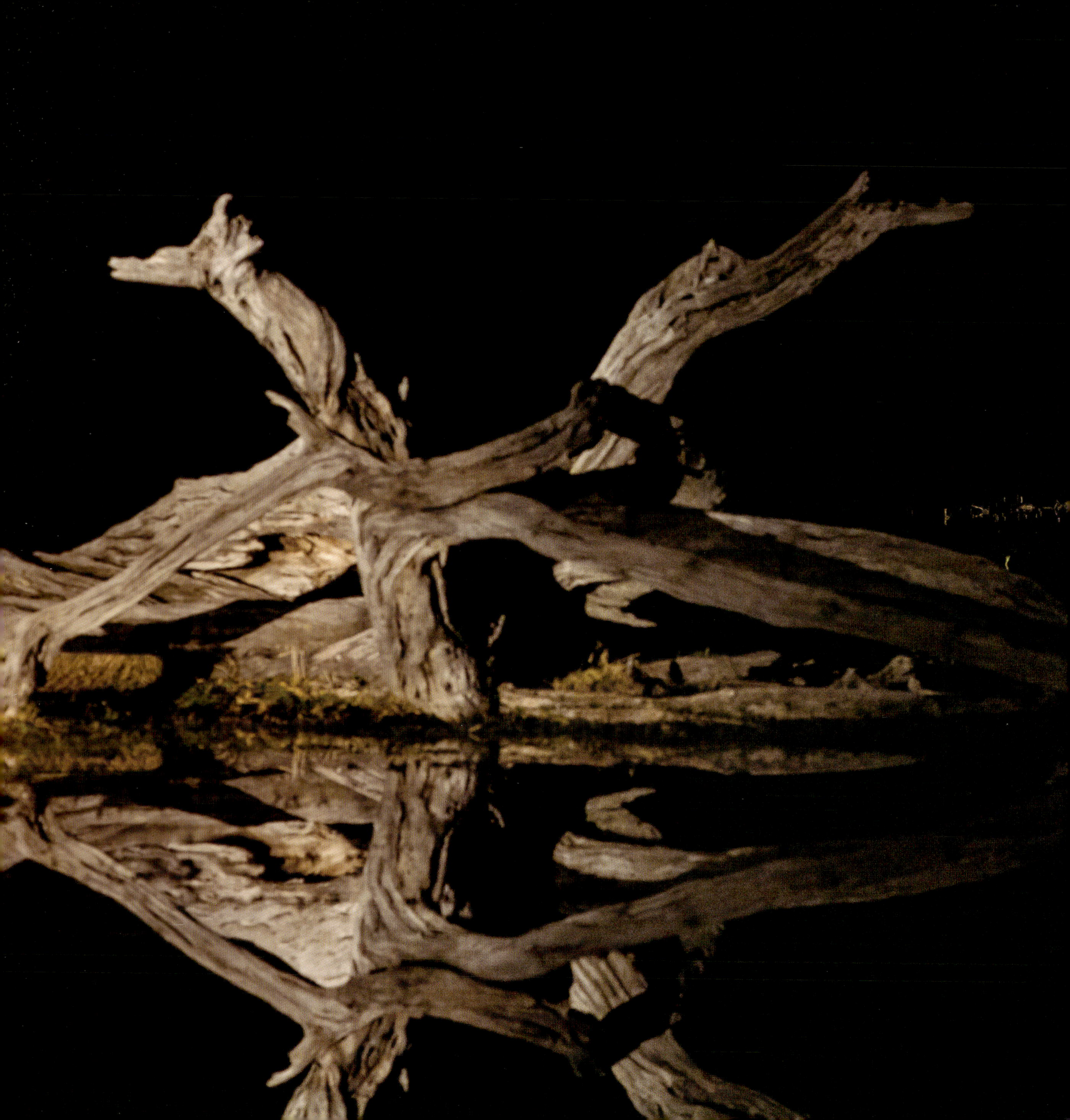

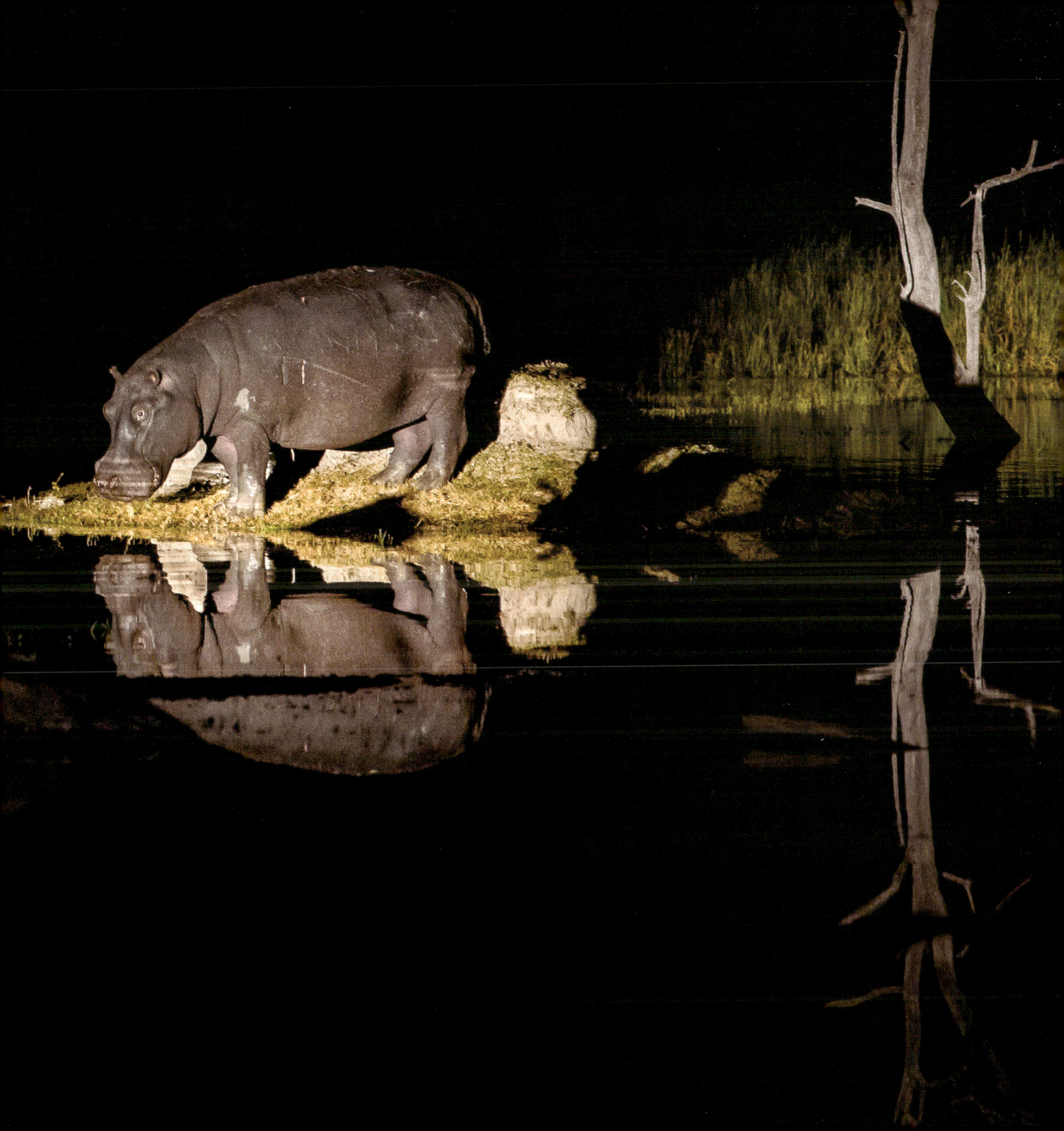

ever underwater surveys of the world below the surface. As we carried dive tanks and compressors to the lakes and built a dive platform out of two mokoros, our activities again attracted salt-hungry bees that arrived in the millions. Research divers were inhaling bees and had to wear hoods and wet suits all day. The weather was hot and humid, and thunderstorms barreled through every afternoon.

Diving in the Cuanavale source lake was like diving on a moon off Saturn: an otherworldly wonderland of wispy algal gardens in crystal clear water, with thousands of fish species uniquely adapted to the acidic conditions. There were crocodile tracks on the bottom and giant tiger fish in the deep water. We mapped out microhabitats for 10 new fish species and took peat cores that helped us consider the relative age, form, and function of the source lakes.

BY AUGUST 2017, we had explored all major rivers and channels in the Okavango Basin—the Okavango Megatransect, 5,178 kilometers (3,217 mi) over 216 days, an early 21st-century scientific baseline created thanks to the partnership between the Wild Bird Trust and the National Geographic Society. By August 2018, we had completed all major tributaries of the Kwando Basin, a sibling river system.

In 2018, the National Geographic Society signed a landmark agreement with the Angolan government that promotes work with local communities to protect the keystone watersheds of the Okavango Delta, bolstered by the Wild Bird Trust, Nkashi Trust in Botswana, Kavango Wilderness Project in Namibia, and Fundação Lisima in Angola. We are now building an endowment to support this important community-based conservation work in perpetuity. In 2018, the DELTA (Defending Economic Livelihoods and Threatened Animals) Act, committing the United States to invest in the mission to protect Africa's last remaining wetland wilderness, was passed. By 2024, we had installed 14 permanent satellite-connected hydrological and meteorological stations across the Okavango and Kwando Basins.

Science without storytelling is lost on most people. In 2018, we premiered *Into the Okavango* at the Tribeca Film Festival. In 2021, our podcast series, *Guardians of the River*, won Best Narrative Nonfiction Podcast at Tribeca and Best Podcast at the Jackson Wild Media Awards. We have also shared our conservation message in short films in local languages. And this is just the beginning.

THE MEGATRANSECT brought the Okavango into the stories of all our lives. Mr. Water, GB, Tom, KG, and Snaps had lived all their lives in the delta, and now they had explored every inch of all major rivers and channels flowing into it. Now it's their job to tell this story to the people of the delta, working through new traditional knowledge centers dedicated to intergenerational knowledge transfer. As we continue to explore and protect these waterways, we will use local knowledge, culture, and language, to support sustainability and conservation goals. The key to success for the entire watershed is community health, education, and prosperity. In the words of Water Setlabosha: "All we want is a better life for all. We want to be free!"

PAGES 138–139: *Hippos were heavily persecuted during the Angolan Civil War, causing them to become aggressive toward people. They remain the most dangerous animal in the Okavango Delta, actively charging mokoros and people living along the Cuando River.*

RIVER GUARDIANS

Elias Ngunga

The Way of the Wild

ELIAS NGUNGA, now in his early 30s, was born in a remote village in Angola's Cuando Cubango Province. Due to the civil war, he had no formal education, yet he is incredibly intelligent. He survived as a young boy by running guns up and down the Cuando River. Today, he is a survivor who prefers to live in the wilderness rather than move to a city. This is where he is free.

I am always struck by Elias's ability to absorb new information and concepts with ease. For hundreds of millennia, humans orally transmitted knowledge crucial for survival. A good memory must have been a useful talent, favored by natural selection. Most of what Elias knows comes from oral tradition and collective memory. He is an expert in how to survive without technology. People like Elias hold the key to our future.

Elias straddles the old and new, traditional and modern, simple and complex. After the trauma of being a child soldier, he went to live in the capital city, Luanda, for nine years before returning to the Lisima Landscape, the remotest part of Angola, to be free. To him, freedom is an axe that can make a house, fish trap, beehive, table, dugout canoe, bark canoe, dinner plate, rope, and almost anything else you need in a place where traditional knowledge, not money, keeps you alive. Elias's life shows us that freedom isn't democracy; it's a close relationship with nature, curated by culture and tradition.

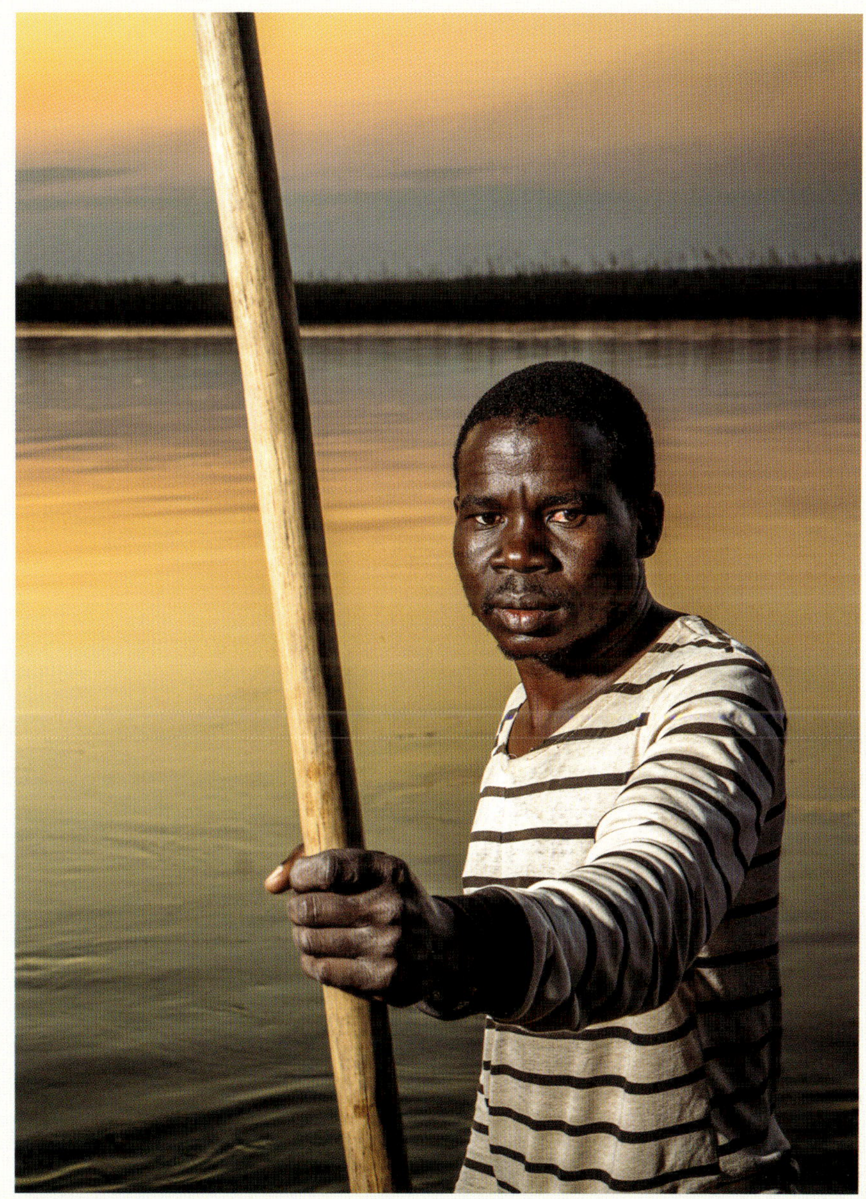

Elias Ngunga lives near the confluence of the Cuando and Quembo Rivers. He is a true man of the Lisima Landscape, spending most of his time in the wild.

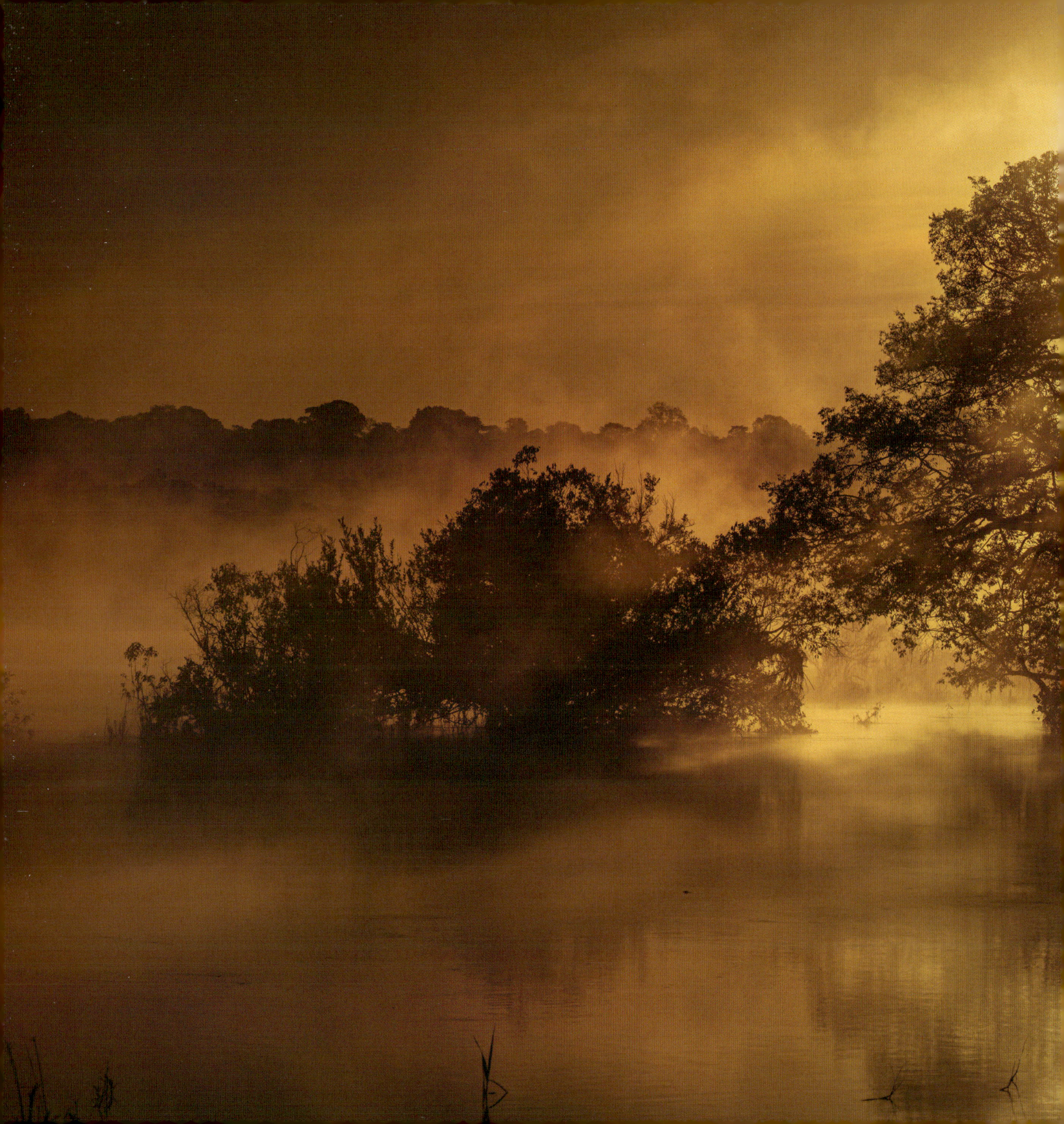

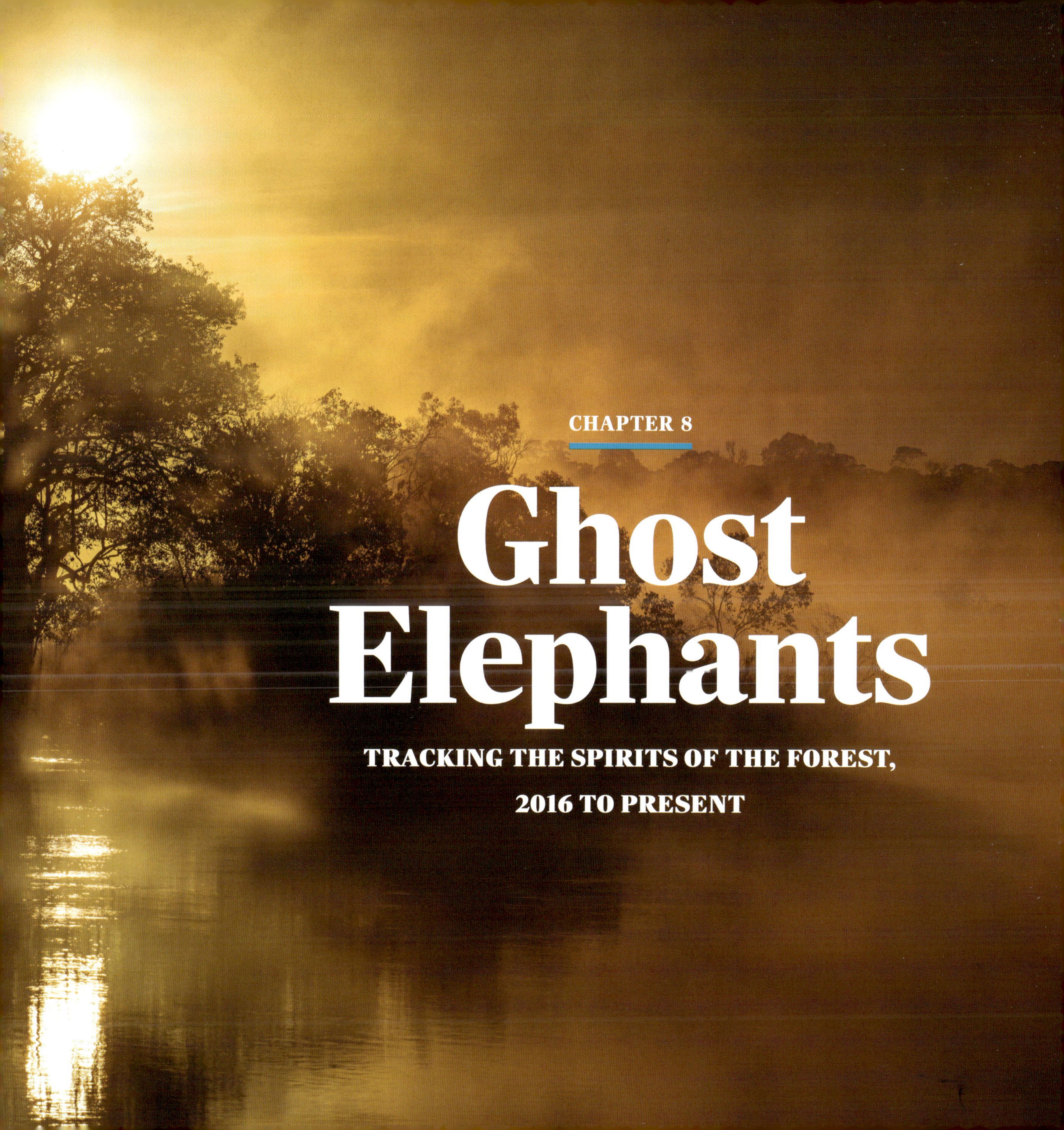

CHAPTER 8

Ghost Elephants

TRACKING THE SPIRITS OF THE FOREST, 2016 TO PRESENT

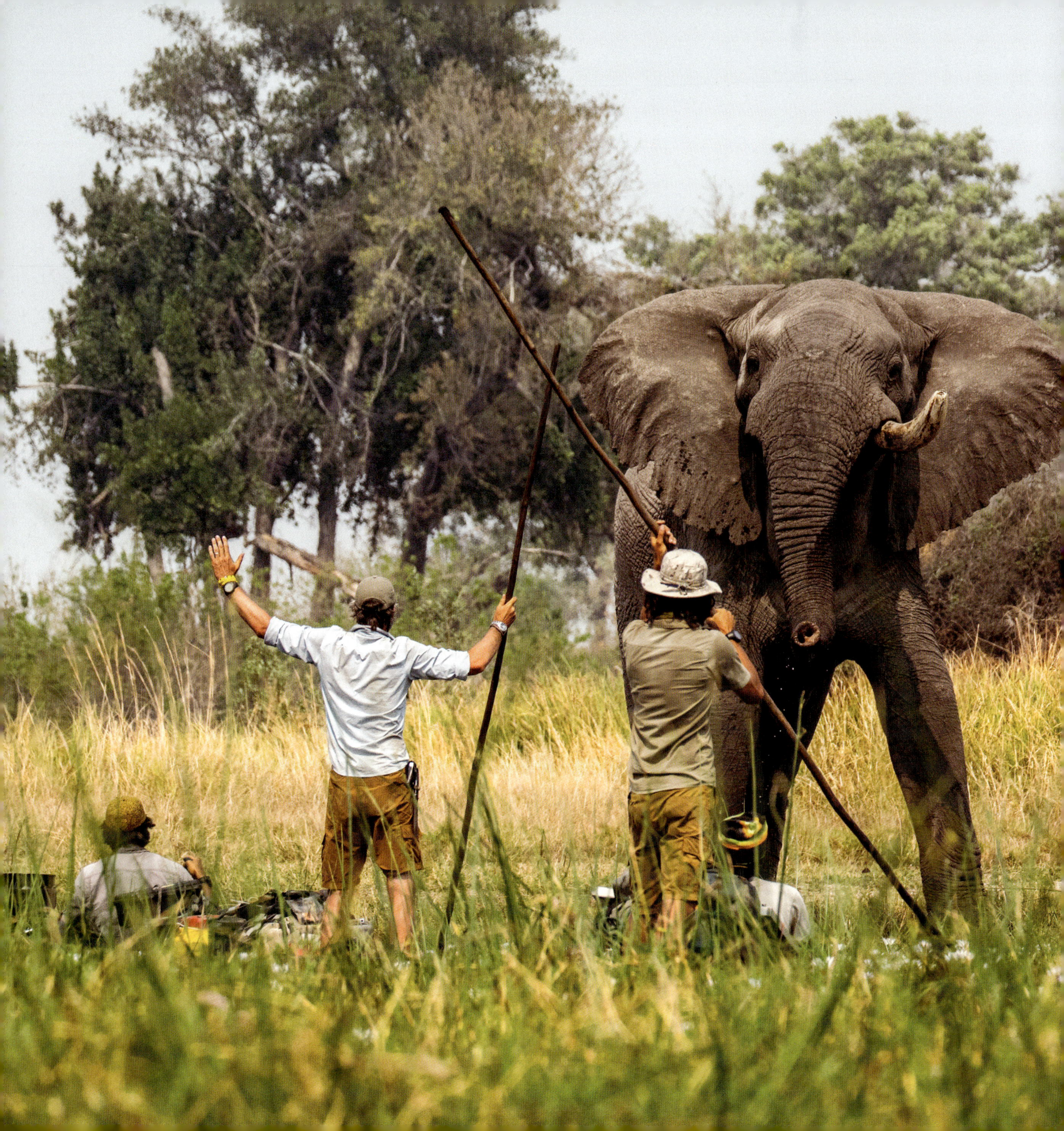

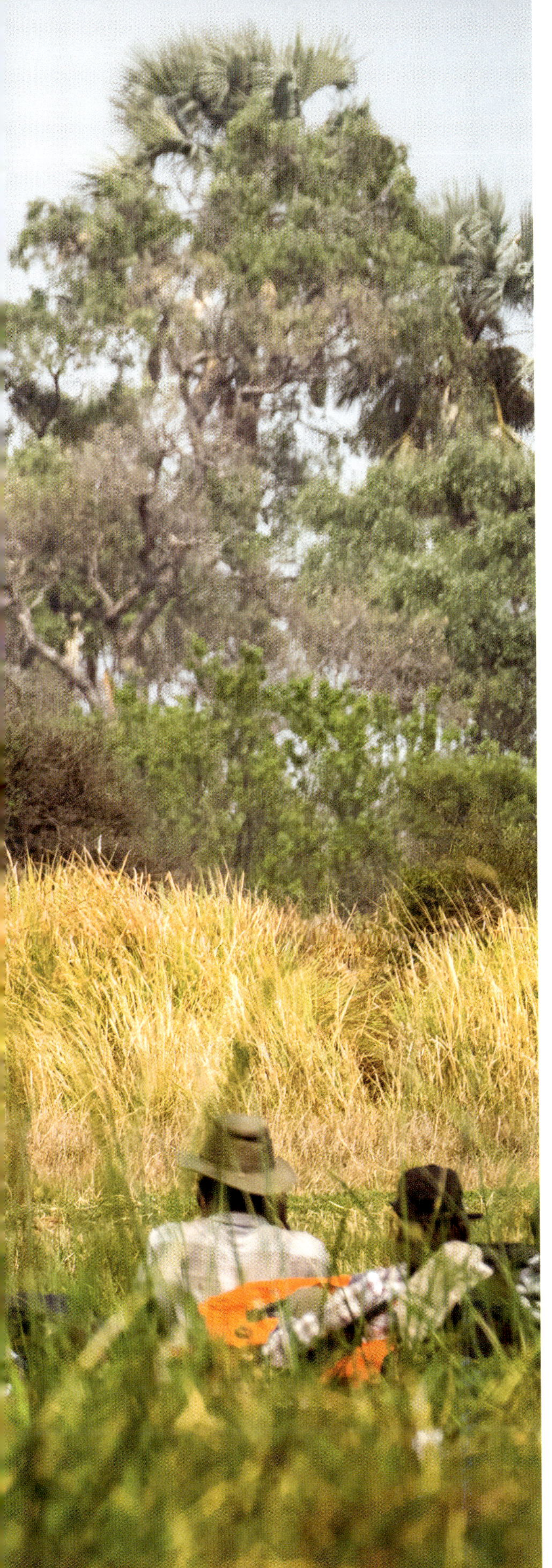

IVORY—A LUXURY PRODUCT created from the tusks of elephants—is a paradox of the beauty and brutality of human endeavor. Deeply embedded in our history, commerce, and spirituality, ivory is part of the story of civilization itself. It was central to the establishment of ancient trade networks extending into Africa and Asia, across the Sahara and down the Silk Road. Now, in the 21st century, our obsession with ivory as a spiritual, artistic, and cultural artifact conflicts with the stark reality of its acquisition.

In the aftermath of the war in Angola, researchers and conservationists presumed that the region's once abundant elephant population was gone. Ivory was one of the currencies of the war, and it's estimated that between 25,000 and 100,000 elephants were killed over the decades of fighting. The elephants that survived were presumably driven out of their ancestral homes into safer lands in neighboring Botswana, Namibia, and Zambia. From the very beginning of our explorations into the Angolan Highlands, we saw signs that the elephants were not gone after all.

In 2012, Angolans had been encouraged to return to their ancestral lands from the cities they had moved to seeking refuge from the fighting. That year, upon returning to a village near the source lakes, the village elders of Saiyoswa described forest elephants with red eyes arriving from the north meeting savanna elephants moving up from the south. If they weren't mistaken, this was an extraordinary story; hybridization between these two species

In August 2018, my brother, Chris (standing at right), and I encountered the largest bull elephant we had ever seen on our expeditions in the Okavango Delta, where they live longer and more happily than anywhere else on Earth. **PAGES 142–143:** *Forgotten in oral history, uncharted on maps, and undocumented in any literature, the landscape of Lisima lya Mwono—the Source of Life—is shrouded in mystery.*

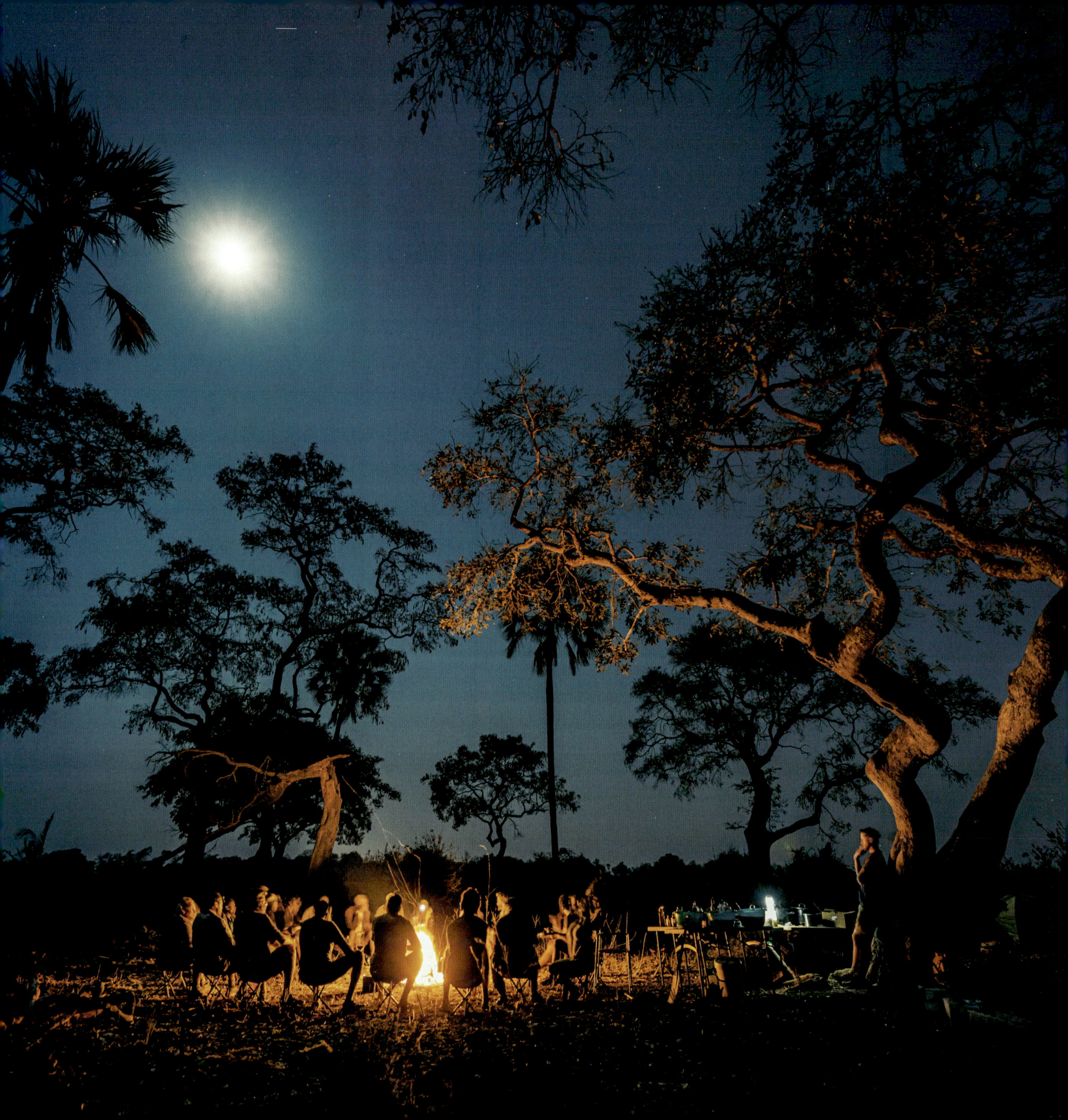

is an unanswered question being studied across tropical Africa.

In 2015, I had found a forest clearing near the Cuito River, below the confluence with the Luapula River in Lisima. It looked just like the sort of clearing I had seen many times in the Okavango: a so-called elephant garden, where a big bull elephant had cleared a nice open area, selected trees to establish themselves over decades on the edge of his woodland, and broken other trees to coppice them into bushes, with lateral branches accessible for young elephants to browse.

I could see that the bull had taken great care in his gardening. He must have been within a few days' walk from us, based on his tracks and dung. I imagined him too old to migrate or go on extensive walkabouts. He tended his gardens in a clearly defined territory between water points, living mostly by scent in the deep forest, moving around silently. He would have been in his 70s or 80s, carefully protecting his last set of teeth. He had witnessed the entire civil war, heard the massacre in Tempue in the 1960s, the Battle of Tempue in the 1980s, and the final standoff between the government and rebel forces in his forests. He had learned to hide from people with guns, motorbikes, and trucks, from forces that had funded the war effort using ivory and diamonds. He knew humans and, like the Luchazi who destroyed bridges to isolate themselves from the armies, had wandered deep into remote forests that couldn't support large numbers of elephants but could support him.

Elephants are sentient beings that live as long as we do. I have witnessed them celebrate birth and mourn death. They show a meandering, rocking, radiating, empathetic, mindful unity, a grandness beyond human conception. It's humbling. Elephants' perception of humans varies individually. They can hear us coming kilometers away and smell us over even greater distances under the right conditions.

They are remarkably sensitive to human presence, exhibiting curiosity, even approaching us in areas where we coexist peacefully. In places where poaching and human-elephant conflict are prevalent, they are extremely wary and sometimes aggressive toward us. If we can learn how to coexist with elephants, we have a model for a sustainable future.

It was much safer now; hunters had surrendered their AK-47s for shotguns and the ivory trade was now illegal in Angola. Yet poachers occasionally came from Zambia. It was wise to stay hidden. He had found a good place to retire.

I sat in his clearing, feeling his presence, hoping to meet him as I had done so many times in the Okavango Delta. I waited until after sunset, but he never arrived.

This was our first introduction to the "ghost elephants" of Lisima. Up in the highlands where the bull had made his gardens, there couldn't be more than 100 elephants left. These bulls wandered the forests like lost spirits, waiting for the breeding herds to arrive in summer on their annual migrations, hoping they would linger long enough for him to find a female in season. Perhaps all he wanted was the company.

By 2018, we had been searching for the ghost elephants for two years. We saw lots of elephant trails along the remote stretch of the Cuando River we were exploring. Much of the conversation that moonlit night around the fire was about these enigmatic "elephantoms."

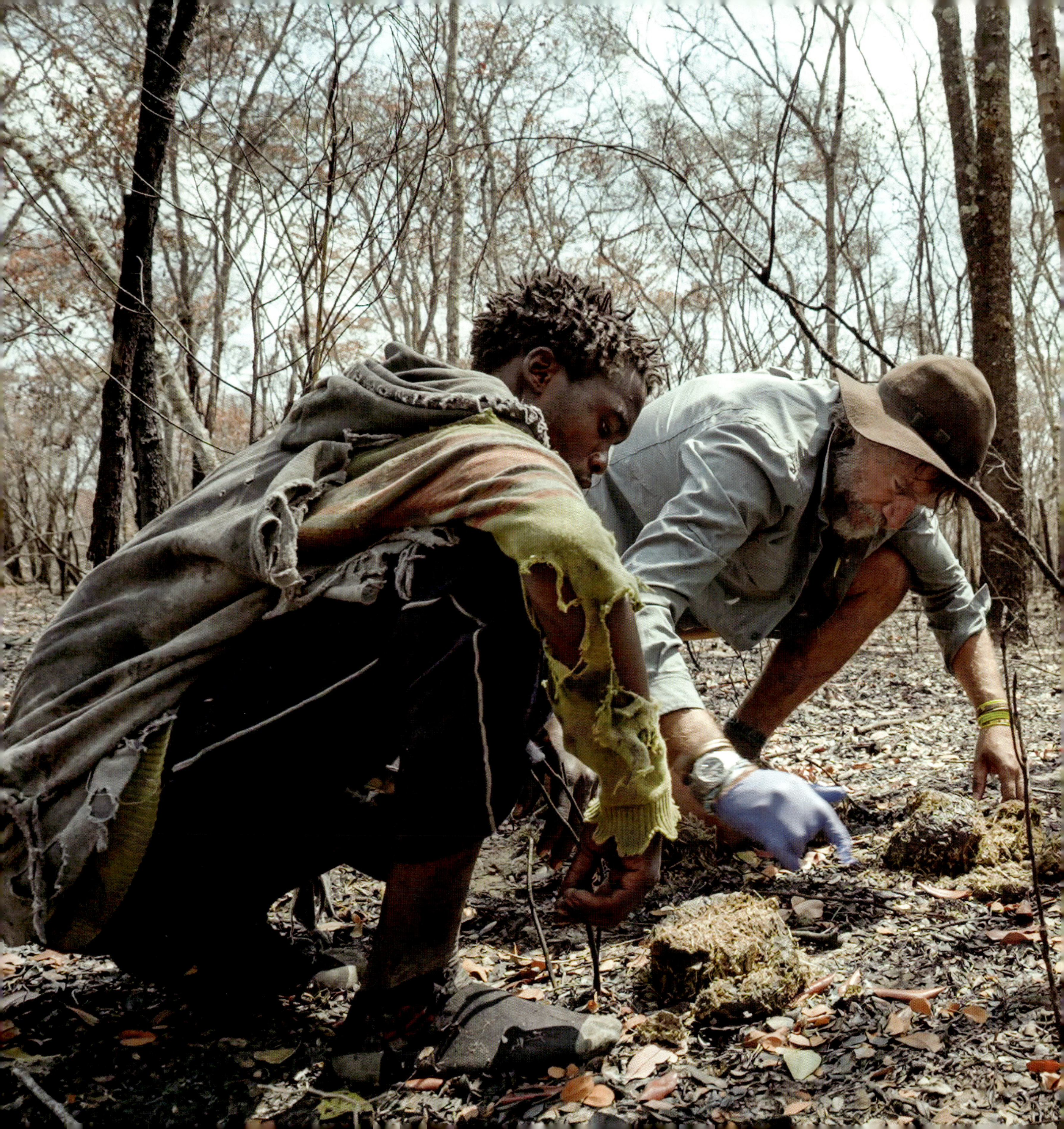

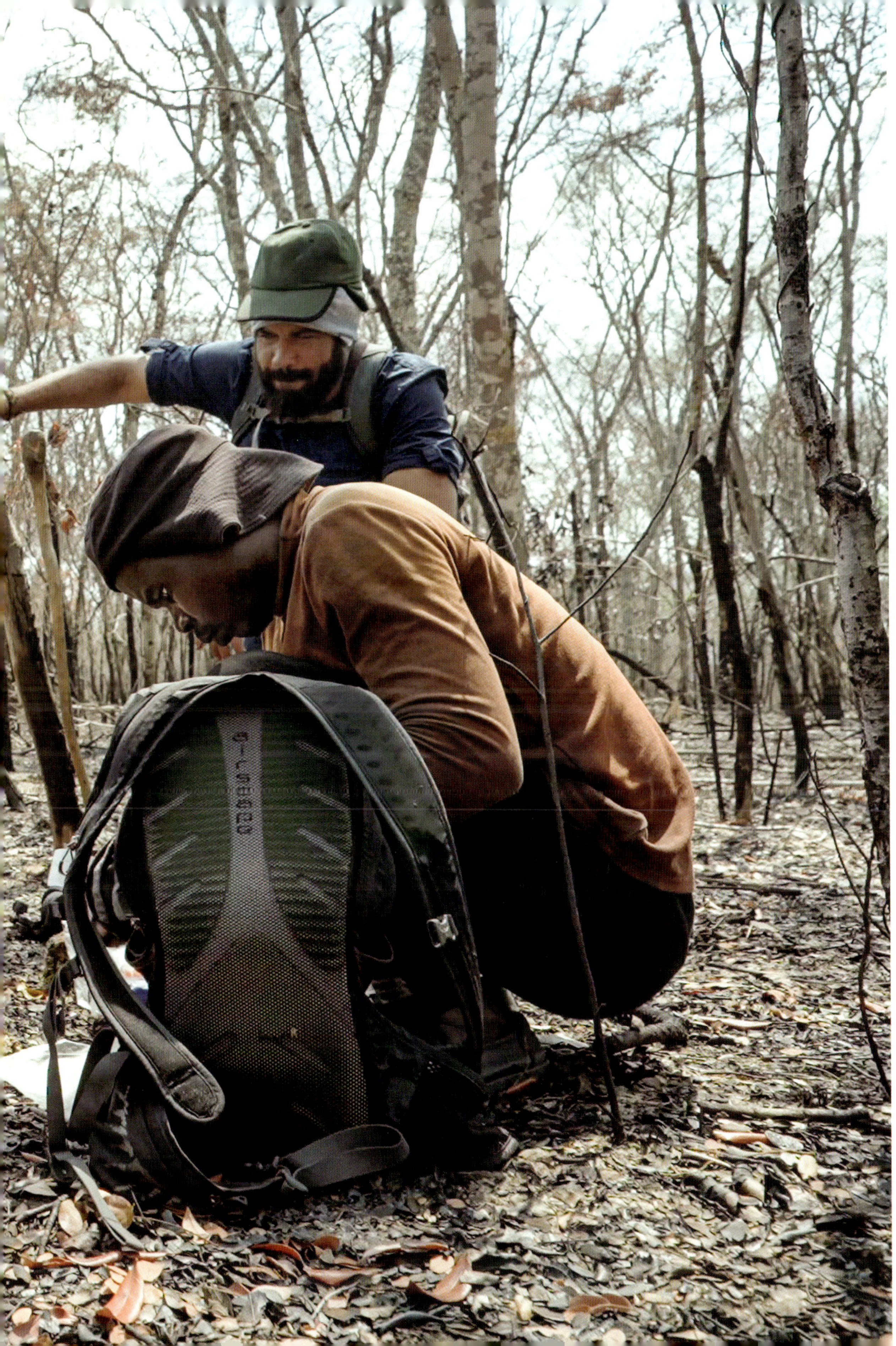

Rodriguez, a local hunter and tracker (far left), found an elephant trail and led (left to right) myself, Elias, and Kerllen to a pile of fresh dung, from which we took DNA samples. These were our first fresh samples from the ghost elephants, whose ancestry they would reveal when compared to the genetics of other regional elephants.

The presence of elephants in these high-altitude forested watersheds was an incredibly interesting discovery. This appeared to be the hiding place of Africa's largest elephants.

IN 2016, we found more signs of elephants in the Lisima Landscape: markings on a tree along the Cuanavale River, just beyond the confluence with the Tempué River. A big bull elephant had rubbed mud more than two meters (6.5 ft) up an old teak tree, perhaps itching his rump. I saw no dung anywhere, strangely, perhaps because of the heightened activity of dung beetles at the end of summer. The presence of elephants in these high-altitude forested watersheds, above 1,250 meters (4,100 ft), was an exciting and incredibly interesting discovery for us. Savanna elephants wouldn't be able to survive in this highly specialized forest habitat at such an altitude. Yet this appeared to be the hiding place of Africa's largest elephants, living mammoths who had shaped an unusual existence in this forested home.

That same year, we also learned about the so-called Fénykövi elephant, a massive bull standing for more than five decades in the main rotunda of the Smithsonian National Museum of Natural History in Washington, D.C. He is affectionately known as "Henry" by museum staff. Visiting the exhibit, we instantly recognized him. He looked like the big bull elephants we had interacted with many times in our mokoros in the Okavango Delta: long, heavy tusks and bulk beyond any other elephants on the African continent. When we looked into the Smithsonian records—Josef J. Fénykövi's hunting journals and logbooks—we discovered something extraordinary. This giant bull elephant was shot in southeast Angola, somewhere in the Lisima Landscape. Were these ghost elephants Henry's descendants that we had been hearing rumors of?

Personal accounts, newspaper interviews, and diaries about the Smithsonian elephant made it clear that this was a Moby Dick tale, as are most of the big elephant hunt stories: The hunter becomes obsessed with felling the big tusker himself. In search of the biggest elephant ever recorded, Fénykövi and his guides ventured deep into the upper reaches of the Cuito, Quembo, and Cuando Rivers. On his second trip to the area, in November 1955, the Hambukushu and Khoisan master trackers who were guiding him found two bull elephants resting under some trees in the dense miombo woodlands characteristic of the eastern Angolan Highlands. On November 13, Fénykövi, along with his three assistants—Mario, Francisco, and Kukuya—tracked down and killed the large bull.

This elephant was remarkable for its sheer size, setting world records with its dimensions. The largest elephant I have ever interacted with in the delta was about three meters (10 ft) at his shoulder; the Fénykövi elephant was over four meters (13 ft) at his shoulder and weighed more than 12 metric tons (13 tn). His skin alone weighed over two metric tons (2.2 tn). The largest elephant alive today weighs no more than eight metric tons (8.8 tn), which is huge. To get this big, he had to have lived a very long time, perhaps 100 years or more.

The Fénykövi elephant had been donated to the Smithsonian Institution in 1958. That year, Fénykövi sent images of the elephant to Remington Kellogg, director of the United

RIVER GUARDIANS

|ui |oa, the Last Master Tracker

The Origin of Us All

In the early 1908s, eight-year-old |ui's father called him to come see something: a massive bull elephant walking past their homestead in Nyae Nyae, a Namibian village. His father was a respected hunter among the Ju|'hoansi San people of northeastern Namibia. In his click language, "Ju|'hoansi" simply means "the People." Genetically, they are our closest link to the small founding population of all modern humans today.

At the time, |ui and his people had never seen an elephant before. Giraffes were then a more common sight in Nyae Nyae; in fact, the Ju|'hoansi did a giraffe dance for healing and spiritual connection to their ancestors. But in the mid-1980s, according to Ju|'hoansi elders, elephants began arriving en masse. They came across the Cubango River from Angola, where a terrible bush war was raging. After the 1987 Battle of Cuíto Cuanavale, Africa's largest tank battle since World War II, a constant flow of elephant refugees poured across the Namibian border into this dry, forgotten part of the Kalahari Desert.

When the elephants arrived, they started tearing down camel thorn trees, a staple in the giraffes' diet. Thousands of elephants tore down these ancient trees, and as a result, the giraffes slowly disappeared. Eventually, the Ju|'hoansi healers adopted an elephant dance, and the elephants came to be considered as sacred as the giraffes once were.

The elephants continue to destroy the trees, however, which damages natural springs that sustained these first people. Wealthy hunters come into the region and cull the elephants, considering giant bull elephants the grandest trophy. When asked, the Ju|'hoansi people of Namibia say they wouldn't miss either the hunters or their prey.

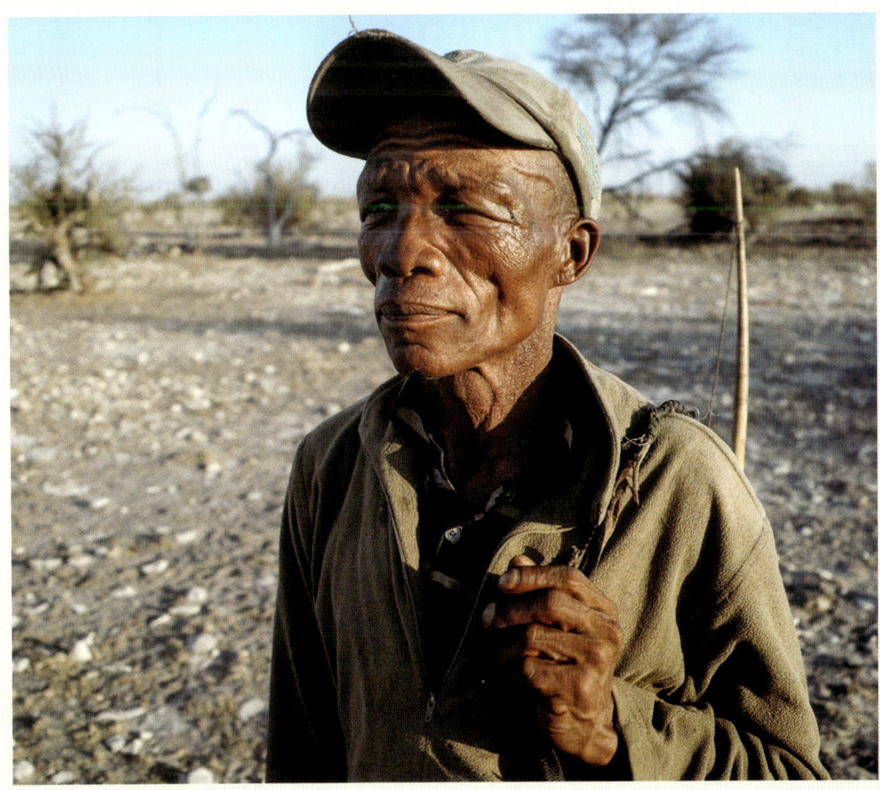

Pictured here, |ui |oa is one of the last Ju|'hoansi San master trackers living in the Nyae Nyae Conservancy in northeastern Namibia.

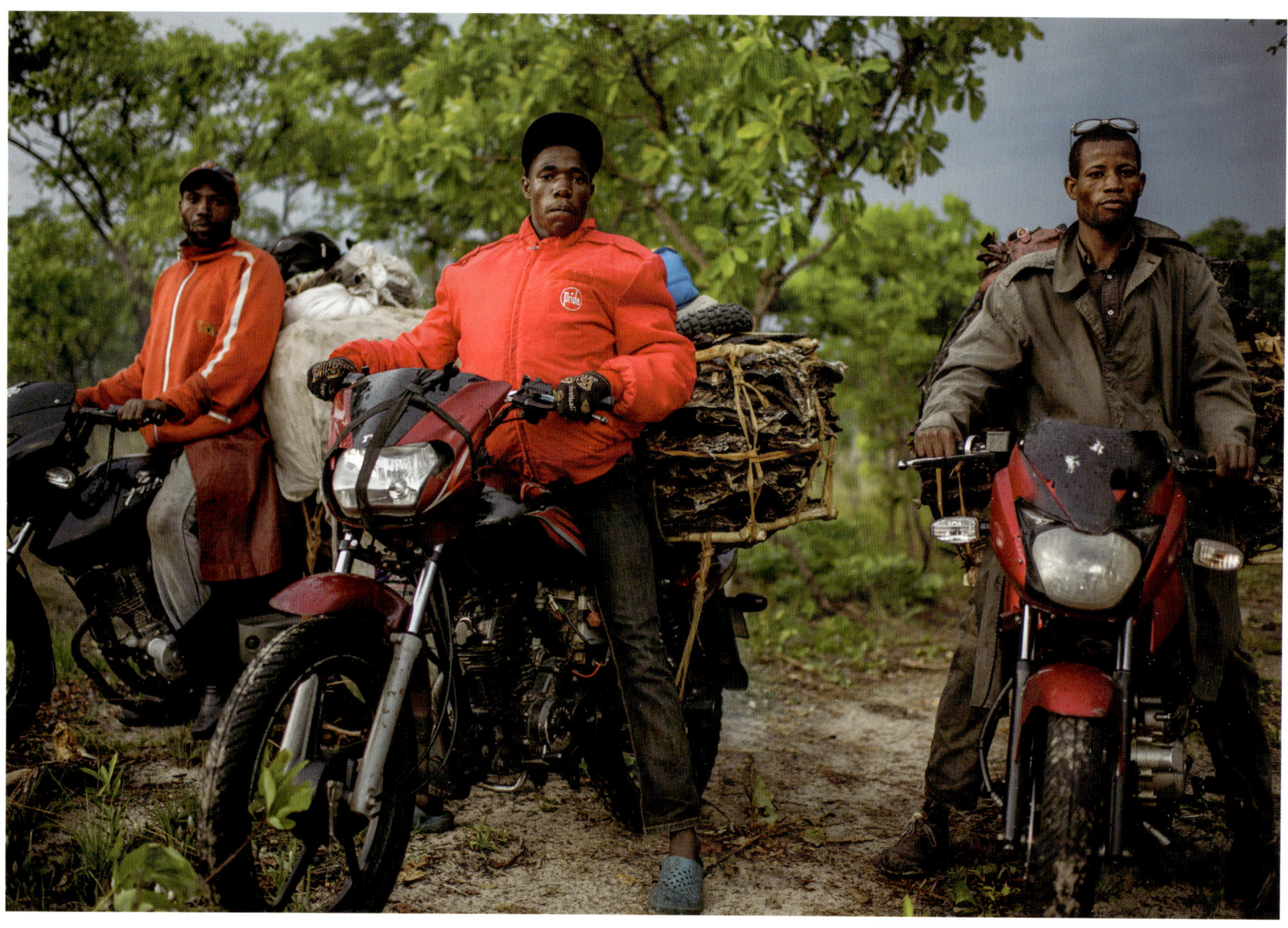

The only way to access hunting camps in the remotest parts of the eastern Angolan Highlands is by motorbike along ancient footpaths. These men once were bushmeat hunters, but they now work with traditional leaders to monitor and regulate that destructive trade in these remote forests.

States National Museum, to help taxidermists William Brown (Smithsonian's chief taxidermist) and Norman Deaton reconstruct the elephant. The museum revealed the restored body to the public on March 6, 1959, giving "Henry" the place of honor in the museum rotunda. He became known as the largest extant land mammal on display in any museum. In 1961, the Fénykövi elephant was publicized in the September issue of *National Geographic* as part of the feature article on Angola. The world was inspired to learn more about the country but was soon locked out as the war of independence began before the year ended.

Recently, the National Geographic Society partnered with the Smithsonian Institution to do a submillimeter 3D scan of the Fénykövi elephant, securing the data necessary to create 3D-printed models of the elephant for visitors. More important, the Okavango Wilderness

Project is creating an augmented reality experience that could be used to digitally repatriate the elephant to Angola, allowing visitors to the museum in Luanda to interact with a holographic version of this lost grandfather of the forest.

WHEN YOU SIT with elephants, you can hear them grumbling and rumbling, snorting and trumpeting. This represents just a fraction of their auditory and seismic senses. They walk silently because they are listening with their feet. These gentle giants, guardians of the forests and grasslands, are in constant communication, touching one another and their environment with their trunks; gesturing and bumping into each other; flapping their ears; freezing, kicking, and charging. They live as long as we do and are awake twice as long, sleeping only a few hours every night. They have good reason to develop a complex, conversational communication system that facilitates ongoing group discussions about migration routes, weather patterns, seasonal foods, parental care, and potential threats. Walking with elephants, you can hear that they are talking about you too.

They also create smellscapes—olfactory profiles of family members, places, events, and objects—that evoke strong emotions and memories, just as scent does for us. I have known blind elephants that lived sustainably for decades. We are just beginning to understand the secrets of elephants.

Elephants grow up in matrilineal family groups and mature at the same rate as we do, so they have ample opportunity for development of culture, language, custom, and traditional knowledge. They care for their young, sick, and elderly, and they grieve and mourn the dead.

Like humans, elephants call out to one another using individual names that they make up. They use a specific harmonically rich, low-frequency vocalization for each individual. They also recognize and react to their own names, while ignoring calls addressed to other elephants. I would like to believe they use the names of the deceased when visiting grave sites. This great elephant's name wasn't Henry, and he certainly wouldn't want to be remembered by the name Fénykövi, the big game hunter who killed him.

Henry's skull is significantly bigger than any other elephant skull in collections around the world, which is why it was never mounted with the rest of Henry's skeleton. Too heavy to support, both skull and tusks have been kept in storage in Maryland. In my opinion, these relics should be considered for repatriation back to Angola to be put on display in the natural history museum in Luanda or returned to where the elephant was shot. Placing the skull amid wild elephants would be a fitting tribute to an extraordinary beast, one who gave us a vision of a grand past.

When you sit with elephants, you can hear them grumbling and rumbling, snorting and trumpeting. They walk silently because they are listening with their feet.

AT THE SAME TIME that we were learning about Henry and his legacy, Adjany Costa and I had started interviewing local Luchazi hunters and village elders about the animals they hunt seasonally and whether they ever saw flagship species, like lions and elephants. Augusto, the soba of Chinjanga, a village on the Cuanavale River, and now a good friend, told us about a beautiful valley that he was protecting: a valley with elephants, lions, lechwe, and all the wildlife we had seen in our camera traps and on our river expeditions. Augusto didn't like people hunting down there because it upsets the lions and elephants, compelling them to come up to his village to kill his pigs and destroy his cassava crops.

He said it would take weeks for us see an elephant in this valley, but that we could find elephant dung within two hours on a motorbike. My brother, Chris, went off on the motorbike while I sat in Augusto's thatched hut, talking about life. Chris came back about three hours later with a bag of elephant dung. He, and the hunter with him, seemed excited.

A year later, in winter, Chris, Götz, and Kerllen Costa, country director for Angola at the Okavango Wilderness Project, cycled into this valley and set camera traps. They said they could smell and sense the presence of elephants in the area. In March 2017, we retrieved our first photographs from the camera traps, after they'd been in place for seven months. We saw new populations of cheetah and wild dog but no elephants. And most showed hunters carrying dead animals.

In 2018, we were starting to get to know the Kwando Basin, a sibling river system to the Okavango with its connection to the delta via the Selinda Spillway. That year, Kerllen and I had explored the full length of the Cuando River, the Kwando Basin's main waterway. Augusto had told us, two years before, about a secretive population of elephants that lived in valleys associated with the Cuando, Cuanavale, Quembo, and Tempué Rivers. Indeed, I found elephant trails all along the Cuando River while walking in the afternoons, and Kerllen found large elephant crossings where breeding herds had cut across into adjacent valleys.

The elephants I was tracking were moving quickly south and, in June, were at least two months ahead of me based on the age of the footprints and dung along the trails. Their tracks and signs signaled stress, as they were clearly moving quickly and were feeding on grass roots, making big excavations along the riverbank. All accessible branches on the trees had been browsed. These elephants were clearly migratory breeding herds coming up to Lisima during summer, then leaving in early April for the floodplains and rivers downstream.

Augusto and the old ivory hunters said the resident ghost elephants moved into the upper reaches of the Lisima Landscape during the dry season, from March to November, when the migratory breeding herds were away. They were big bull elephants, he said. The old elephant hunters said the best chance of finding one of these big tuskers was in September: When fire season was over, the landscape was at its driest, and these almost mythical elephants used small springs known to the hunters in remote sections of forest.

We were intrigued. We set camera traps along elephant trails in the valleys that Augusto and other hunters identified. However, each time we went back, the cameras

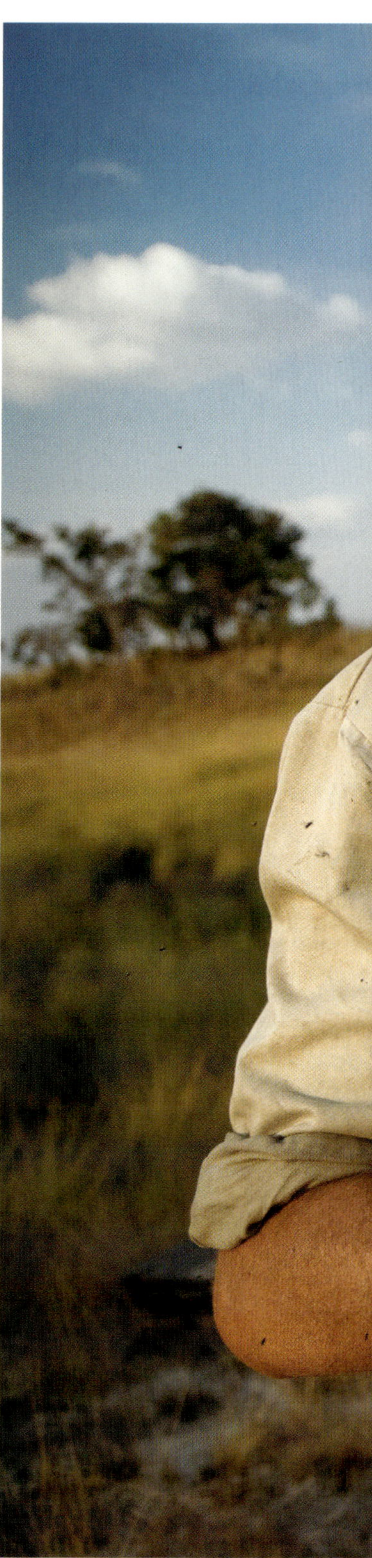

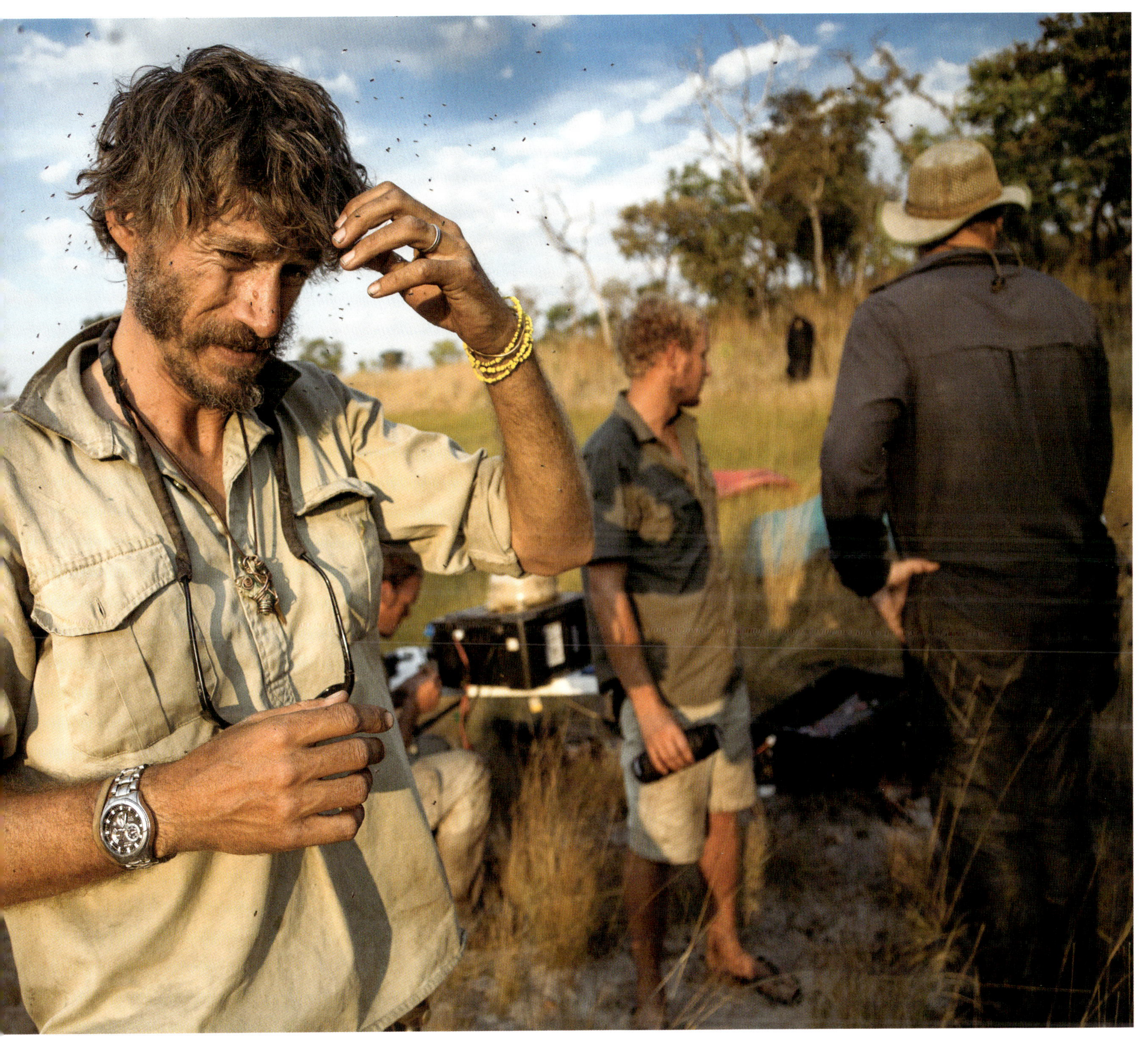

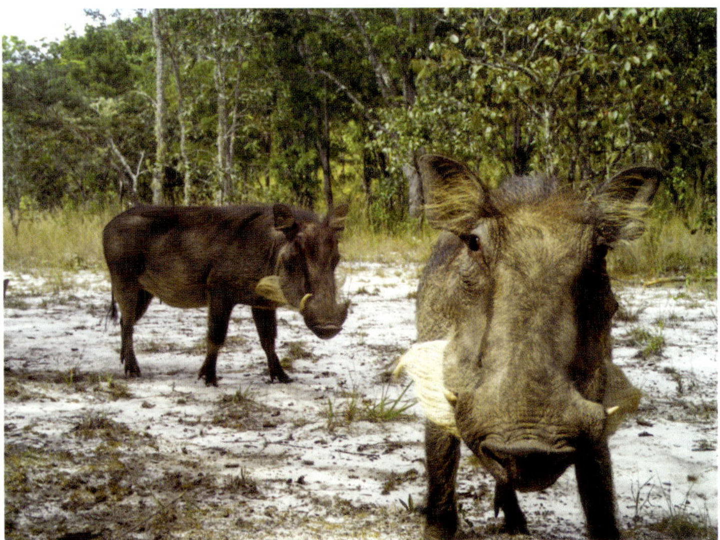

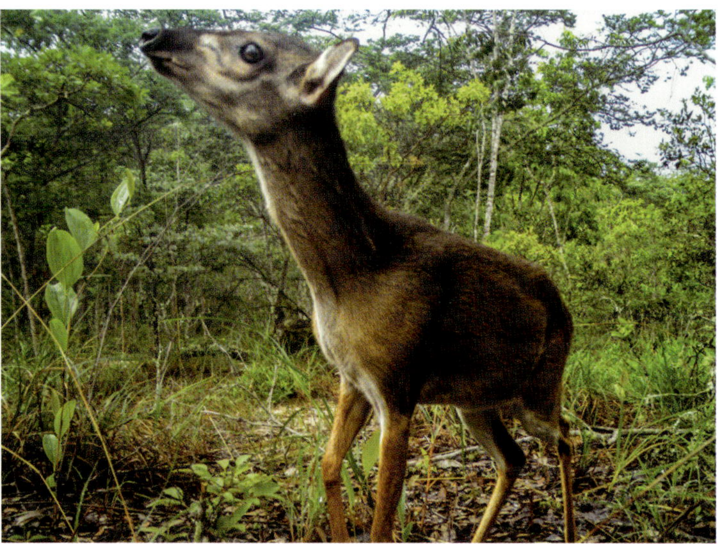

Cameras set in the Angolan Highlands captured photographs of (above, left to right) warthogs, a blue duiker, a vervet monkey, and a leopard. **PAGES 154–155:** *Sweat bees are a constant bother during the summer months in the Lisima Landscape. Stingless, they protect their hives by swarming around intruders, especially buzzing about a person's mouth and eyes.*

were either missing or destroyed. We thought perhaps the ghost elephants could be treading on the cameras, as we saw signs of elephants around them. Between 2016 and 2022, we captured photographs of leopards, duiker, monkeys, bushpigs, and porcupines, but not much else. We put signboards up saying the camera traps belonged to us. That just made the vandalism worse.

By 2019, we had discovered that mostly hunters were disturbing the camera traps. They weren't concerned that we were looking for them or the elephants; they thought we were looking for Lukhisi-khisi, a mythical spirit that wandered the forests, protecting the hunters and imbuing them with powers. They believed we were trying to capture this creature and tame the forests to bend them to our will. So we changed our strategy. We stopped cycling into these valleys on our fat-tire mountain bikes and asked local, trusted hunters to deploy the camera traps and retrieve the memory cards. Only then did we start getting results.

In November and December 2022, we got our first photographs of elephants: beautiful breeding herds with the matriarch posing for selfies before parading her family of sisters, daughters, and their calves past the cameras. A few ghostly images could have been of lone bulls following them, spirits in the darkness. Those elephants moved only at night. We caught no daytime photographs.

TWENTY-SIX MILLION elephants once roamed precolonial Africa. By 1970, the population numbered 1.3 million, and today there are only about 400,000 elephants left on the continent. Imagine how most elephants experience us: the *tap tap tap* of distant gunfire and then the rumble of trucks, smell of fumes, odor of elephant blood. There's no escaping the banging that comes from everywhere—no escape from the knowledge that you may be tracked to the

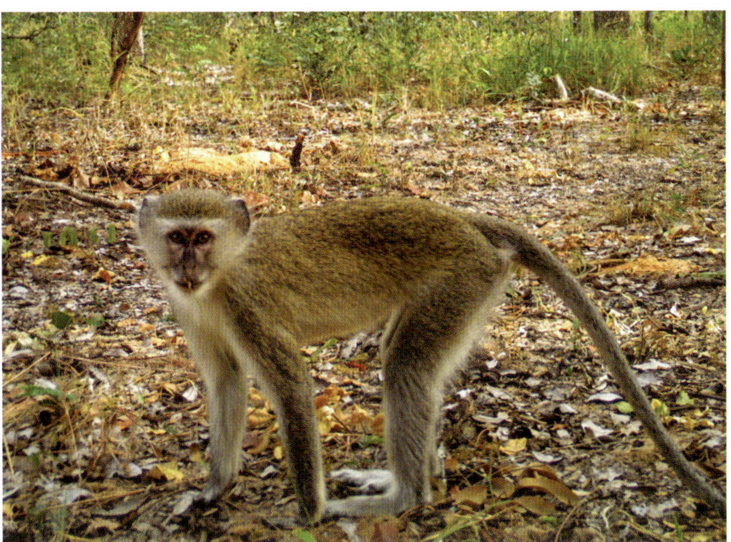

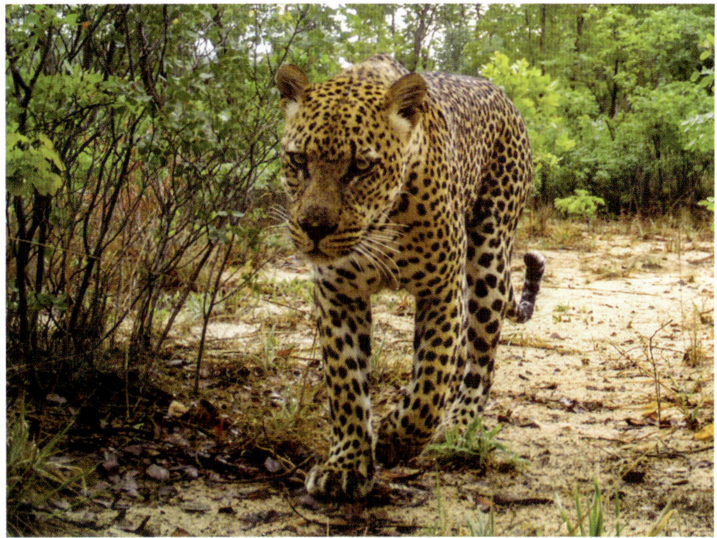

very end. In a war fought over you, all you want is to disappear into wildness, become a quiet heartbeat at sunset, a lost set of footprints through uncharted territory.

The ghost elephants of Lisima must remain lost in freedom, hidden from harm. They must be revered and allowed to protect themselves as they have done for thousands of years, before the Portuguese colonizers and the terrible wars. Yet I wonder at the consequences of seeking them out. Like Fénykövi before us, we looked for Henry's descendants with the same persistence that we had explored all the major rivers and discovered all the new and endangered species. Augusto said that the last white people to come to Chinjanga, his village, were Portuguese ivory hunters in the late 1950s. The last time the villages along the main track to Tempue saw white people was when a Portuguese colonel arrived in the early 1970s and massacred all the inhabitants who didn't flee. Local people hide the stories and places they care most about from outsiders, fearing that, like every other time, these things will be taken away from them. Maybe looking for the elephants was a step too far?

I get this feeling when it comes to the Fénykövi elephant. There must have been many stories of this massive elephant before it was taken from its homeland. Maybe it's better not to have heard them, not to enshrine them in our modern museums. Perhaps some stories are best told in the local language, shared around a fire.

When I visit Henry at the Smithsonian, I think about extinction: about trying to find hope in time that is always running out. We lose an elephant every 30 minutes in Africa with that *tap tap tap*, a sound that rangers trying to protect the elephants are often helpless to stop.

Our search for Henry's descendants is urgent, as the last megafauna slowly disappear into the memories that museum exhibits preserve. The knowledge we gain will help ensure

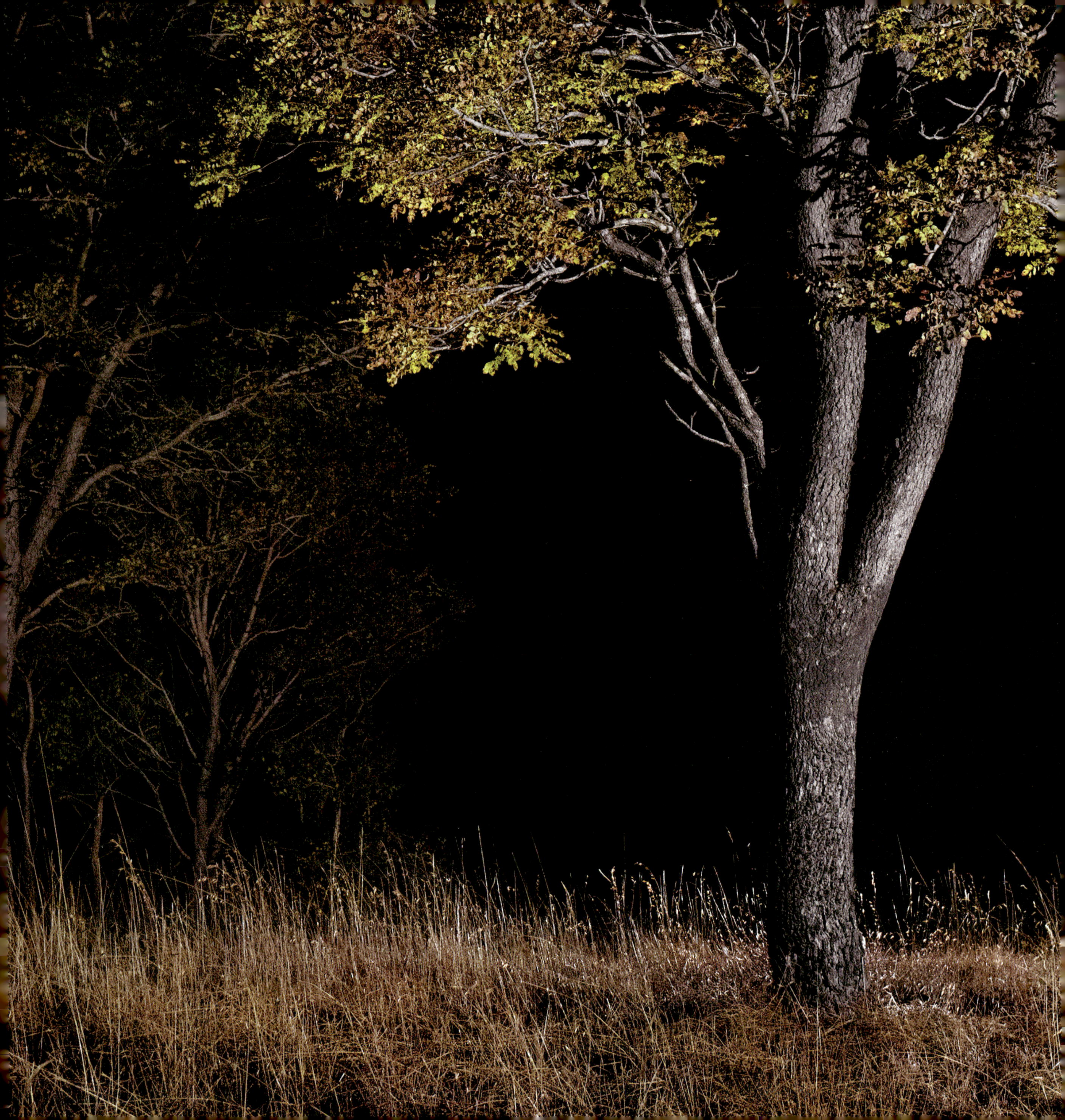

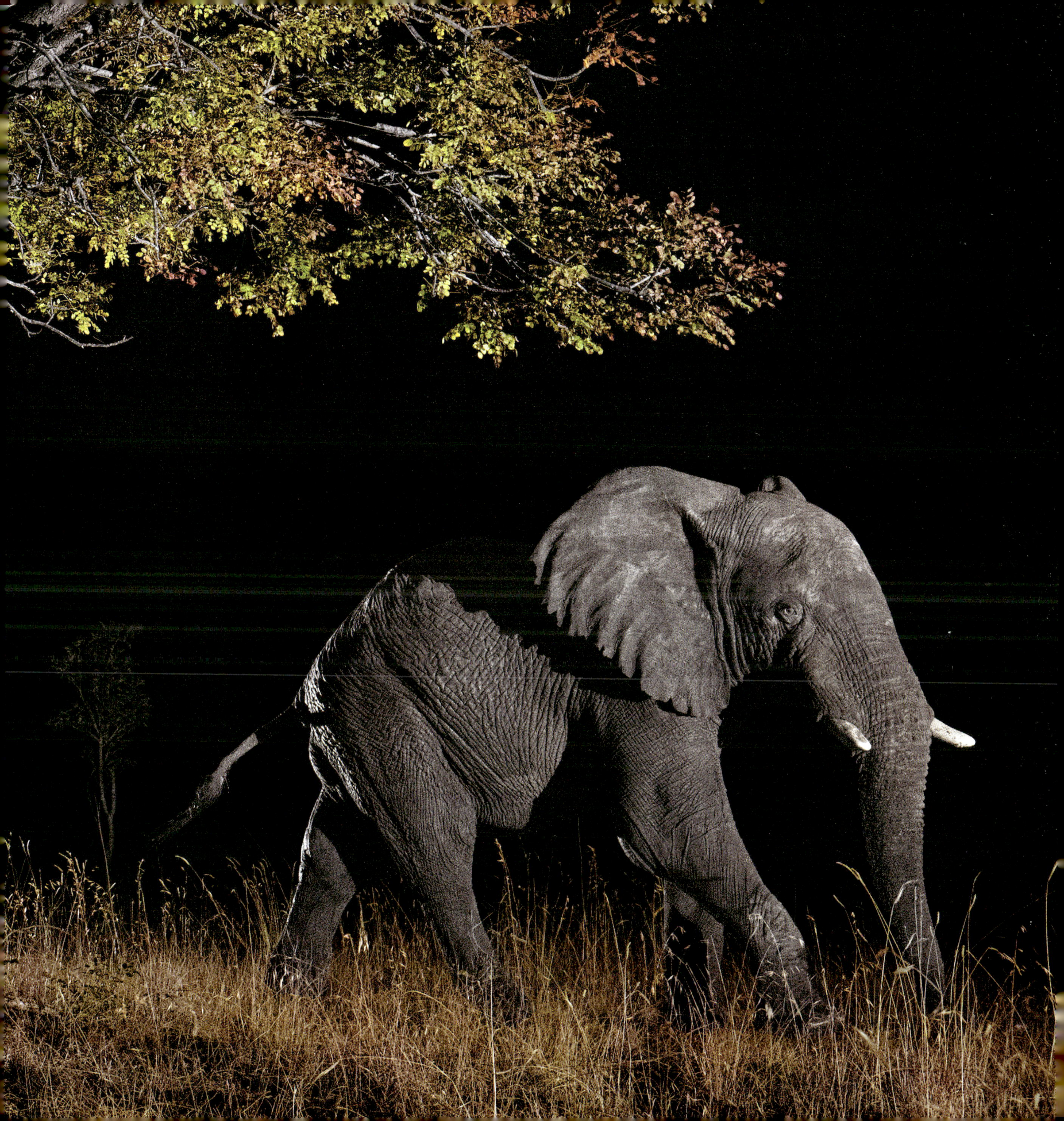

their survival. We have deployed an array of acoustic sensors to listen to the ghost elephants as they talk to each other, passively monitoring the seasonal movements of these secretive forest spirits. Hundreds of camera traps will continue to watch them indirectly, and every now and then we will visit to see whether they trust us enough to reveal themselves. Curiosity is at the root of exploration and human endeavor. These ghost elephants have captured our imaginations, and we are determined to help keep them and their unique habitat alive and thriving. Our world needs places like this, an Eden where we can begin again.

ARE AFRICA'S last free-roaming elephants destined to be ghosts? Spectral beings roaming an Earth scarred by our desires, surviving on the world's fringes as a regret? The ghost elephants challenge us, beckon us across an existential divide with a silent trumpeting in the language of trees. They ask us to look beyond what is seen, to feel the pulse of wildness that still beats beneath the concrete and steel of our creations.

Their specters a pale reflection against the canvas of modernity, ghost elephants are fading whispers in the wilds of our imagination. Though they walk silently, their colossal forms shrouded in the dust of memory charge loudly into a global consciousness struggling to remember their majesty. Their footprints are everywhere, but they remain unseen. Their footfalls, soft as whispers, are echoes of Earth's ancient heartbeat. The ghost elephants are memories of the world as it once was: pure and untroubled, wild and unbridled. They tread an eternal path through this twilight of existence. Ethereal. When they return, there will be thunder.

As we ponder their fate, we must ask ourselves if the future of all wildlife is to follow these ghosts into obscurity: there but hidden from sight, phantoms in darkness. Architects of landscapes, shapers of ecosystems, they have the power to create and destroy. Their all-powerful trunks wield the power to topple trees and excavate rivers, but they also have the tenderness of a mother's touch, caressing the fragility of life.

A war is raging, a battle for their survival, perhaps for the very soul of Earth. This isn't just a fight against extinction. It's a struggle for human nature as well. We face a choice between a world where spirits roam free and one where they are but ghosts, haunting the edges of human consciousness, a collective memory of humanity, a reminder of what once was.

The natural state of both humans and elephants is peaceful. We used to coexist by living alongside one another, trusting that the one wouldn't surprise the other. Like all siblings, we had our disagreements and conflicts, but we always resolved them. We both have the capacity for unprovoked rage and anger, but these feelings pass. But unlike elephants, humans have stewarded forces that are driving us to destroy the natural world.

Fostering a more harmonious relationship goes far beyond protecting wild landscapes, creating corridors for elephant migration, and reducing human-elephant conflict. It isn't just about closer interaction; it's about acceptance. It's about equal rights to exist and a shared destiny.

PAGES 158–159: *Since 2016, we have deployed hundreds of cameras across the remote eastern Angolan Highlands, seeking evidence of the region's ghost elephants. In 2025, we finally captured definitive photographs. Those who live here believe these secretive breeding herds have magical powers.*

RIVER GUARDIANS

Vice-King Sachindamba of the Luchazi

King of the Forest

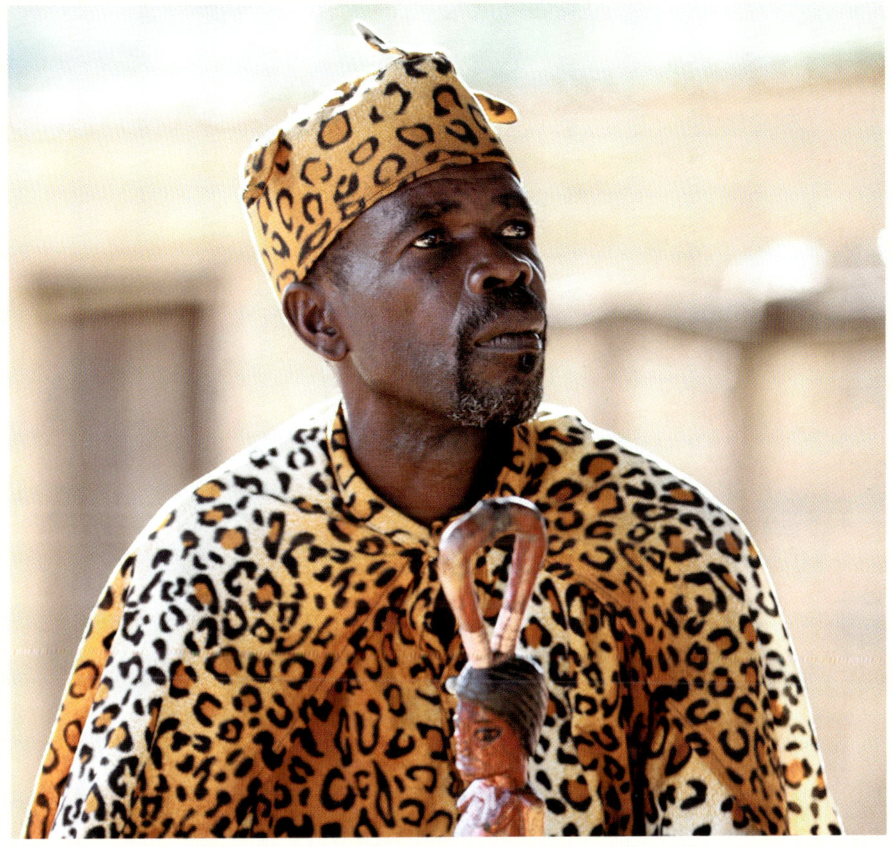

Vice-King Sachindamba of the Luchazi has a kingdom that spans much of the Lisima Landscape. He upholds the traditions of his people to protect natural resources.

Vice-King Sachindamba lives in the Lisima lya Mwono Landscape near the source of the Lungué-Bungo River, the true source of the mighty Zambezi River. He presides over a large kingdom spanning some of the wildest parts of this Source of Life landscape. He enforces hunting seasons to protect the wildlife populations and the natural heritage of his people. He only allows traditional bark canoes on the rivers—no bridges or mokoros—to discourage commercial bushmeat hunters. He is the protector of the last ghost elephants.

Muito floresta—a lot of forest—is his pride. The unending miombo woodlands, globally important peatlands, and sacred source lakes aren't just wilderness to him. They are dynamic characters imbued with conscious indifference, all-powerful over humans yet innately us. These untamed, wild forests challenge and sustain the Luchazi and Chokwe villages, such as Tempue. Source lakes, rivers, and forests are sacred, protected by mythical spirits in communication with their ancestors. The people living here burn forests to optimize them for honey production, to create pastures and grassland corridors for wildlife, to clear moribund grass, and to purify the rivers and streams. Sachindamba's kingdom has been taking practical steps to preserve its environment for generations. The outside world should celebrate this example.

Traditional cultures such as that of the Luchazi people live by an intricate decision-making system based on knowledge and wisdom that takes a lifetime to understand. A fluid boundary between this world and the spiritual world has been one of the primary mechanisms of protection in the Luchazi culture for thousands of years. While great civilizations have collapsed into ruin, traditionally living communities—if not destroyed—have persisted. We are the existential threat. We must protect their freedom.

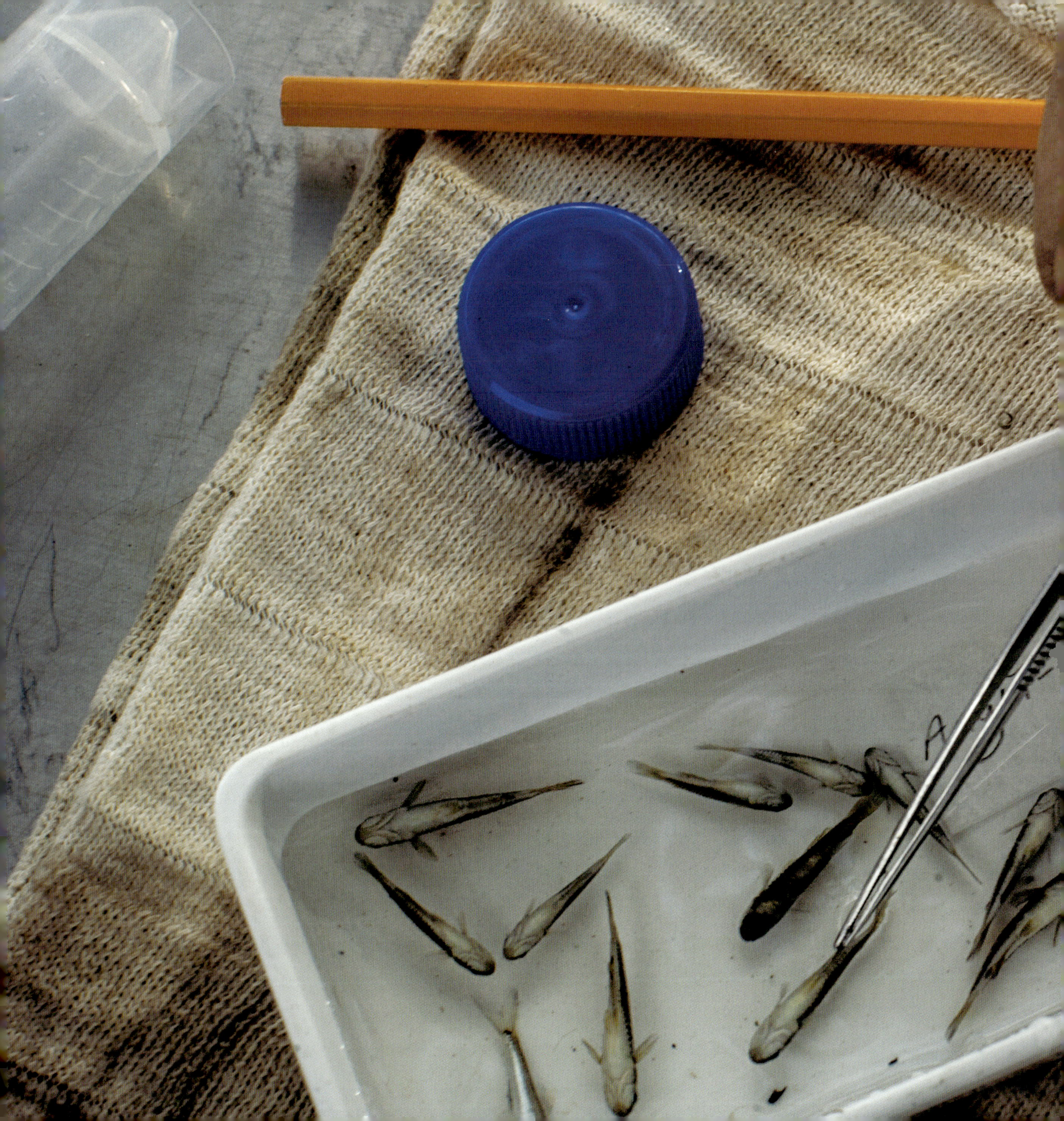

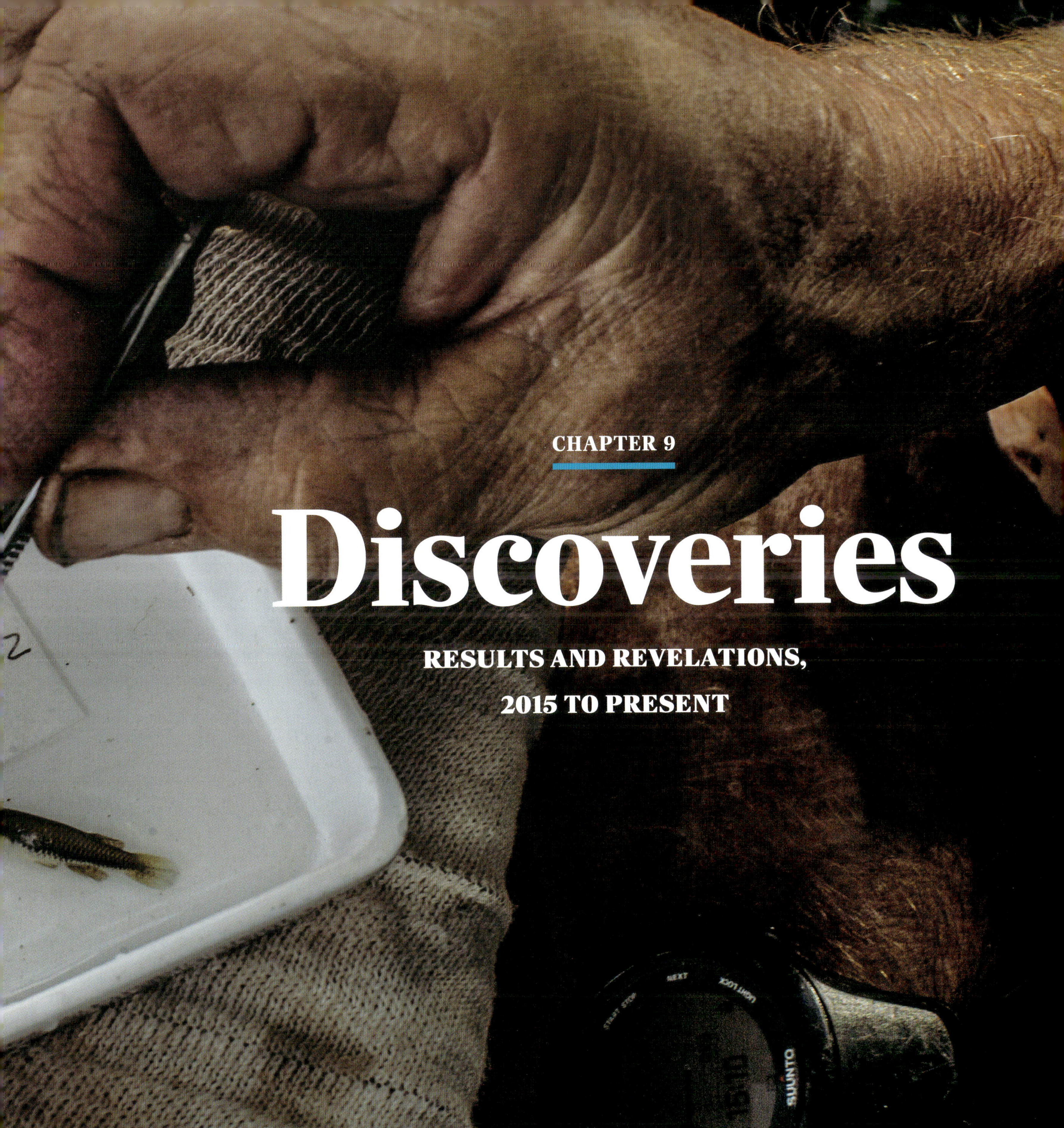

CHAPTER 9

Discoveries

RESULTS AND REVELATIONS, 2015 TO PRESENT

LIVING DAY AFTER DAY in wildness, noticing every sign of life or danger, has changed how I see science. It becomes clear that there's no grand theory of everything; we're trying to explain things that are often beyond description. We can't help but think about the world around us in diagrams and graphs, systems to be understood. We've become convinced that nature is mechanical and that we can predict what's going to happen.

I'm that ecologist; I often give talks about the Okavango Delta as if I know its every intricacy, every nuance. Yet I know that the nature of nature is unpredictable. We're desperately trying to understand and control it because we're scared. We've realized that life will go on without us, and yet we know that we are inherently linked with these ecosystems—we reach out and yearn to be part of them. Nature isn't a puzzle to be solved: We are. Words and numbers were invented to explain us to ourselves.

WINTER NIGHTS in the Lisima Landscape are freezing cold and damp; during the day, the sky is hazy and discolored by hundreds of fires burning. It's almost entirely silent. There are no signs of wildlife or people, no birds calling or insects buzzing, no movement beyond the constant gusts of wind in the trees.

Yet as we travel between the source lakes, we get the sense that we're being watched. Our motion-sensing camera traps showed us that eyes were indeed watching and ears listening. They were hiding: not from us but to conserve energy and to outlast the dry season.

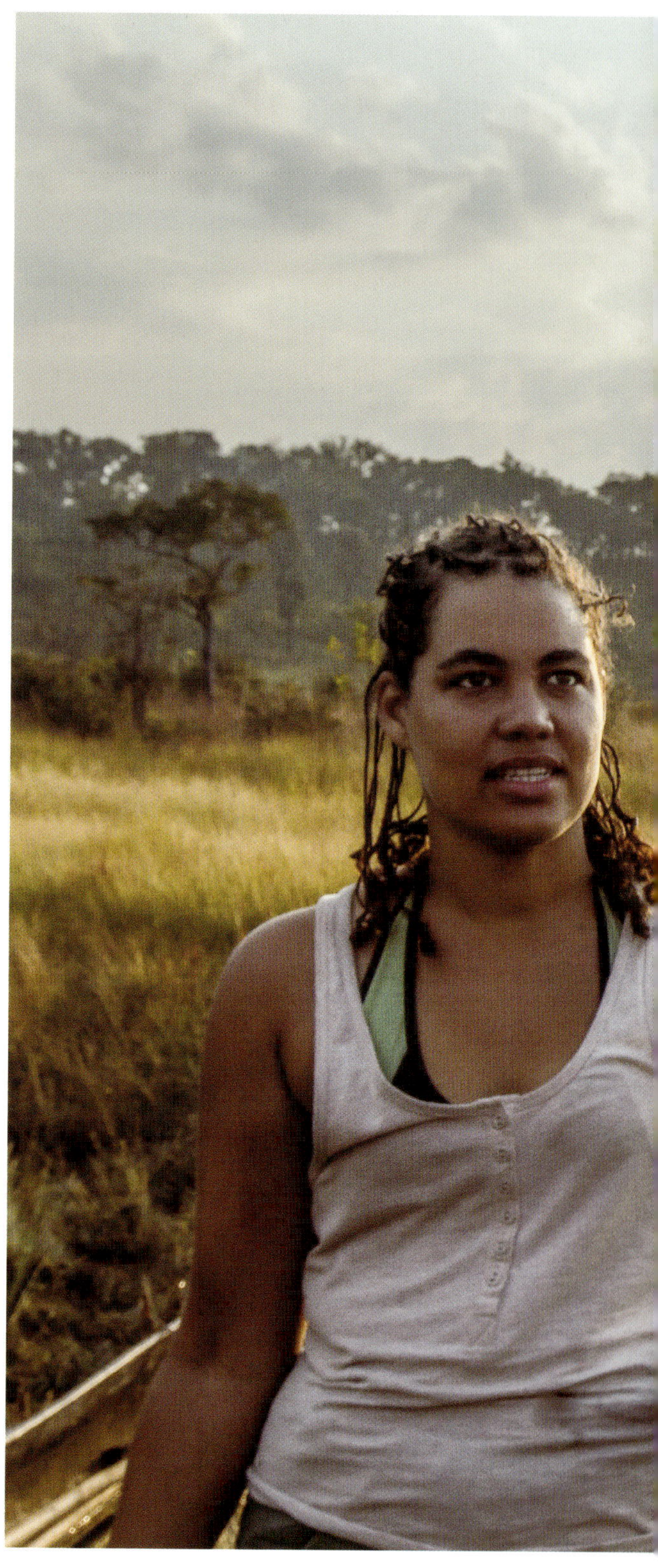

Adjany Costa (left) and Paul Skelton examine a small fish caught near the source of the Cuito River before photographing it for the expedition inventory. **PAGES 162–163:** *Every morning on expedition, we check our nets to see what small fish were trapped overnight. We use a few drops of clove oil to put them to sleep, photograph them, and take fin clippings for DNA sampling. Then we release them back into the river.*

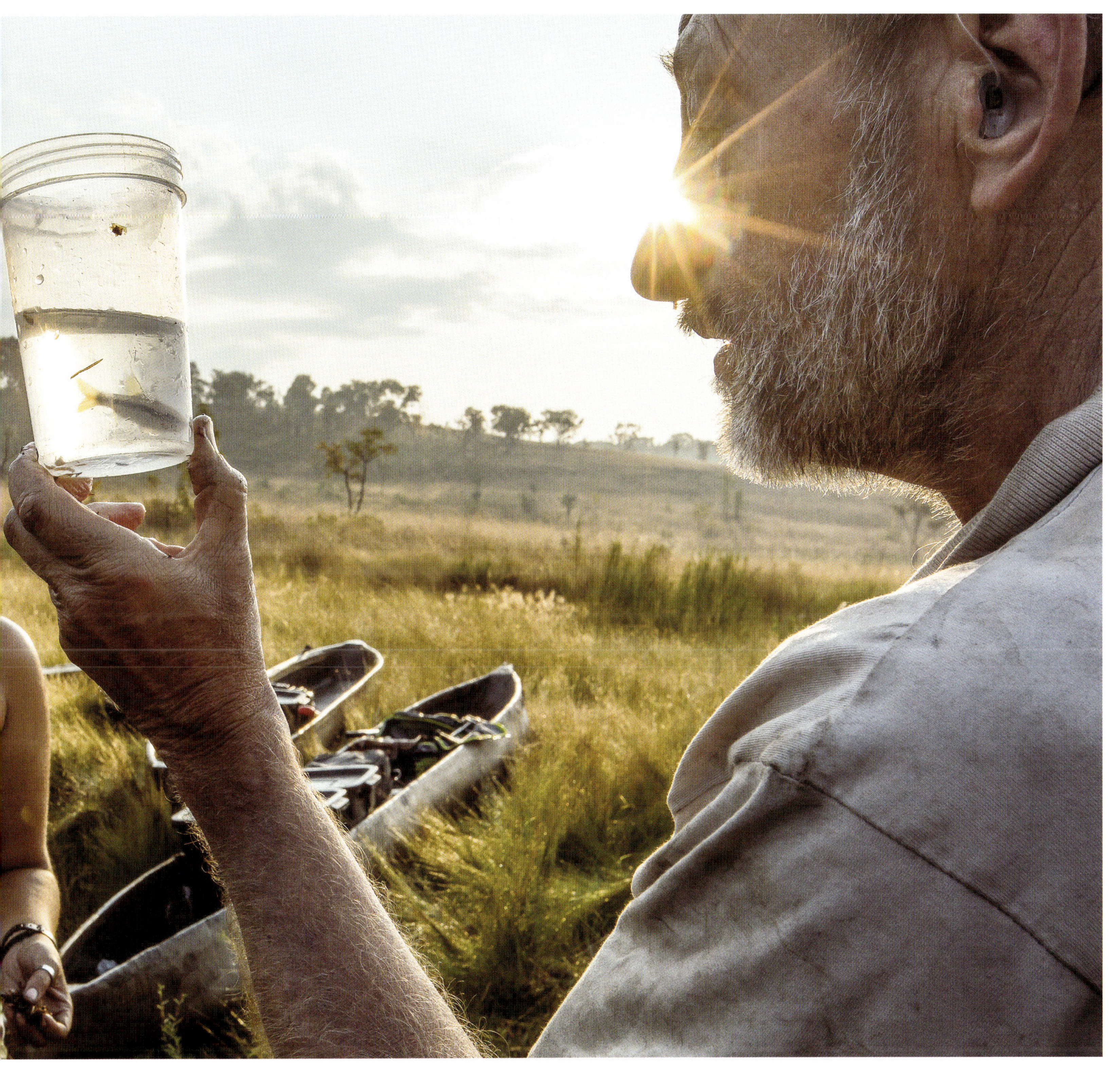

During the winter months the local people start lighting fires, the grasslands desiccate in the sun, and trees drop their leaves. Winter is a defining characteristic of this unique wilderness, and one that seems to give an opportunity to rest and reset before summer's return.

In summer, towering thunderstorms gather every day to drench the Angolan Highlands with moisture rising from the Congo Basin. What follows is an emergence of life on a scale I have never witnessed in other subtropical forests.

If you are traveling through Lisima forest in midsummer and you notice that the leaves on the trees are all gone, beware: The bark on those trees isn't bark but rather moths that can literally choke the air and clog any light source. A tent in the moonlight will be covered in moths. If not moths, then it will be millions of beetles on and inside the tent. Marching swarms of red stinging ants and large black army ants swarm through camp, forming dense mats that behave like a single animal, reaching up onto tabletops and biting holes in your tent. I once woke to a strange rustling sound and found one such swarm of giant black ants using my body as a shelter from the rain. Once disturbed, they moved off, making a nightmarish rattling sound.

When you're having a private moment in the bushes, dung beetles range in from the surroundings, crashing into you even before you have time to finish your business. Near permanent water, fluttering butterflies and the collective, unnerving buzz of millions of honeybees and sweat bees can fill the air. Such abundance, though nightmarish to some, speaks to this landscape's health: tarantulas the size of my hand, colonial nesting spiders with webs the size of cars, millipedes more than a foot long, centipedes with massive poison claws like pliers, so-called hippo flies that can bite through leather, longhorn beetles the size of a smartphone, and praying mantises of every possible size and description. Bugs are simply inescapable in the Lisima Landscape: Insects crawl on or over you all the time.

Tropical forests the world around are typically places of abundance, with the highest concentrations of species, but the oscillation between boom and bust in the eastern Angolan Highlands is unrivaled. It's like spring is wound up slowly during winter and then enthusiastically released.

It took us several years and 57 scientists passionately documenting the secrets of a little-known landscape to discover for ourselves what the local people knew all along: This is truly the source of life for those who live here and a source of hope for us all. The Lisima Landscape is a biodiversity hot spot, a Key Biodiversity Area (a designation approved by an international partnership of conservation organizations), and one of the world's largest globally important wetland ecosystems (named a Ramsar site, an international designation identifying key wetlands deserving protection). This landscape is the primary water source for the world's largest transfrontier conservation area, the Kavango-Zambezi, as well as for almost two-thirds of Africa's remaining savanna elephants. It is a living ark, precious for its regional biodiversity and for the Indigenous

We were able to paddle the upper reaches of the Cubango River. No one has ever navigated the full length of these source rivers, so everything we document is new to science, contributing to a deeper understanding of these complex ecosystems.

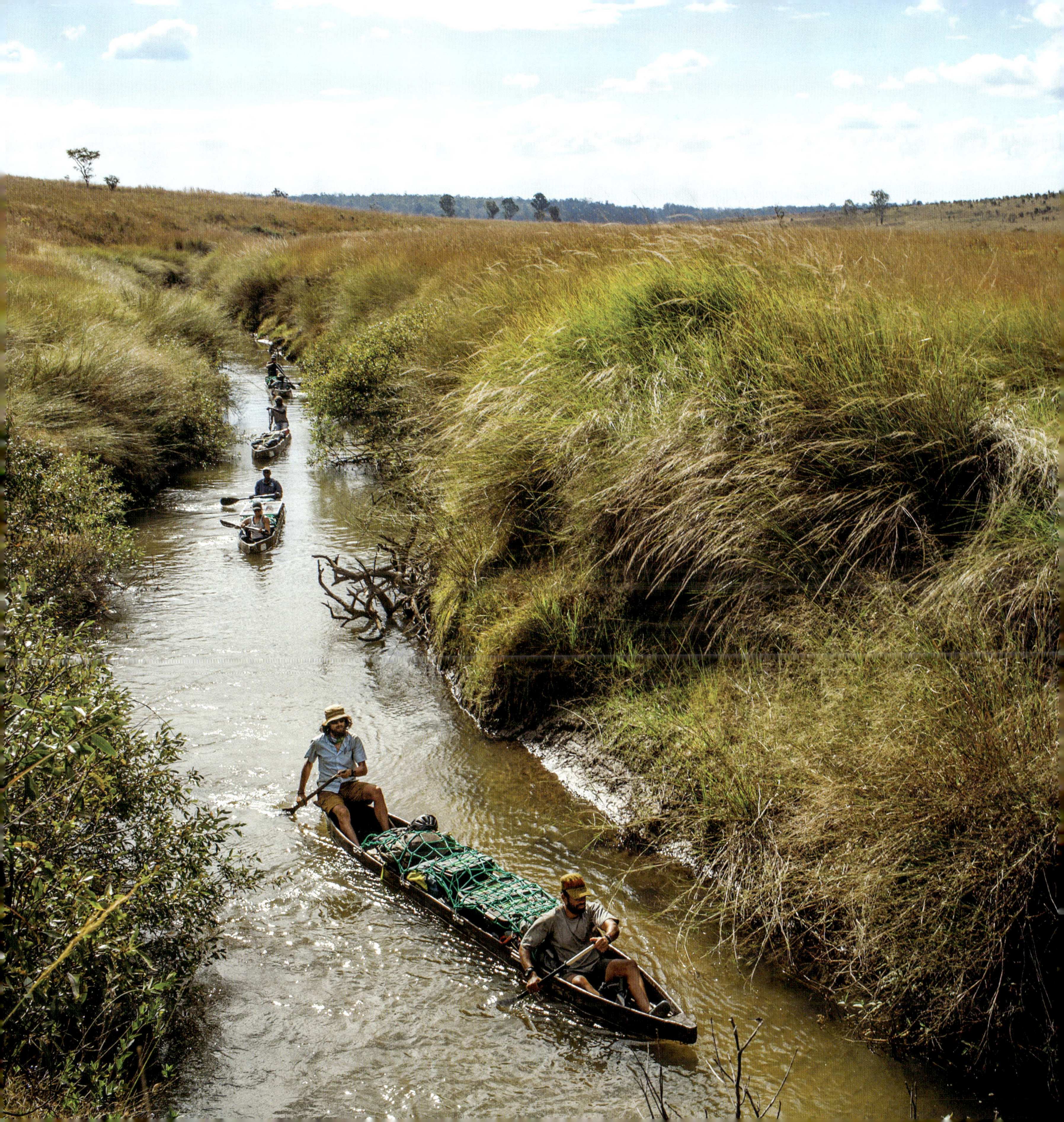

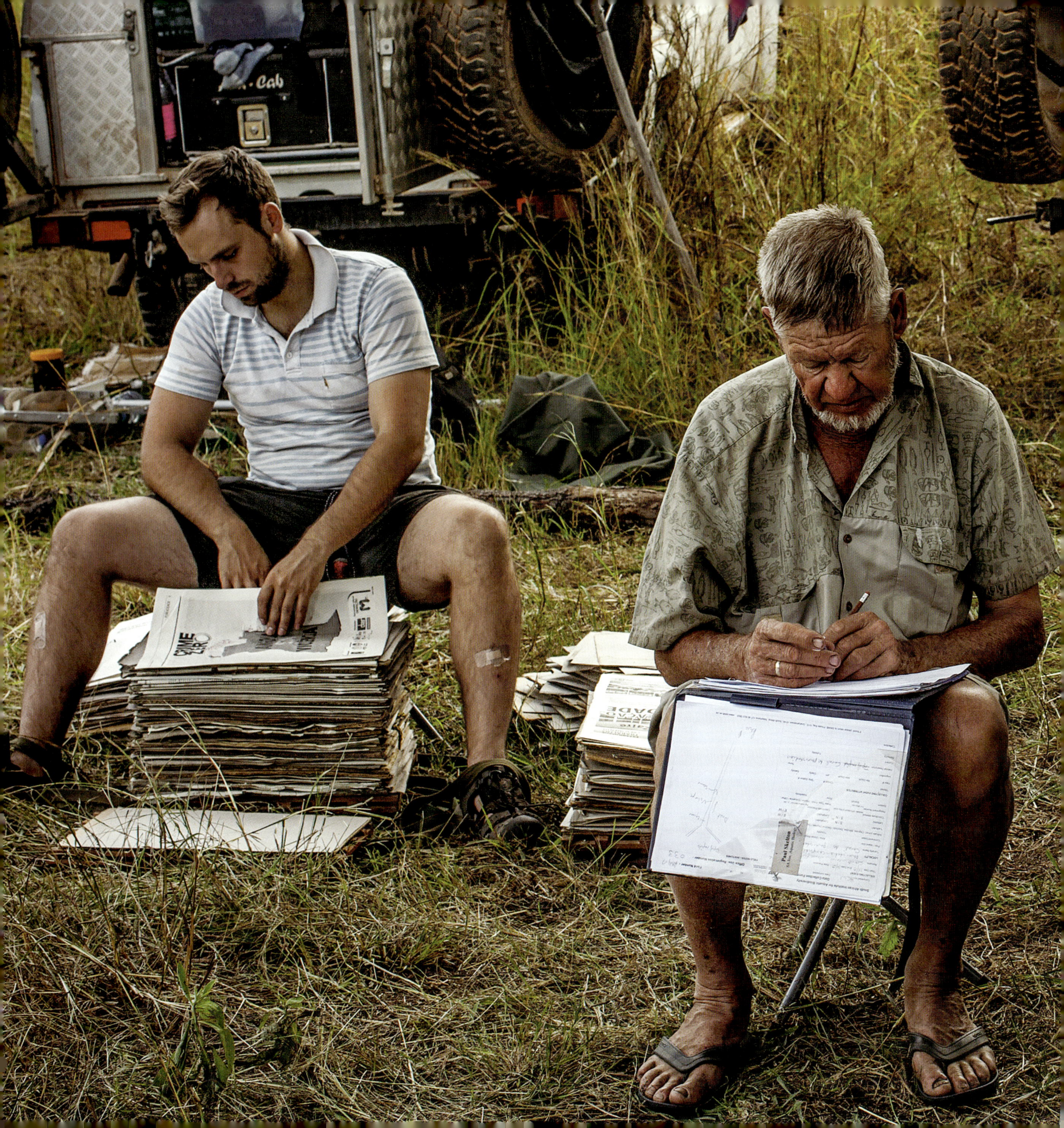

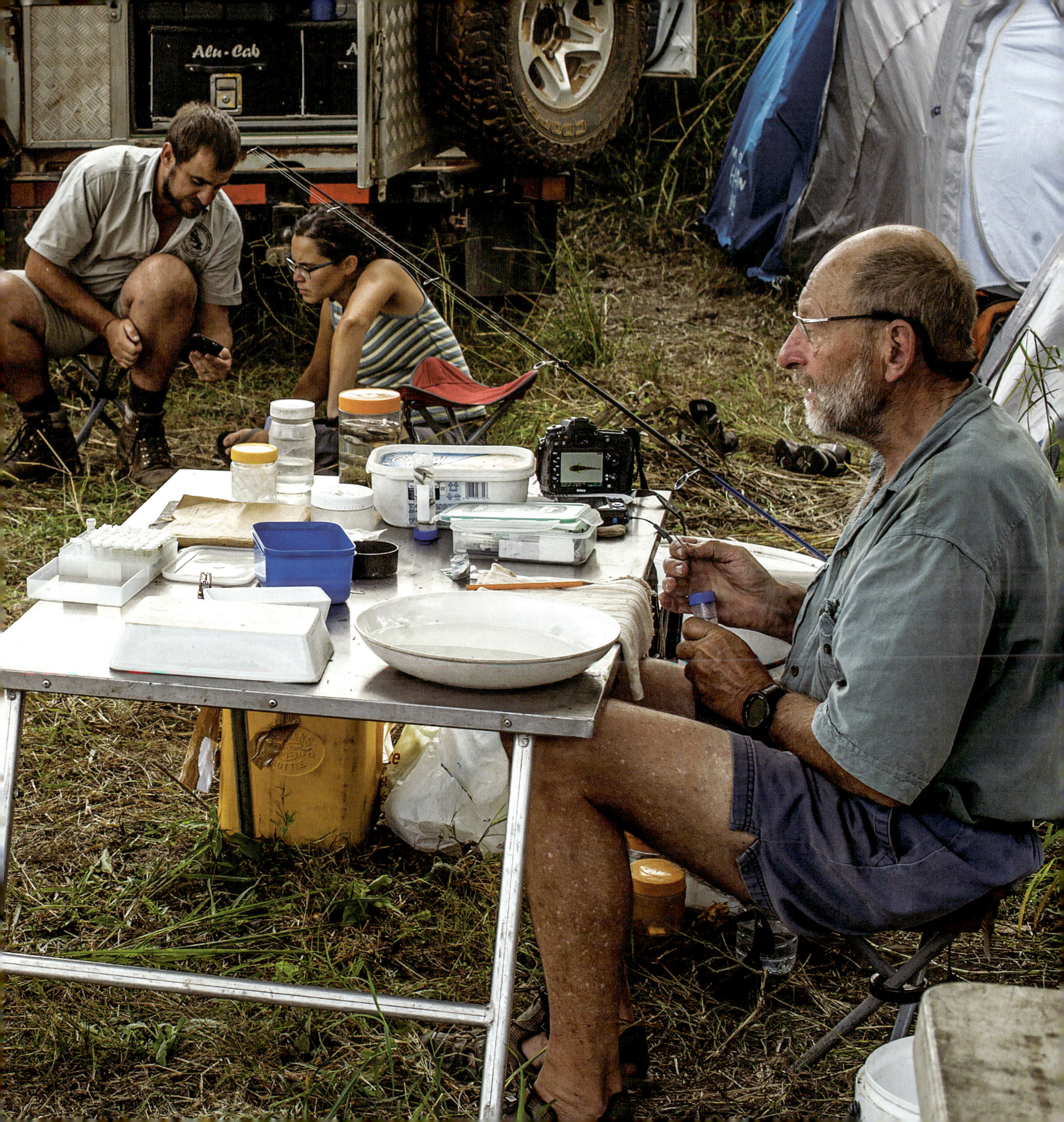
Alu-Cab

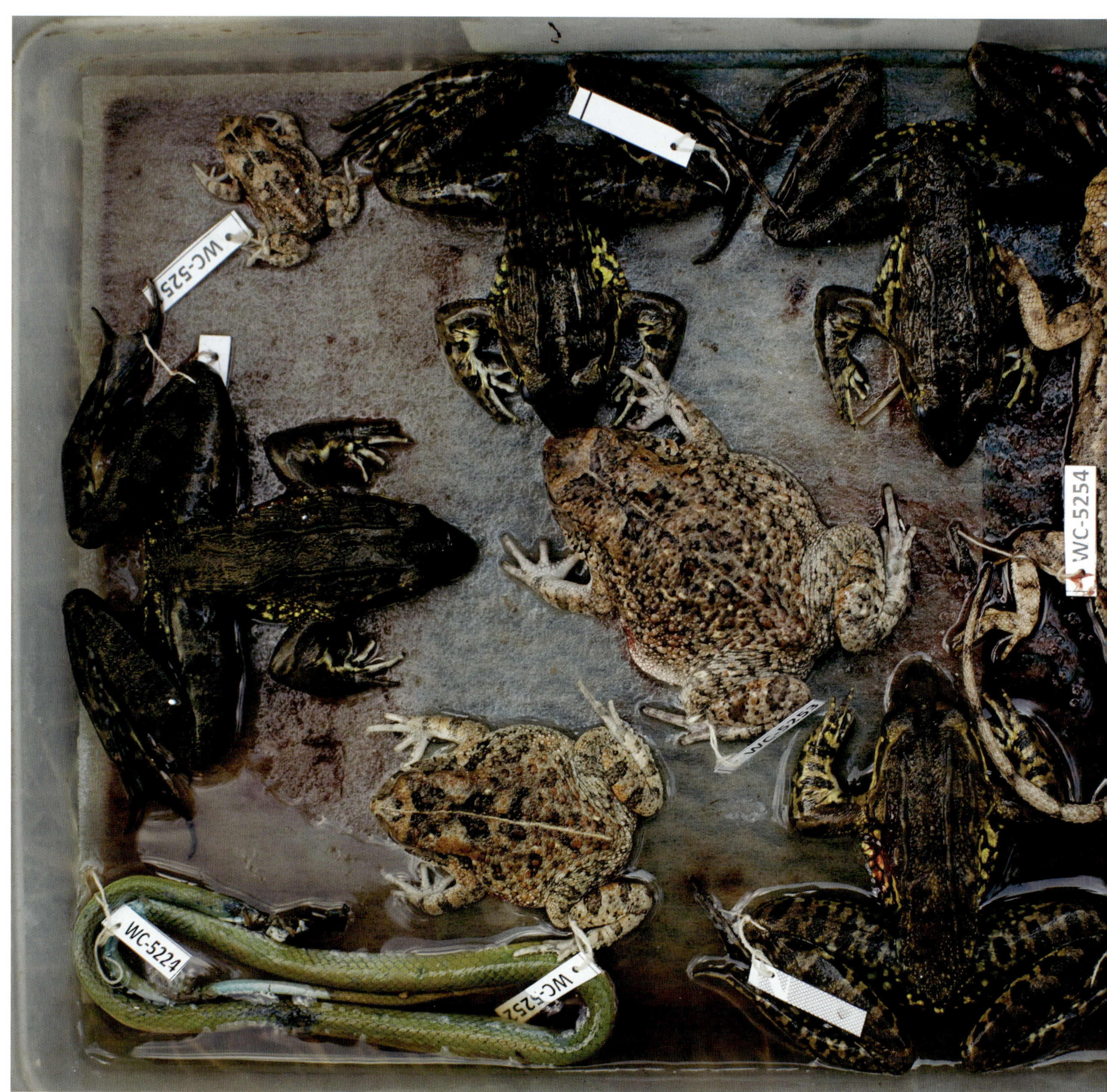
WC-5254
WC-5224

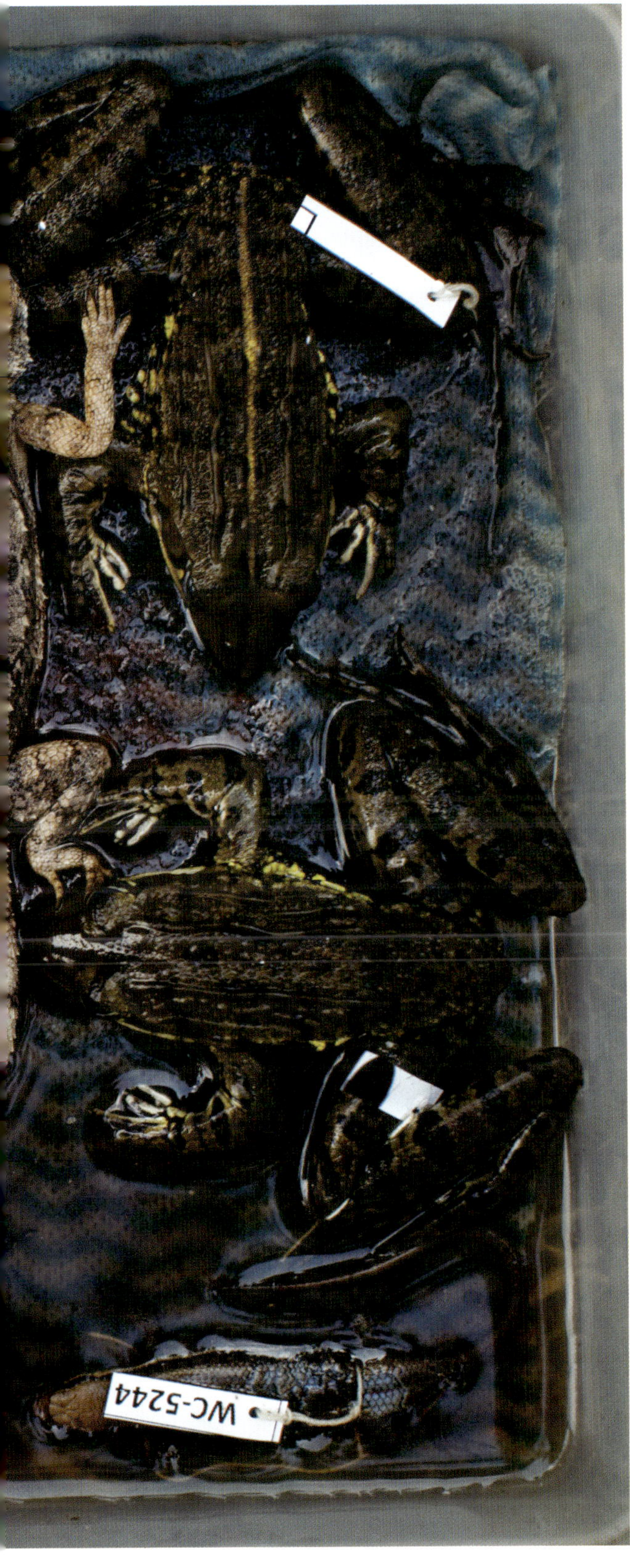

conservation knowledge held by its Luchazi stewards—a living textbook for sustainable forest management and biodiversity conservation.

FROM 2016 to 2023, the initial Okavango Wilderness Project expeditions focused on biodiversity and wildlife populations. Our team has documented more than 150 species new to science, as well as hundreds of species previously unknown in southeast Angola. These are our top-level research findings so far—more are likely to come.

The eastern Angolan Highlands represented the biggest gap in scientific knowledge of African flora until we started our detailed surveys with David Goyder from Britain's Kew Gardens. Of the 1,600 botanicals collected, the team has discovered 92 new plant families, including 450 species. Of these, 21 species were new to Angola, three are confirmed to be new to science, and 10 more are potentially new and in the process of being described. Undoubtedly, we will continue finding new plant species for decades to come.

The mycological collection now totals 458 specimens of fungi, representing 65 genera and more than 100 species. So far, two species have been new to science, with more than 10 others potentially new as well. The misty conditions and thick mats of moss support the growth of mushrooms of every color. If you go out at night with a fluorescent lamp, the forest floor lights up with glowing mushrooms near the source lakes.

Cataloged entomological collections now total 14,141 insect specimens, including at least

Unfortunately, specimens need to be preserved for detailed description and DNA analysis, so new species can be described.
PAGES 168–169: *Our land-based biodiversity surveys are large undertakings with up to 50 scientists and experts setting traps, stalking around the forests, and patrolling the rivers looking for new and interesting species to document.*

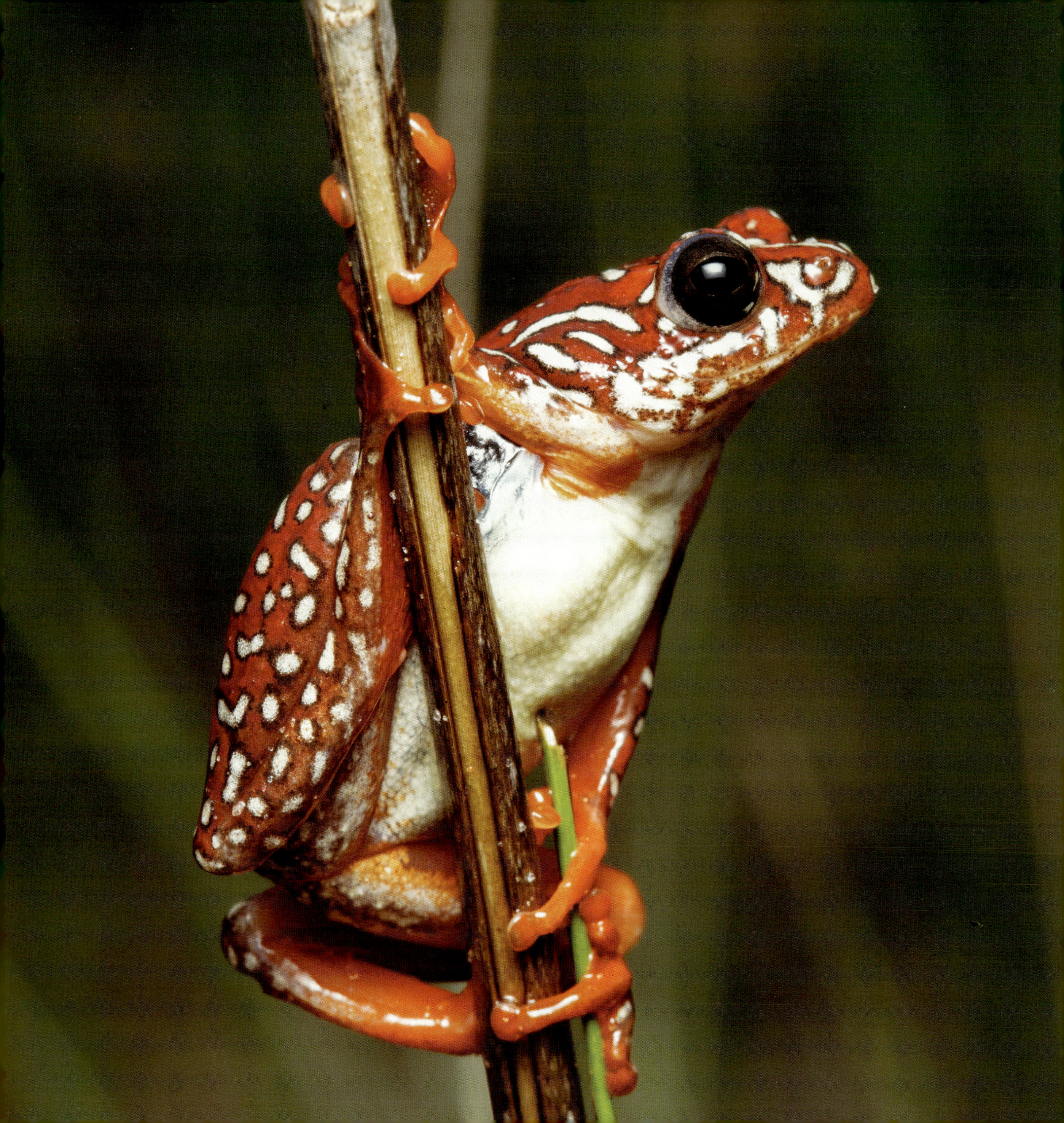

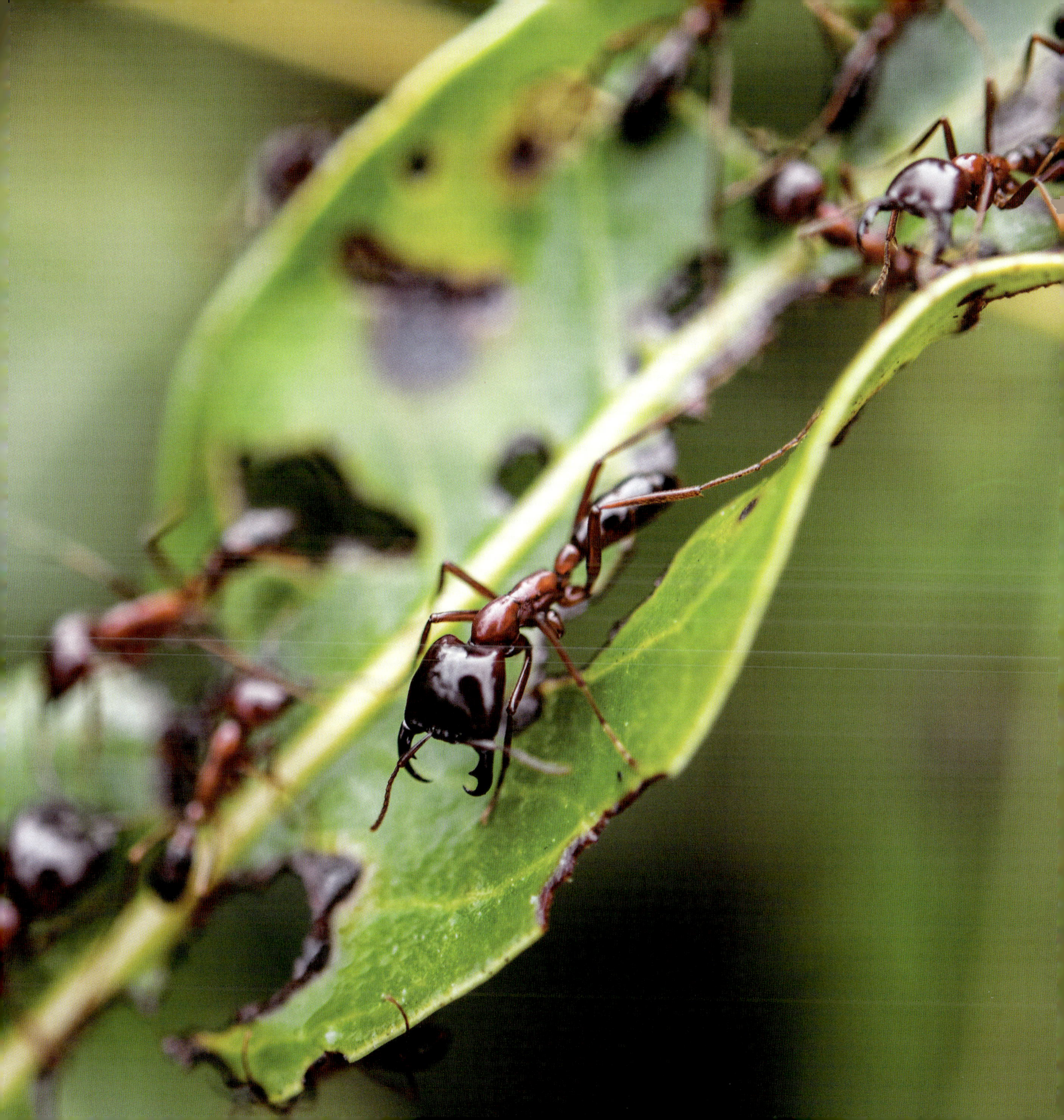

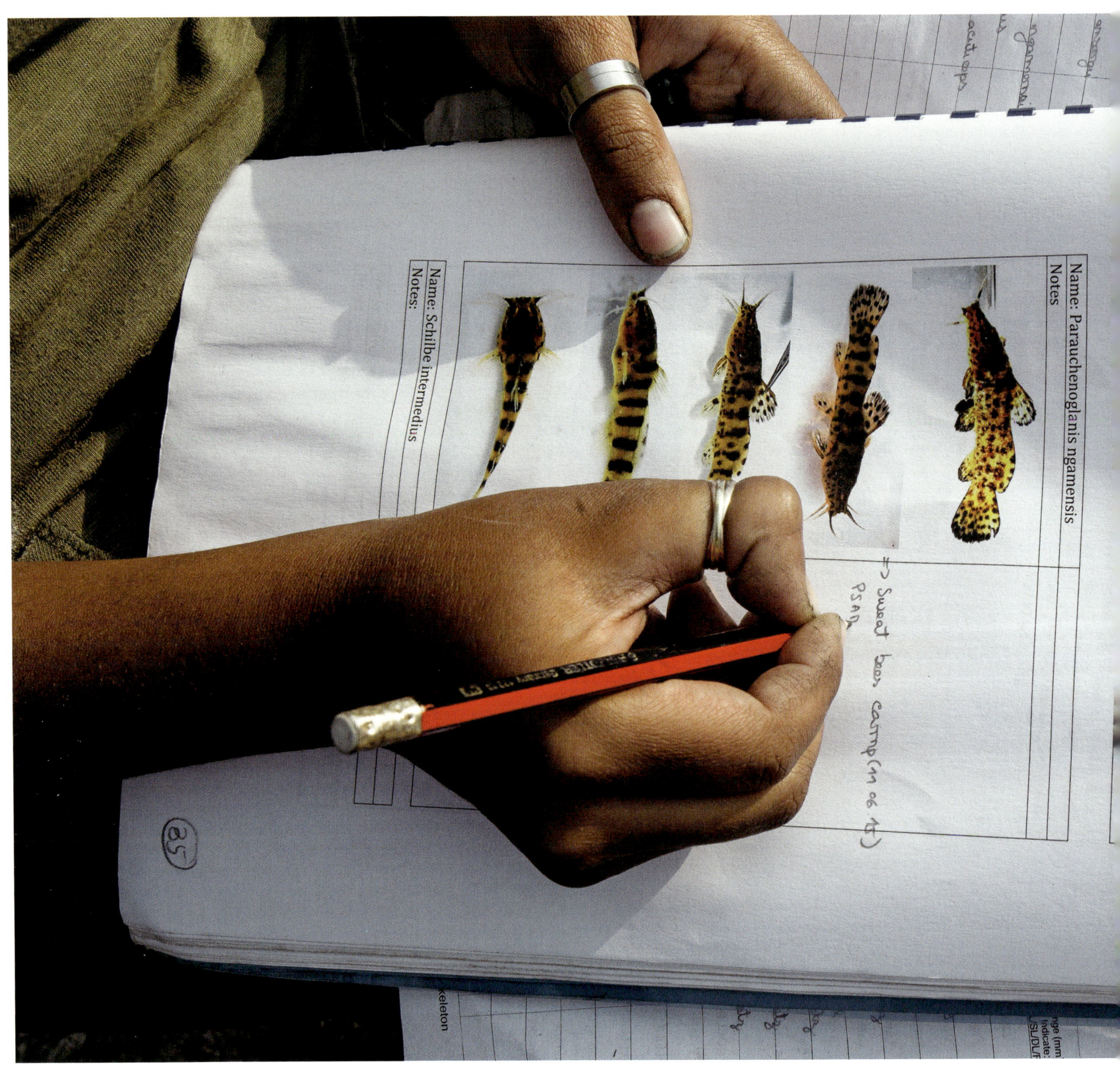
Name: Parauchenoglanis ngamensis
Notes
Name: Schilbe intermedius
Notes:

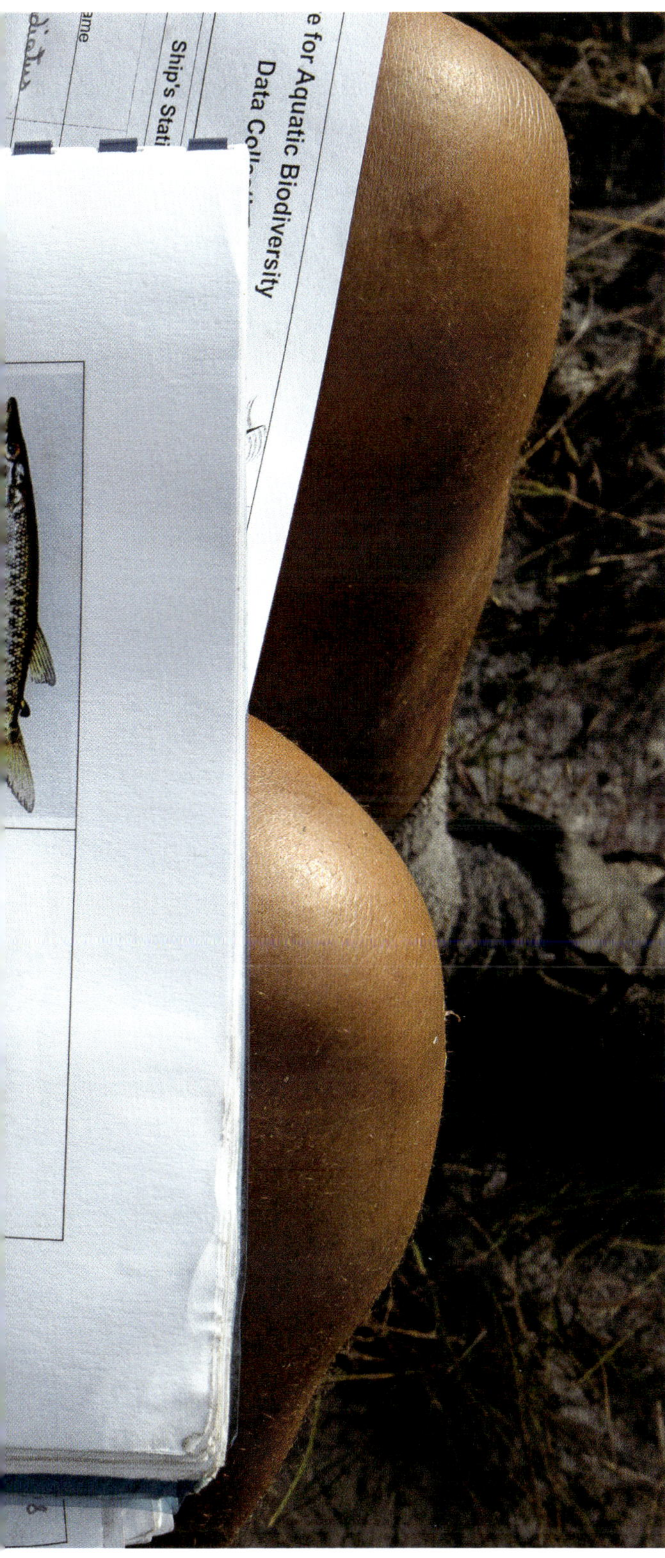

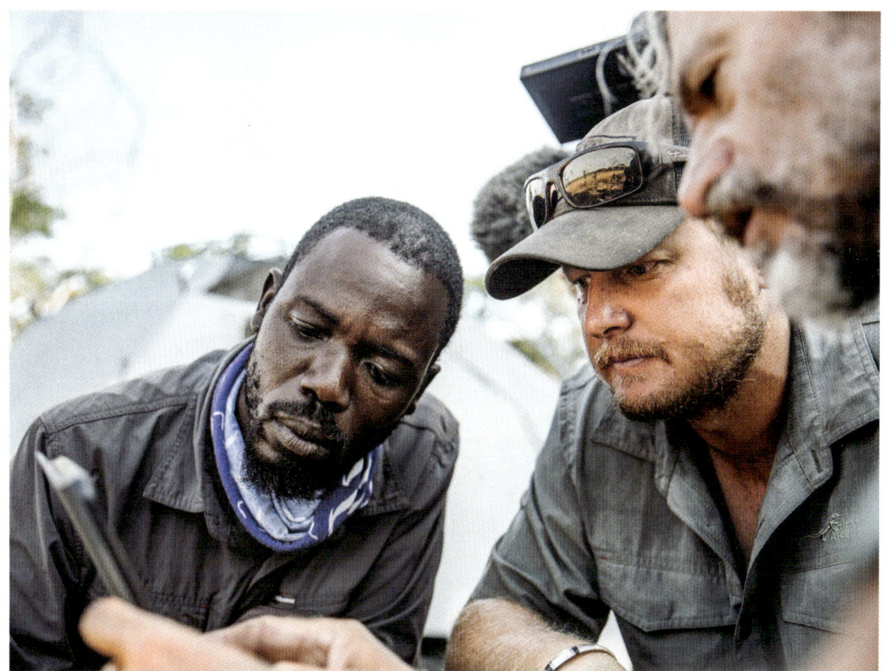

33 species new to science. The three most exciting finds are Africa's smallest dragonfly, proposed to be named lilliput prickleleg; an alarmingly hairy mite; and a large tarantula with an impressive horn on its back—the largest ever recorded on a spider—that appears to function like a camel's hump, storing water and nutrients. The research team is assessing 50 more insect species potentially new to science, including 12 dragonflies. One species emerges in large numbers at sunset at the outlet of the Cuanavale source lake looking so much like little halos flying around that we have proposed to name it *"haloensis"* after the HALO Trust.

The ichthyological collections now total 7,668 fish specimens from 16 families and 113 species. We described five species new to science; four more are under assessment. In the

Antonio Cangamba (left) and Götz Neef help me work out where to camp along the Cassai River. **OPPOSITE:** *Adjany Costa makes notes about the fish species found in the Cuito River.* **PAGE 172:** *Angolan painted reed frogs in the millions call from these great floodplains.* **PAGE 173:** *Harvester ants can overwhelm the forest floor during the summer months, making camping almost impossible.*

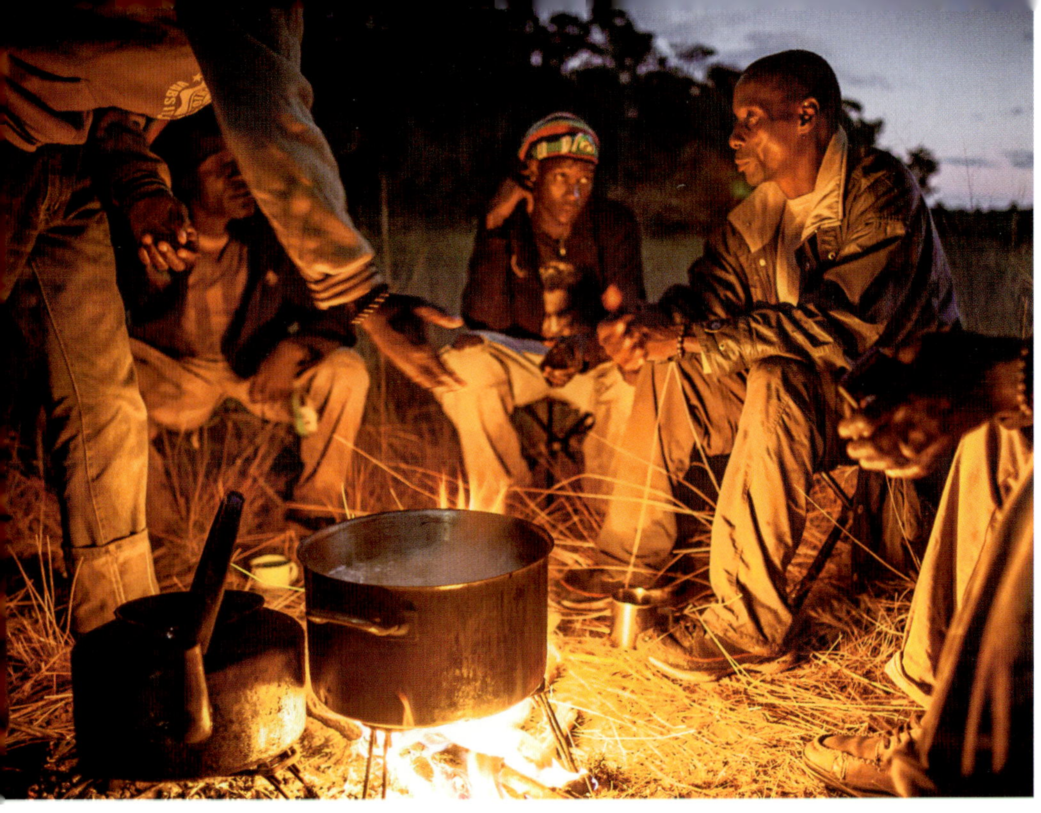

These local Wayeyi polers came with us to explore the sources of the Okavango Delta in the Angolan Highlands. After exploring all the major rivers and channels and meeting the Angolans and Namibians living along these tributaries, they will take the stories of these personal discoveries back home with them.

Cubango-Okavango and Cuando River systems, we also found seven species never documented in Angola, as well as a ghost stonebasher—a species that sends out electric pulses to detect smaller fish to prey on at night. This odd-looking fish—cited as having the highest brain-to-body weight ratio of any fish—had been seen here only once before, half a century ago, and offers new information about the historical connections between these river systems.

Paul Skelton, our ichthyologist during the first expedition, told me that one of the great privileges of his long career was exploring these waters and watching the microhabitats change as we progressed downstream. Almost six years later, Paul sent me an issue of *Ichthyological Exploration of Freshwaters*, which included his article describing the first new fish we had found in the Cuito source lake and calling it *Microctenopoma steveboyesi*. You haven't quite made it until someone names a small African fish after you.

The herpetological collections now total 1,284 specimens of reptiles and amphibians, including 29 families and 115 species: three new to science, nine potentially new to science, and seven new for Angola. In 2016, we found a blind, 7.6-centimeter-long (3 in), red-brown tadpole. We kept it in a tank for nine months, wondering when it would transform into its adult form. For a time, we thought it may be a species that didn't metamorphose, something specially adapted as an eel-shaped tadpole in the source lakes' peaty fringes. Eventually, though, it turned into a Bocage's tree frog, a species poorly known and hardly ever seen.

In 2017, on a walk with a bushmeat hunter showing us a system of more than 200 two-meter-deep (6 ft) pitfall traps, we found a massive Gaboon viper, the first recorded appearance of this highly venomous snake in Angola. In 2018, we pulled at a dead branch for firewood and half the tree came crashing down. Flying down with it was a massive black mamba with a small green variegated bush snake wrapped around it. The tree fell on the mamba, only the second ever reported in Angola. Its back was broken, but it was still squirting venom at us—venom that can kill a person in 10 minutes. We managed to subdue and euthanize the mamba, preserving its skin for the Bayworld museum in South Africa. The green bush snake, meanwhile, disappeared in the chaos.

The ornithological team recorded 421 species in more than 30,000 bird distribution

RIVER GUARDIANS

Gobonamang "GB" Kgetho

A Simple Life Worth Living

LIFE IS ABOUT DECISIONS. Do I stay or do I go? Do I stop or do I carry on? Gobonamang "GB" Kgetho, my good friend for many years, has never struggled with making decisions. To him, it's always been clear: Respect your elders, live your culture, and honor your traditions. GB trusts in his ancestors and lives according to traditional values—a simple life worth living.

GB decided to stay, to live a traditional life as a fisherman, to follow his father and grandfather into the wild, into the freedom of the present moment. He doesn't live in the hope that at some point in time, in the distant future but not now, he will achieve happiness, security, and peace. His freedom is natural, and it's now. Tradition is our living connection to this freedom.

GB's father, Kgeto Kgetho, in his calm, reflective, stress-free manner, conveys that there's no future, no time, in nature. Likewise, GB has shown me that we can only reflect upon past rhythms and cycles as they echo into the present, that it's best neither to dwell in the past nor to predict what is going to happen. In Wayeyi culture, nature is more musical than purposeful when you get to know it. GB taught me to pause and listen.

Immersed in the present moment as he sets his fishing nets at sunset, GB's existence is a harmonious melody. Reality becomes an artwork. His senses grasp natural wonder. There's no anxiety, nothing to worry about, in nature's present. He isn't lost in the future we can't seem to escape from. Tradition and living memory taught GB that the universe doesn't exist in the future; it exists now, as a reflection of the past.

Gobonamang "GB" Kgetho joined me on our first mokoro crossing of the Okavango Delta in 2010: He is, effectively, a founder of this work. We both had no idea what we were getting ourselves into.

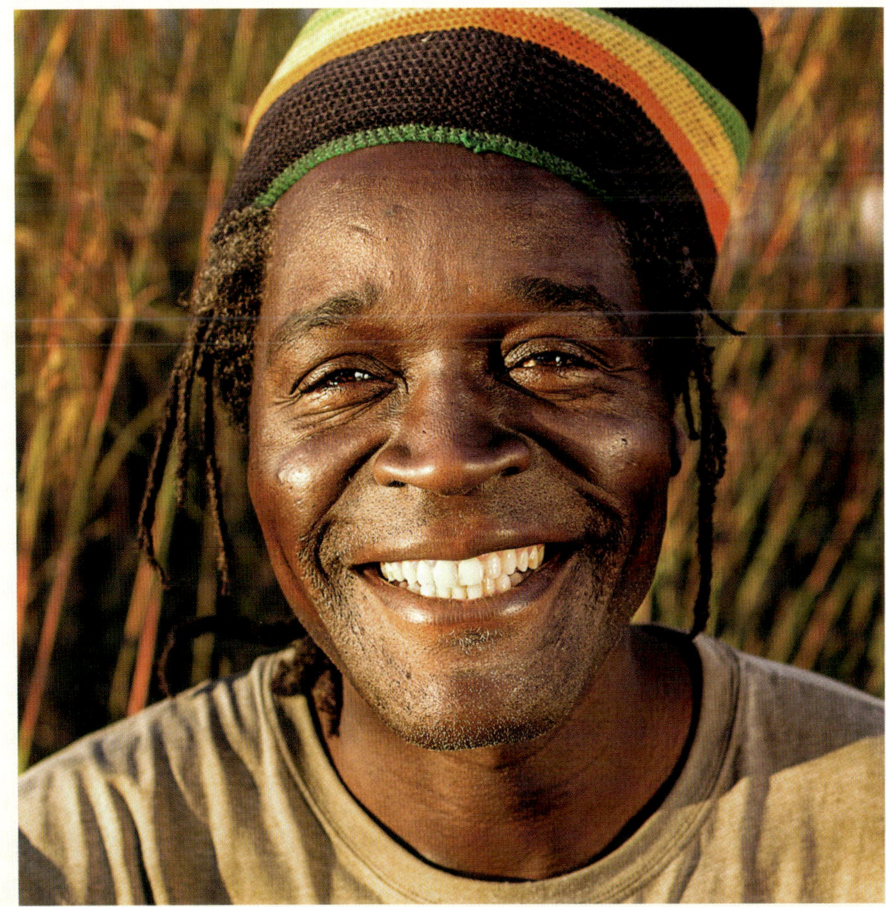

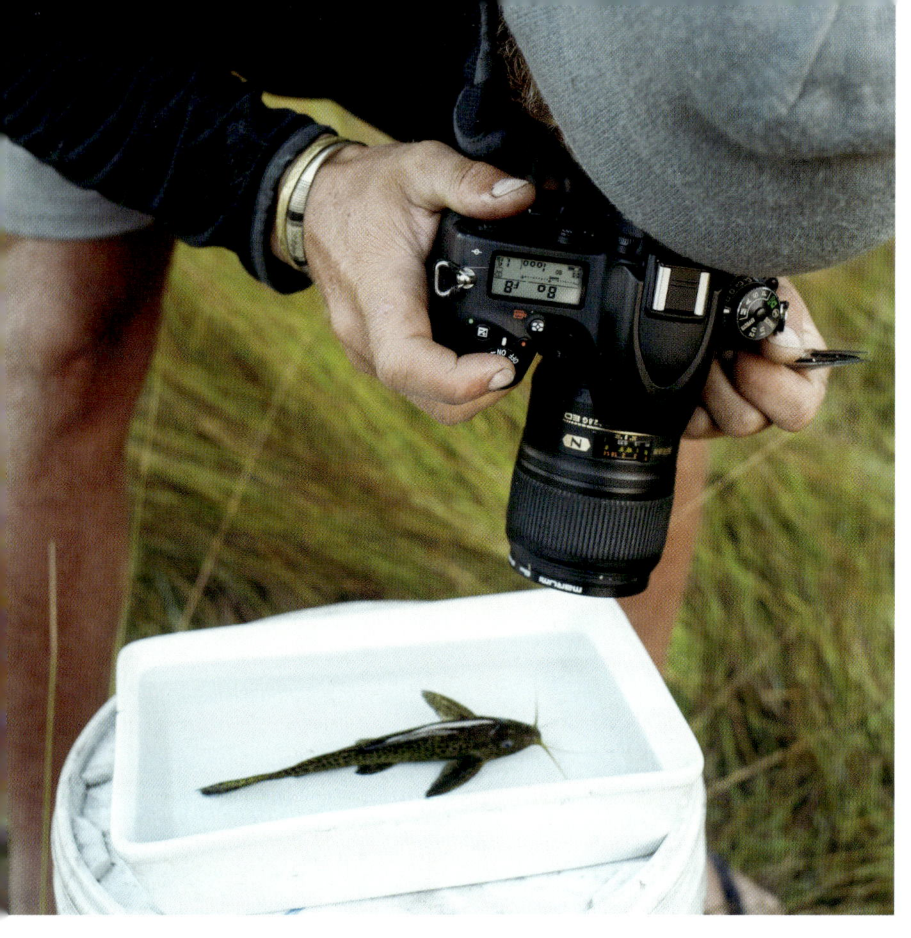

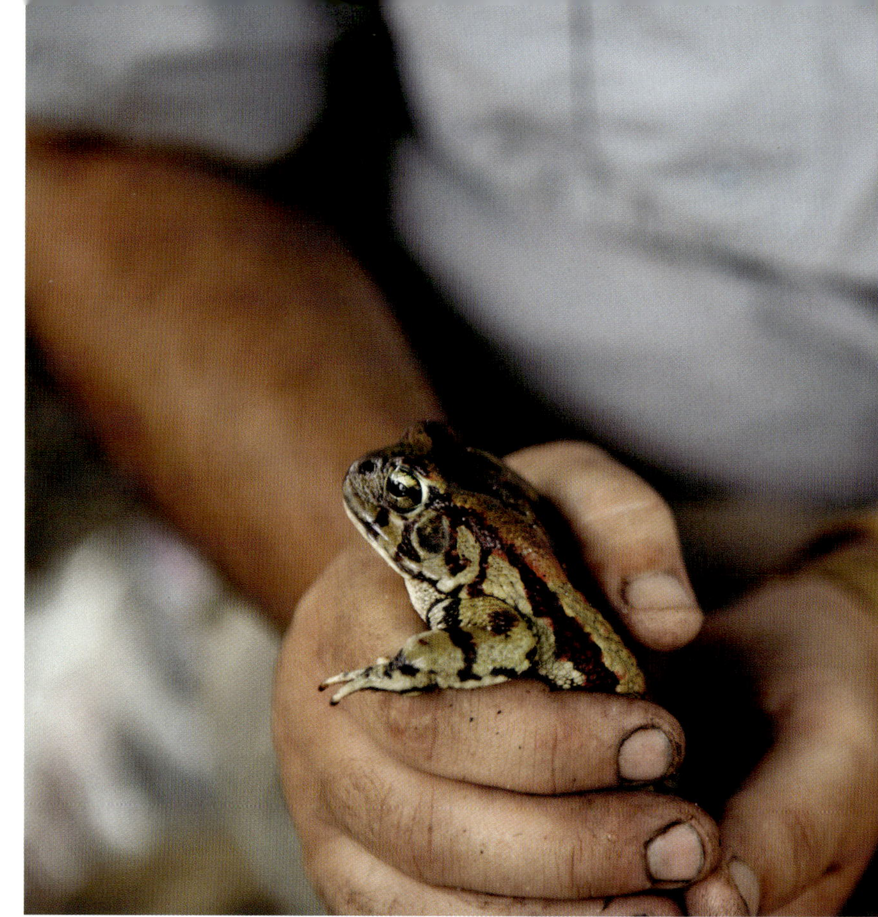

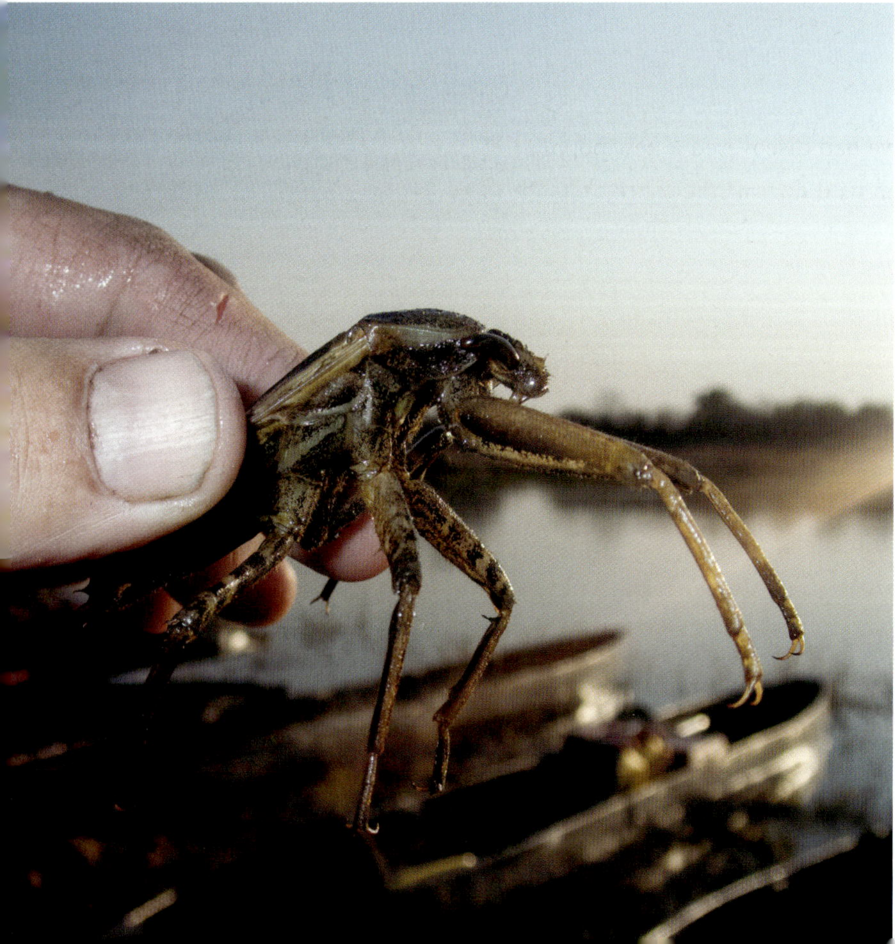

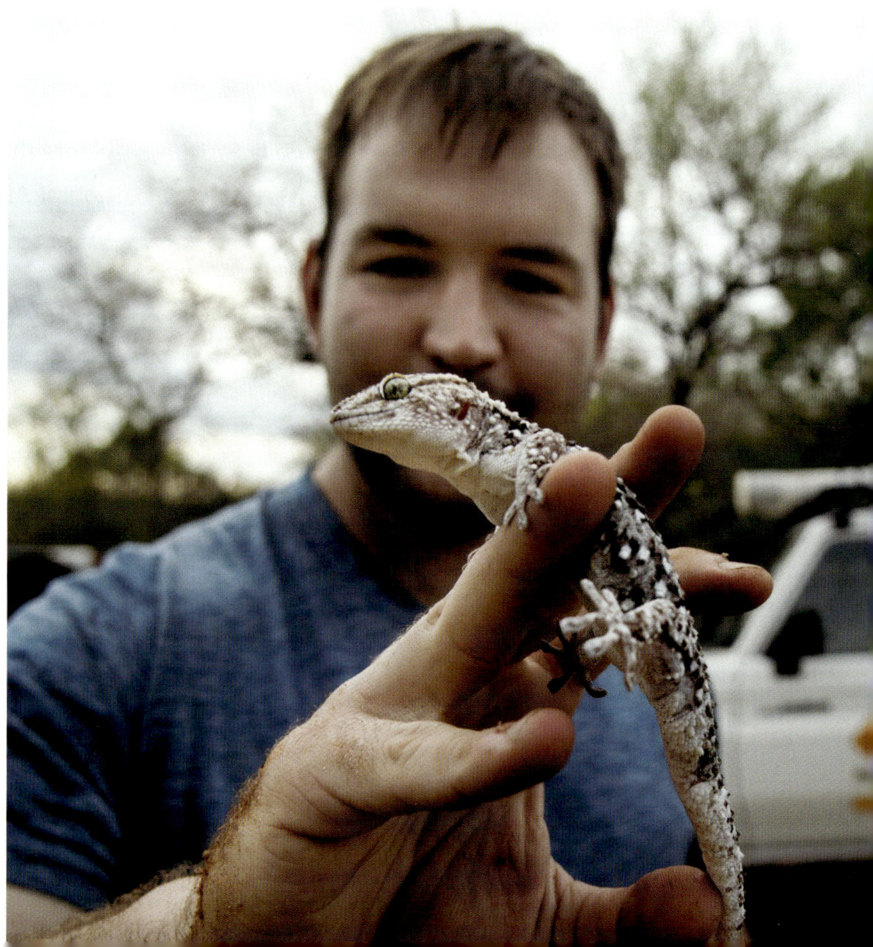

records. Miombo woodlands are notoriously difficult for birders, given their dense vegetation, but you can easily find yourself surrounded by a menagerie of species feeding together. The ornithologists found 10 species outside their previously documented range and two new species for Angola.

The mammal collections now total 223 groups of small mammals, with eight families and 35 species, including three bat species and a striped mouse believed to be new to science. Some of Africa's rarest bats appear to be abundant in the Lisima Landscape. In summer, small rodents and shrews scurry around in the grass. They have chewed into my tent and started cleaning my beard a few times. The most fascinating sighting was a Gambian sun squirrel, a giant creature with a banded tail, whose species tends to live in savanna woodlands across tropical Africa but had rarely been seen in northern Angola.

The camera trap team has completed 17,876 trap days across several dozen cameras, sighting 2,979 large mammals from 38 species, including flagship species such as cheetahs, wild dogs, lions, leopards, spotted hyenas, sable antelope, red lechwe, and yellow-backed duikers—the first recorded outside of the Congo Basin. We had hoped to capture photographs of Angola's giant sable antelope, the national animal, a horse-size subspecies with horns arching over several meters to their rump. We didn't find them, but we did photograph a common sable hundreds of kilometers north of the species' established range. There were also two photos of what appears to be a new subspecies of grysbok, a tiny deer smaller than a pet dog.

In early 2017, I was at the main TED Conference in Vancouver when I got a call at midnight from Kerllen to tell me that he had just downloaded the first photographs of cheetah in the Lisima Landscape. I was astonished that we had found a grassland species, a large predator, at such a high altitude. Spurred by my excitement and, admittedly, a few cocktails at dinner—while trying to be heard over the poor satellite phone connection—I didn't realize how loudly I was speaking until hotel security asked me to go to my room for the conversation. I spent the next 30 minutes standing in the shower stall, where I got the best reception, getting all the details of the photographs.

Surprises kept coming. The week before, Kerllen had picked up a memory card with photographs of the first black-maned male lion seen in the Lisima Landscape. A few weeks before that, we had discovered a new population of African wild dogs at the source lake of the Cuito River. The dogs posed for photographs before devouring a small forest antelope right in front of us. The next day, our camera trap took photos of a white-headed vulture and martial eagle at the kill site, both birds classified as vulnerable—at a high risk of extinction. We also saw hippo, sitatunga, red lechwe, baboon, spotted-necked otter, African clawless otter, and pangolin during land surveys—all species classified as threatened.

The most common mammal species photographed by the camera traps, in order of their relative abundance to each other, included

The biodiversity discoveries made on our expeditions have been extraordinary: more than 150 species new to science and hundreds of species never before found in Angola and the Okavango Basin. Here, biodiversity experts are studying (clockwise from top left) a species of grunter that makes grinding sounds, a Power's toad, a Fischer's thick-toed gecko, and a great water bug.

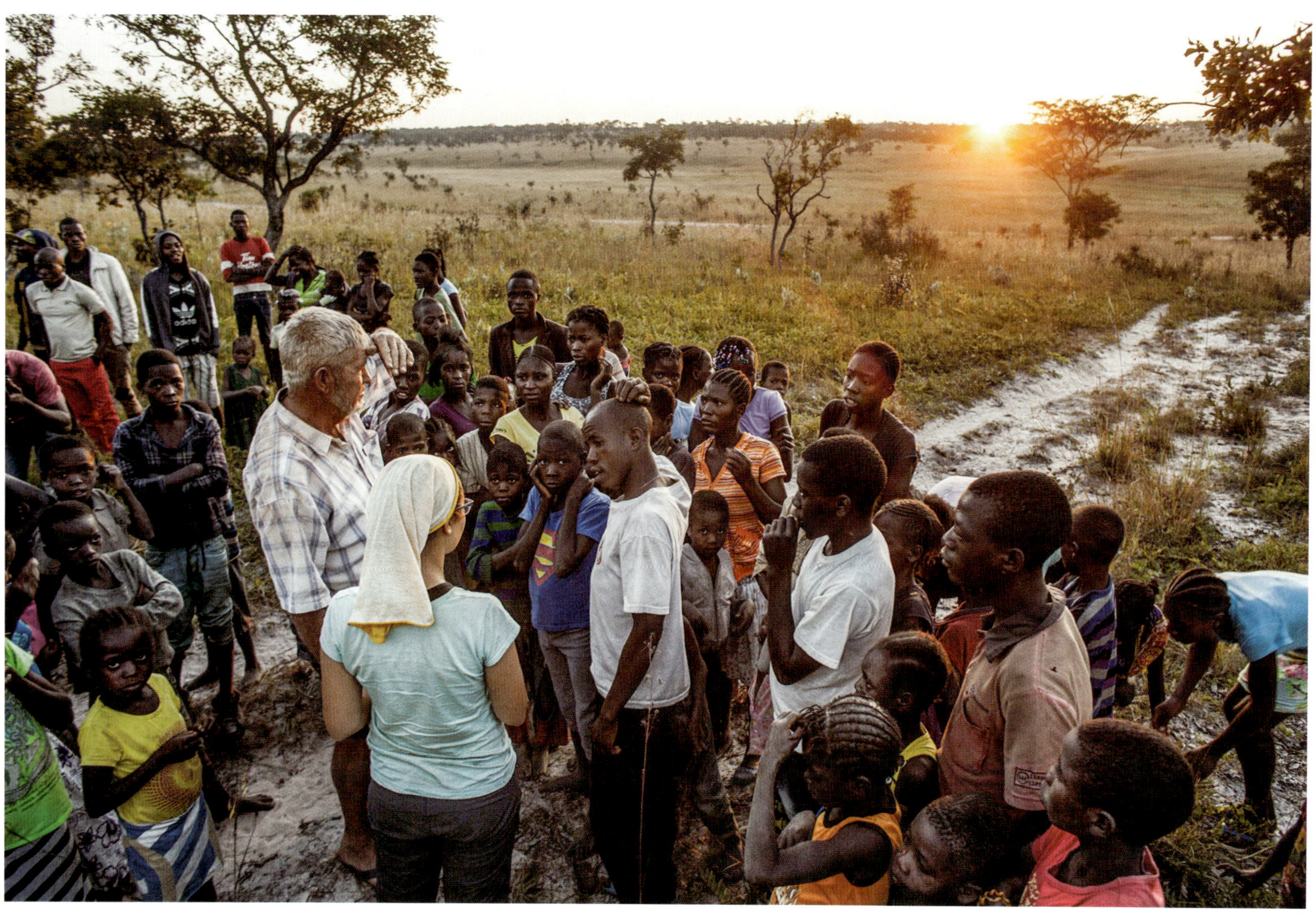

Traditional knowledge shared with us by local communities was essential, enabling our research teams to gain access to source lakes, valleys, confluences, peatlands, and remote forests using cryptic footpaths and motorbike tracks. Local guides led the way for the most important discoveries, documented in their own backyards.

common duikers, blue duikers, side-striped jackals, Cape porcupines, and scrub hares. In the highest-altitude camera trap transects, the duiker species were the most abundant, followed by leopards. Five of the species spotted were of conservation concern: African wild dog, cheetah, lion, leopard, and yellow-backed duiker. Seven species were outside of their known range: sable antelope, Gambian sun squirrel, bat-eared fox, African wild dog, yellow-backed duiker, bushpig, and steenbok. In truth, everything we found was surprising: Nothing was known here, and the habitat was so unique.

Notably, we did not document some species: black rhino, Cape buffalo, giraffe, plains zebra, Lichtenstein's hartebeest, eland, blue wildebeest, sassaby, and bushbuck. All of these species historically lived in this area but appear to be locally extinct due to the Angolan Civil War.

Humans were by far the most frequently photographed species: 44 percent of the photographs showed hunters carrying rifles, bows and arrows, snares, and carcasses, suggesting that hunting levels were high. Miombo woodlands typically have a low carrying capacity (the number of animals its resources can sustain) for medium and large mammals like duikers and roan antelope. This makes them particularly vulnerable to overhunting and the commercial bushmeat trade, which can have a significant impact on potential wildlife population recovery. Fires occurred at 19 percent of the camera trap stations, most likely set by humans to aid hunting activities. Yet local people from nearby villages also helped with our camera trap surveys, and the cameras they deployed had a much higher success rate.

We will spend the next few decades documenting the wildlife of the Lisima lya Mwono Landscape. Every time we look in these vast forests, we find something new and extraordinary. Hunters told us that the Lisima Landscape still hosts black rhinos, a subspecies considered extinct that disappeared in Botswana and Namibia in the 1980s. We never spotted them, but maybe one day we will find them.

THE ROLE of the Angolan Highlands as a source of life goes beyond the wildlife. It supports the landscape, thanks to its high water storage capacity. Though they sit more than 500 kilometers (310 mi) from the major rivers they sustain, these vital watersheds provide a delayed-release lifeline to downstream systems during the dry season. Our research and ground truthing have revealed that these mossy, high-altitude forests have more than 12,000 square kilometers (4,600 sq mi)—perhaps up to 25,000 square kilometers (9,600 sq mi)—of peatlands knitted between them. Deep reserves of water rest belowground, sustained in a layer cake of sediments like a sponge. The features of this landscape represent the largest forested surface expression of a water tower structure on the continent, if not on the planet. As such, the Angolan Highlands water towers are essential to the security of tens of millions of people, regional wildlife and biodiversity, and the ecosystems that sustain them.

In 2015, when we first drove up to 1,400 meters (4,600 ft) to get to the source of the Cuito River, we found ourselves atop an anomalous hump of sand, several hundred kilometers wide, in the eastern part of the Angolan Highlands water tower. The hill straddles the Kalahari sand basin, the longest continuous piece of sand on Earth, which extends from South Africa to the coastline of Gabon. We were struggling to drive our off-road vehicles and trucks through the endless ridges of deep sand and valleys of peat—relict sand dunes, like stranded sections of the Sahara. Hardpans must have formed under the source lakes over thousands of years, keeping water from disappearing into the sand as organic matter accumulated and compressed.

Experts suggest that when this region formed, geological uplift triggered swarms of small but powerful volcanic episodes called kimberlite eruptions, when tube-shaped rocks called kimberlites punch through the earth and allow magma to erupt. This phenomenon

The Luchazi have an oral history going back more than 600 years, but their ancestors have been living in these remote watersheds for some 2,500 years.

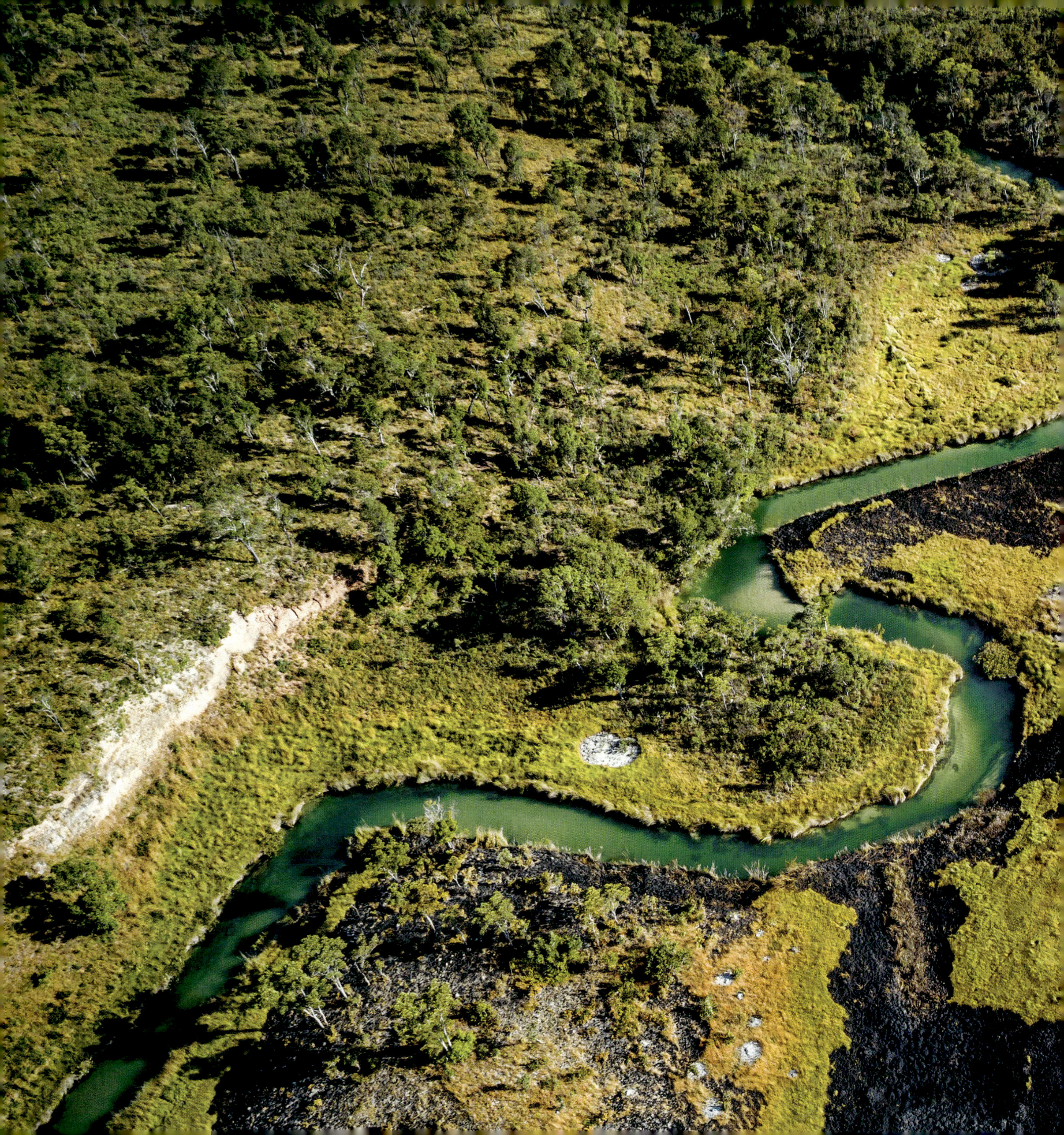

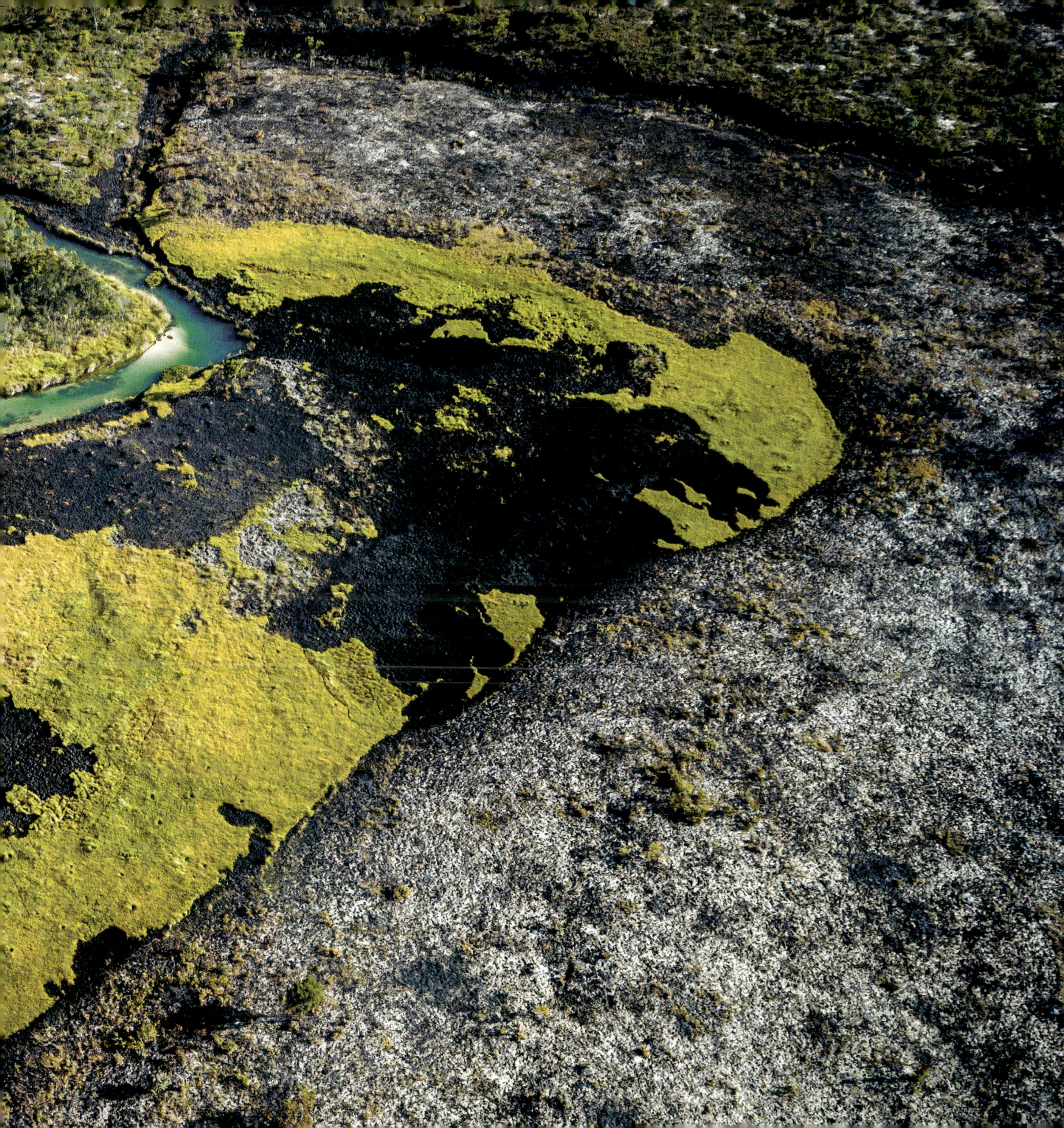

caused the ancient sand dunes to slip and slump, diverting rivers, making them change direction, and enabling the Congo Basin to capture the Zambezi's Chavuma Falls, along the northern divide with the Okavango Basin, to form the Cassai River. When we explored the Cassai in 2023, our surveys found Zambezi fish species, evidence of this ancient river piracy. We found strange yellow crocodiles and oddly small hippos living along the river. The kimberlite eruptions brought diamonds to the surface, too, and they are now mined intensively in the upper Cassai and upper Cuanza and along the Kwando headwaters. Likely, more strange discoveries will be gleaned from the Cassai, the remotest river I have ever explored.

REMOTE SENSING DATA from satellites over the past 25 years demonstrate no natural fire-related forest loss in the Lisima Landscape, but logging and charcoal production elsewhere have devastated forest cover. We can attribute this difference to the Luchazi people who call these forests home, whose every activity is shaped by a need to protect them. When asked what is the most important aspect of the landscape, they say, "*Muito floresta!*—A lot of forest!"

The Luchazi have an oral history going back more than 600 years, but their ancestors have been living in these remote watersheds for some 2,500 years. To the Luchazi, these unending miombo woodlands, source lakes, and peatlands aren't just wilderness; they are dynamic characters in their lives, all-powerful over humans yet innately part of us. These people

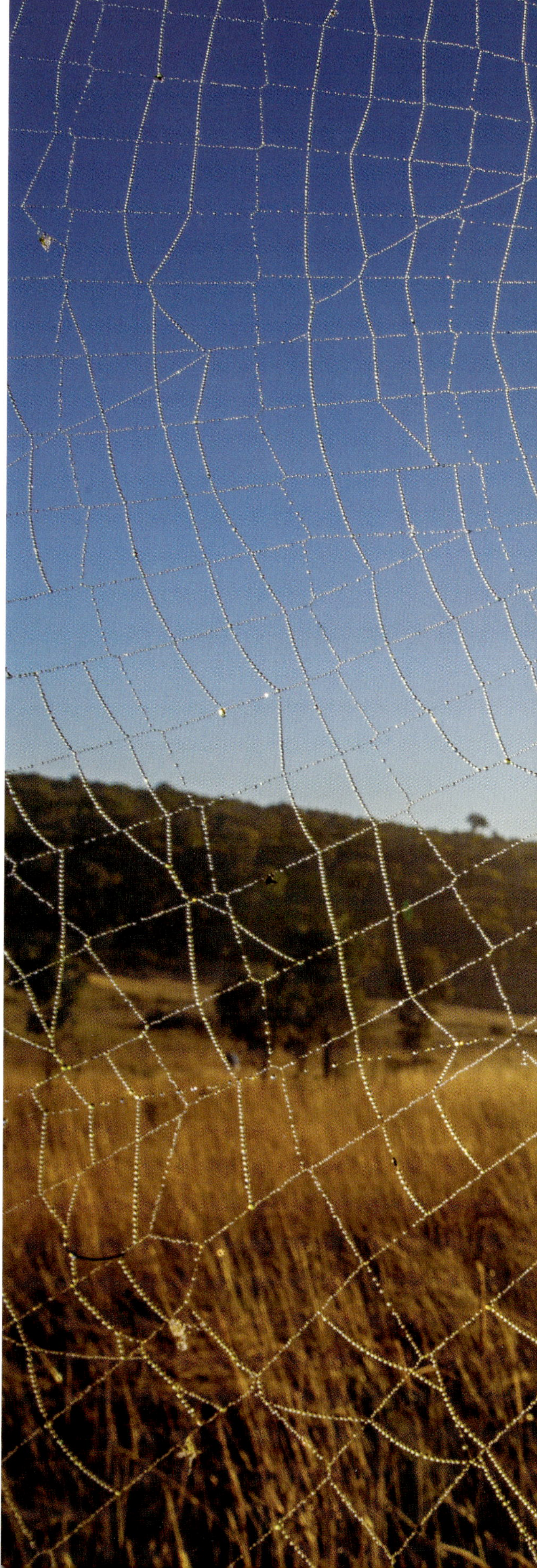

Intricate webs built by golden orb weaver spiders are common on the edge of the forests along the hundreds of rivers and streams of the Okavango Basin. **PAGES 182–183:** *Sharp lines of demarcation separate the forest growth from burned land. The Luchazi and Chokwe peoples have been using fire to manage this land for more than 2,000 years. Now an integral part of the ecosystem, these seasonal fires are anticipated by plants and animals alike.*

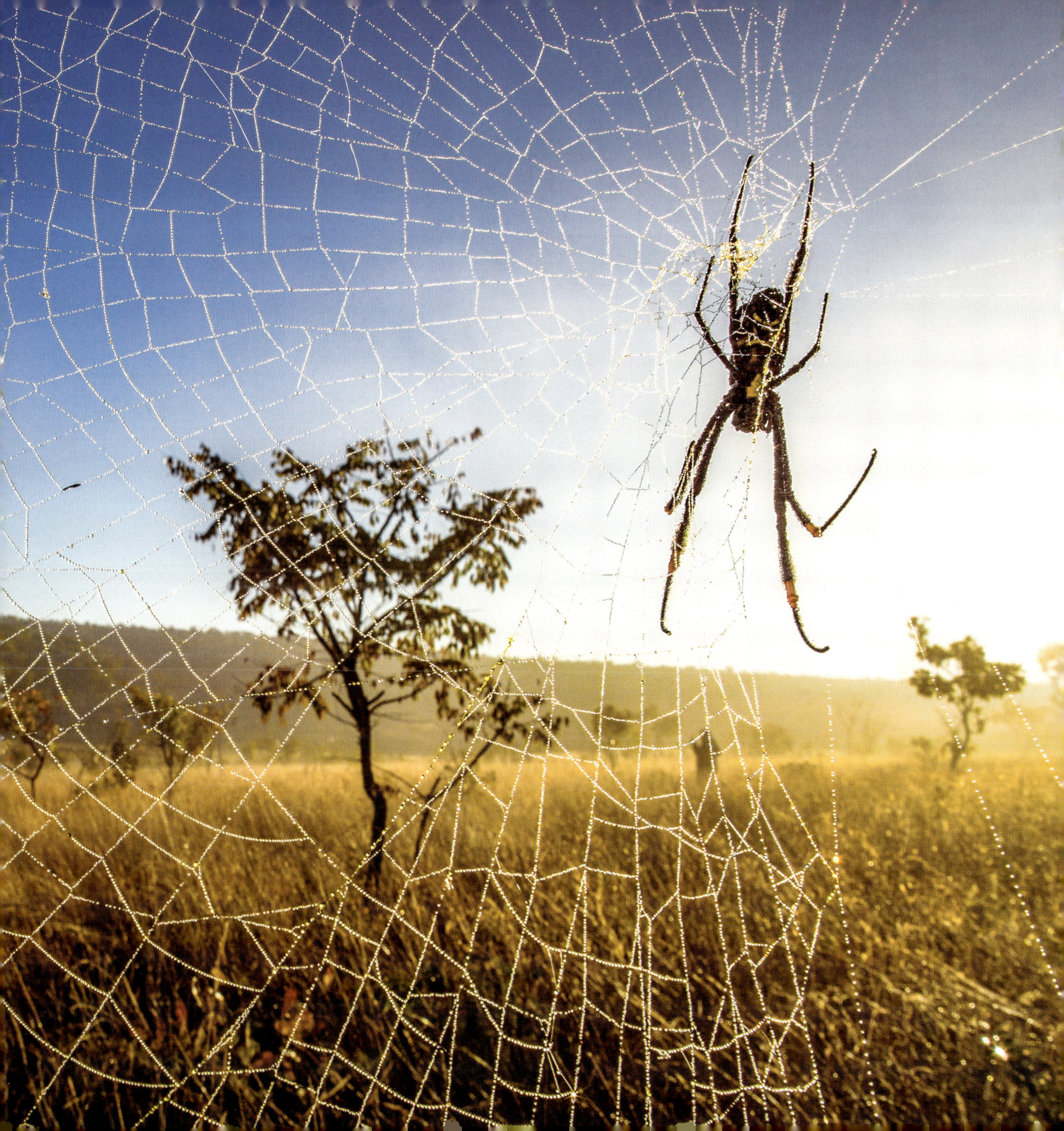

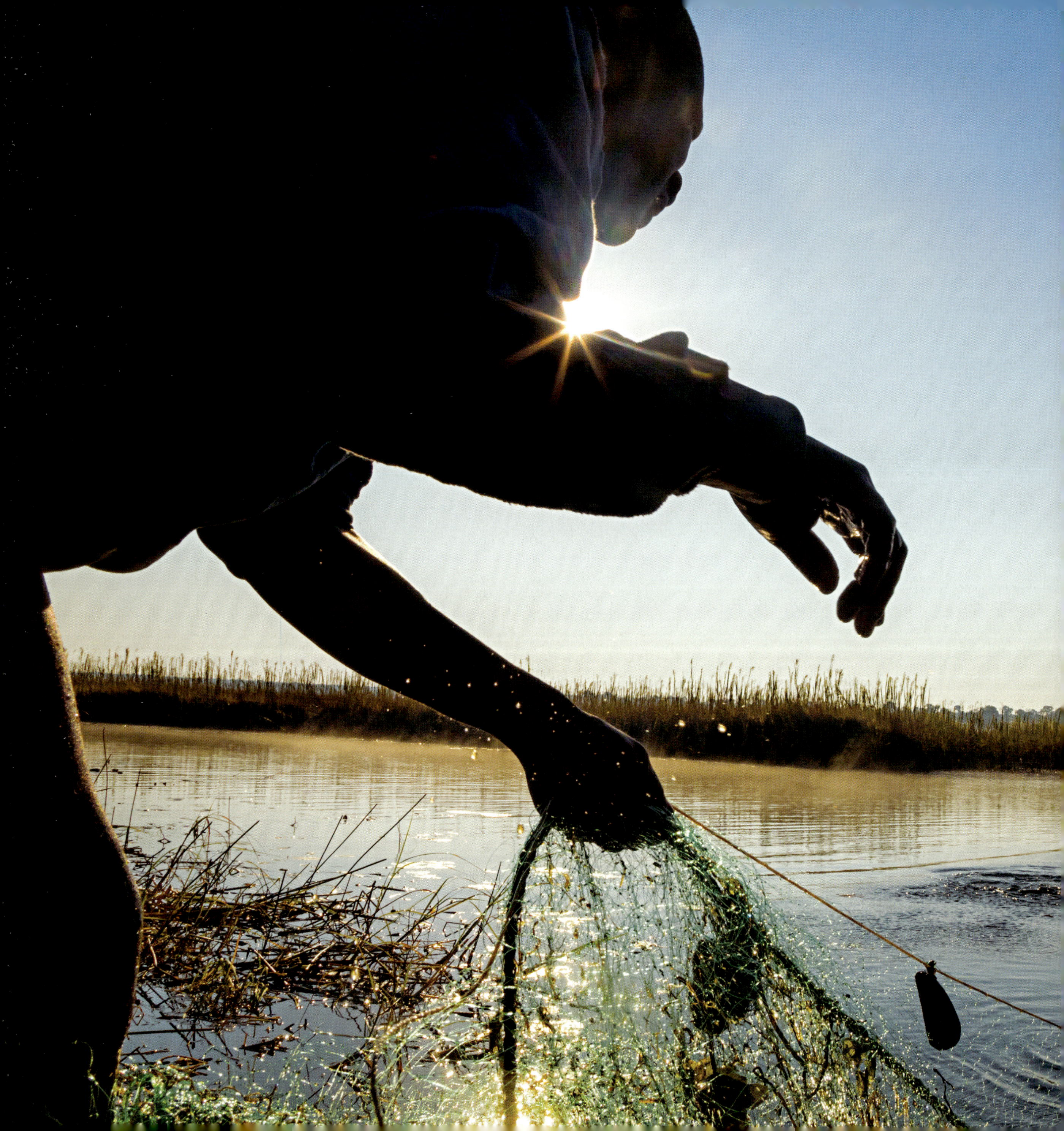

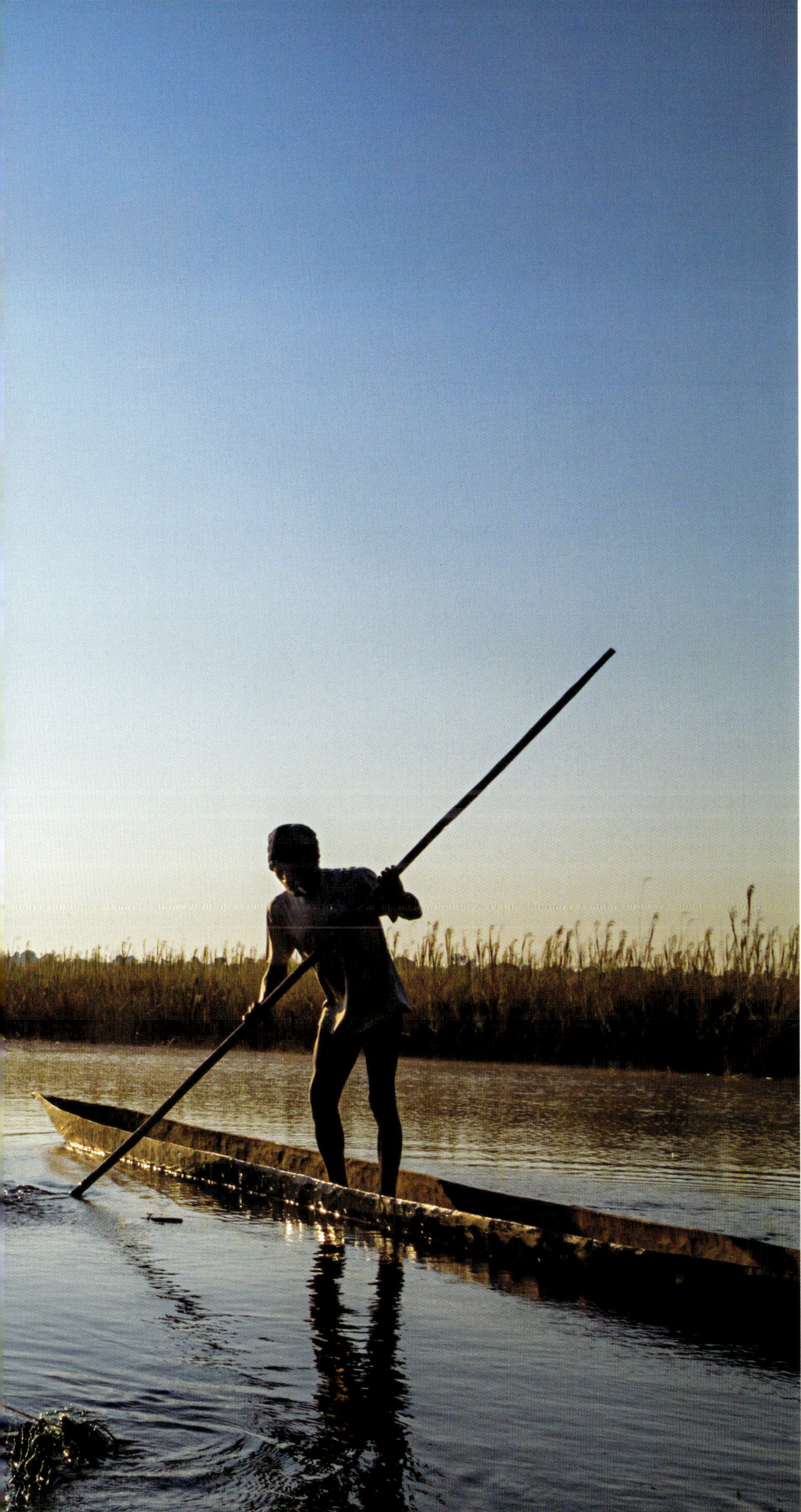

have muito floresta because they protect the forests. They burn forests carefully to optimize them for honey production and wildlife, to create grassland corridors, and even to purify rivers and streams. The Luchazi have been in the business of clean air and water for millennia.

Over the past decade, I have had many all-day meetings with the religious and traditional leaders of the Lisima Landscape. We talk about conservation, sustainable livelihoods, the rivers and forest, the wildlife and biodiversity, and plans for the future. In one of the first meetings, with about 100 sobas (chiefs) and regidora (chiefs of chiefs), our agenda was to teach them, but we got schooled on conservation instead. We tried to explain the concept of a "conservation area," a place set aside for wildlife with no human settlement or activity. The most powerful Luchazi chief had one simple question: "How can you possibly protect a place where you can't be?" They were natural experts in conservation and had been so through millennia by being part of their landscape. Protected areas didn't make sense to them.

Our botanists and GIS technicians could identify only three dominant forest types based on satellite imagery, remote sensing data, and ground surveys, an assessment that cost us more than $50,000. The Luchazi farmers, beekeepers, and hunters could identify seven different kinds of forest, detailing the specific characteristics, uses, values, strengths, and vulnerabilities that differentiated them.

We explained the importance of the source lakes and rivers to millions of people living downstream. They told us they were upset that we'd used soap when washing ourselves in the river, a contamination that was forbidden in their culture. When some Luchazi elders visited a biodiversity survey, they were concerned by all the small mammals, reptiles, fish, insects, and bats that were being killed as specimens for species description. We showed them how the specimens were humanely euthanized and explained where they would be stored. They agreed this was important because the goal was to demonstrate the importance of the Source of Life. With their help, we became far more successful at finding species. The lakes, rivers, and forests were not only familiar; they were sacred.

In Samanunga, Marisa Rodrigues, a honeybee researcher, wanted to sample bees using tweezers and a sampling jar. She reached into a hive and grabbed a worker bee. The Luchazi beekeeper, Mario, stopped her and told her to put the bee back. "That's my bee," he said. "She works for me." He had more than 400 beehives spread across the landscape, and this is how he felt about all his bees. They fed his children, provided money for medicine and schoolbooks, and brought their own small pleasures—such as his mead.

From the smallest bee to the largest tree, the Luchazi feel pride and ownership in this remote wilderness. They are guardians of these watersheds, and their traditional knowledge systems are the key to long-term sustainability and resilience in the face of climate change. They see their ancestors incarnate in the wild.

This fluid boundary with the spiritual worlds has been one of their primary means of protection for thousands of years. Source lakes, rivers, and forests are protected by mythical spirits communicating with their ancestors. The foundation of their lifestyle is freedom—to make mistakes and change according to nature's rhythm. Communities like theirs constitute some of our planet's oldest surviving cultures, and modern society, in the form of globalization and unregulated resource extraction, represents an existential threat.

We must protect the freedoms of Indigenous communities like the Luchazi around the world. The human experience in the wild is a powerful force connected to all time. These are the people who stayed when everyone else left to colonize the planet. They are the custodians of humanity's true, untapped superpower: the ability to be in nature and truly understand our role within it. I believe that their traditional knowledge systems are the key to our long-term survival as a species.

Indigenous culture and beliefs are rooted in generations of trial and error focused on long-term survival. The concept of sustainability, generation to generation, is embedded in all traditional knowledge systems. Yet oral histories and traditional knowledge systems are threatened by cultural homogenization and the digital age, both of which have historically devalued Indigenous languages and dialects. There are now 573 known extinct languages, half of which have disappeared since 1950. A third of all languages have fewer than 1,000 speakers. A lot, as they say, is lost in translation.

An African wild dog stares straight into one of our camera traps, which have revealed previously unknown populations of these canines, as well as cheetahs. Both are endangered species facing extinction. **PAGES 186–187:** *As much as we could, we removed abandoned monofilament gill nets from the rivers. These nets kill any creature that gets tangled in them, exterminating local fish populations.*

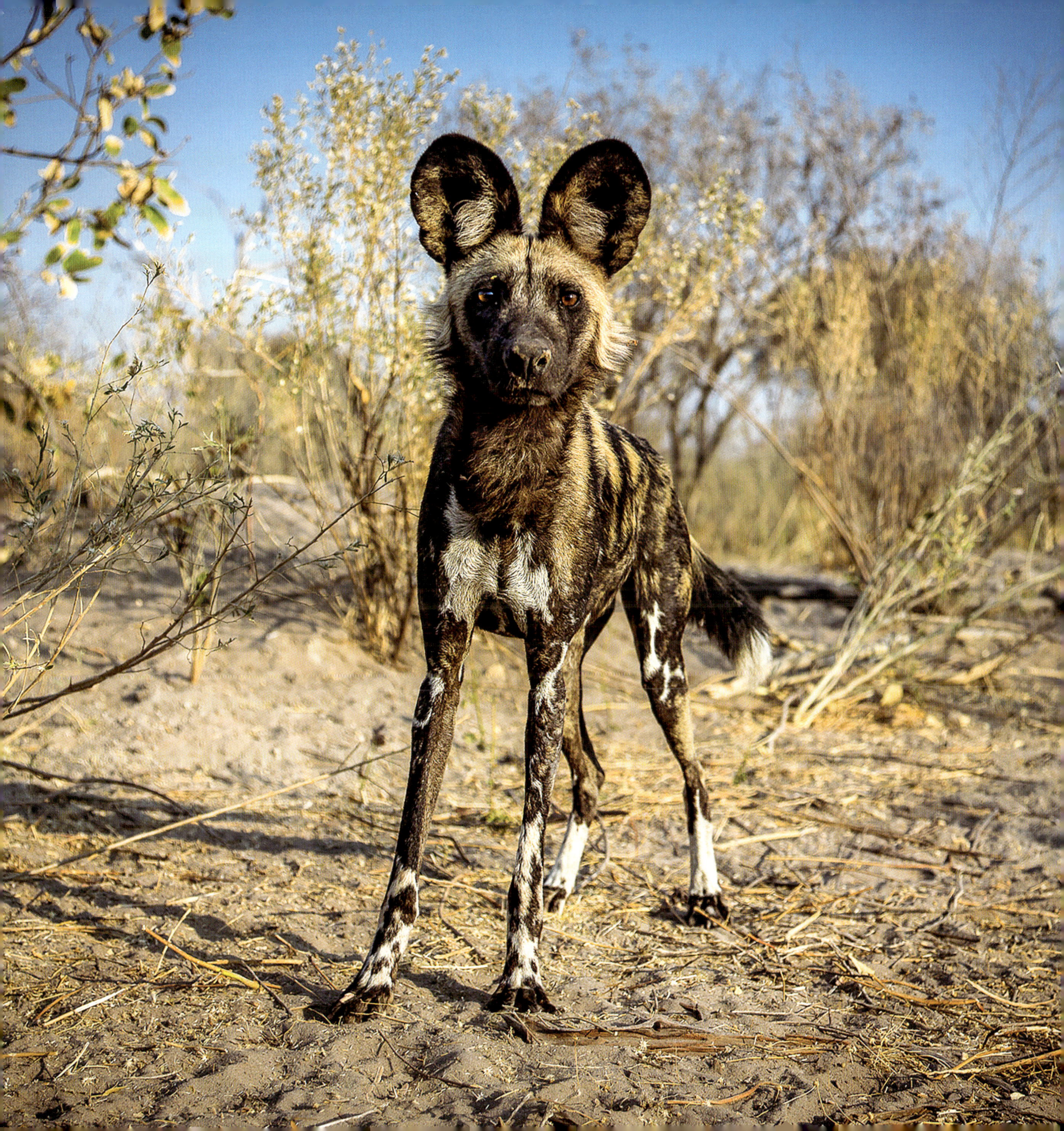

The Okavango Basin has almost no light pollution, so wherever you are, if the moon has not yet risen, you can see the Milky Way rotating above you. It is so bright that local tribes use it to navigate after dark and tell many origin stories linking the Milky Way to all creation.

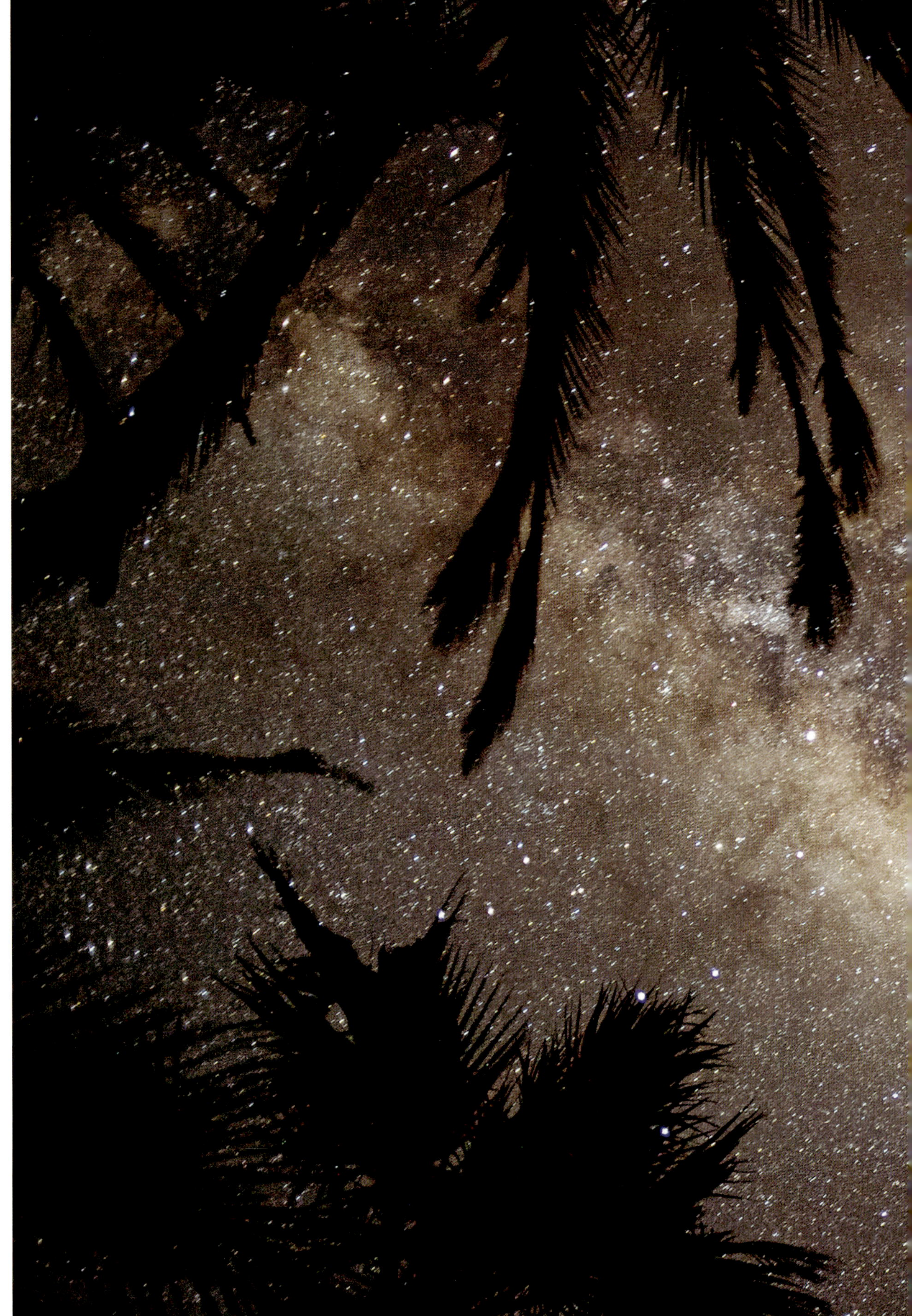

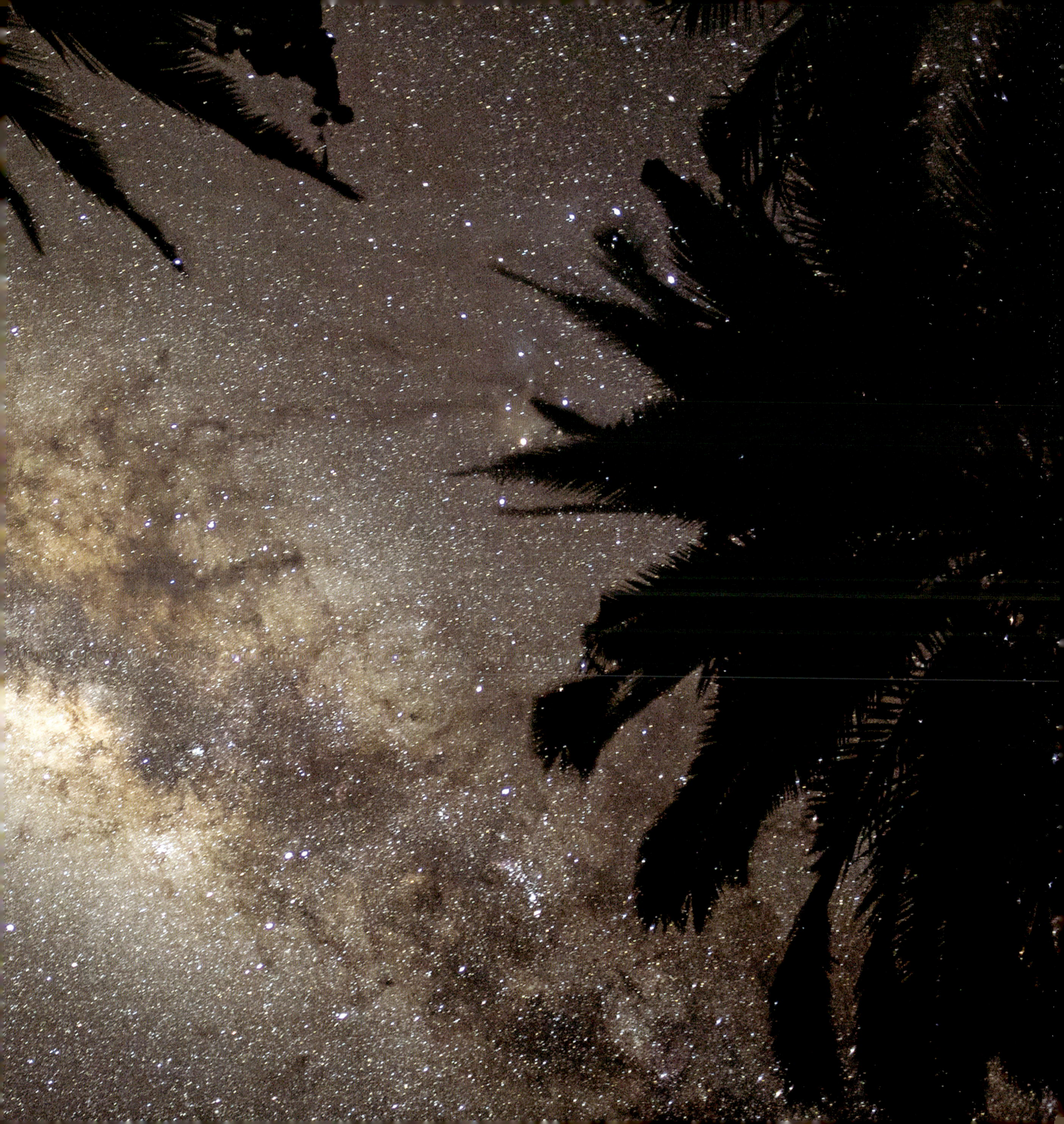

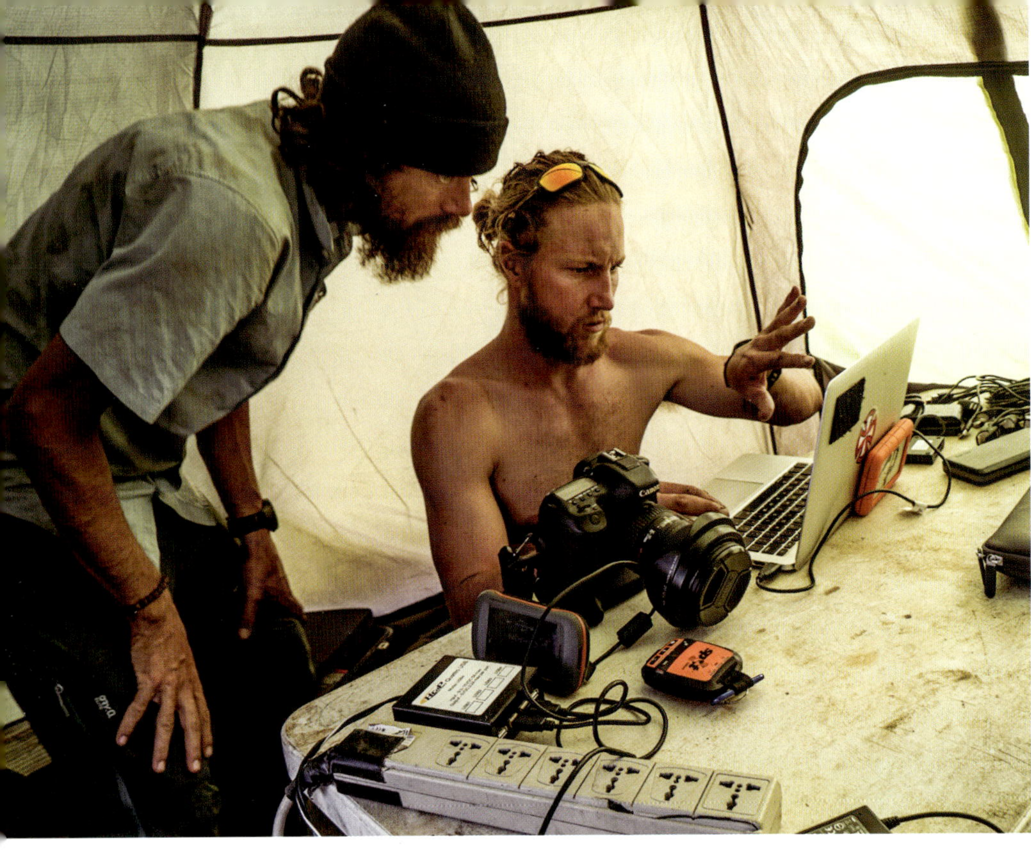

In science, we seek truth and understanding. In storytelling, we seek impact and meaning. Scientific data without explanatory narrative holds little meaning for most people. Here, Kyle Gordon (left) and Chris Boyes talk about the photographs to be shared on social media via satellite that night.

TODAY WE HAVE an unprecedented opportunity to protect this place where we all come from and the Source of Life that feeds it. Our Okavango Megatransect and more than 30 subsequent research expeditions into the eastern Angolan Highlands represent true 21st-century exploration: enabled by the latest technologies and undertaken with the intention of securing a more sustainable, prosperous, and abundant future.

Yet these discoveries are far from complete. We have set a conservative estimate that scientists will identify more than 250 new species in the highlands by 2030. There are portions of this watershed that we have visited only once or not even seen. We will continue to work with the Luchazi to better understand and respect their ancestral knowledge, ensuring that it forms the center of any conservation policies affecting the land and wildlife they have protected for so long.

After four decades of war and instability in Angola, these highlands are still not protected. By 2027, the National Geographic Okavango Wilderness Project aims to establish community-owned, traditionally protected reserves spanning the Lisima Landscape, an area larger than the whole of England. This dynamic, ever changing watershed has avoided exploitation, prewar and postwar, due to its inaccessibility and harsh climatic oscillations. This ancient water tower is a 21st-century sanctuary for biodiversity, wildlife, and ancient ways. We need places on Earth where people can choose to live traditionally while benefiting from modern technologies: satellite internet, solar power, vaccines, and artificial intelligence, all of which have the power to bolster traditional values and intact ecosystems.

Far away from modern society, at the ends of the earth, is the Source of Life, a place and a way of life that overcomes adversity and conquers cataclysm because it fits in with the natural rhythms of nature. We didn't discover anything new that year; we found a place we all used to inhabit, a people like we all used to be. Ten thousand years ago, our entire world was wilderness. Today, wilderness is all that remains of that world now long gone. Ten thousand years ago, we were, as we are today, a modern dreaming intelligence, living amid nature that has taught us to speak, to craft objects—fire and stone, bow and arrow, medicine and poison—and to depend on one another and all life around us. We are these last wildernesses, every one of us.

RIVER GUARDIANS

Bogolo Joy Kenewondo

A Woman Rising

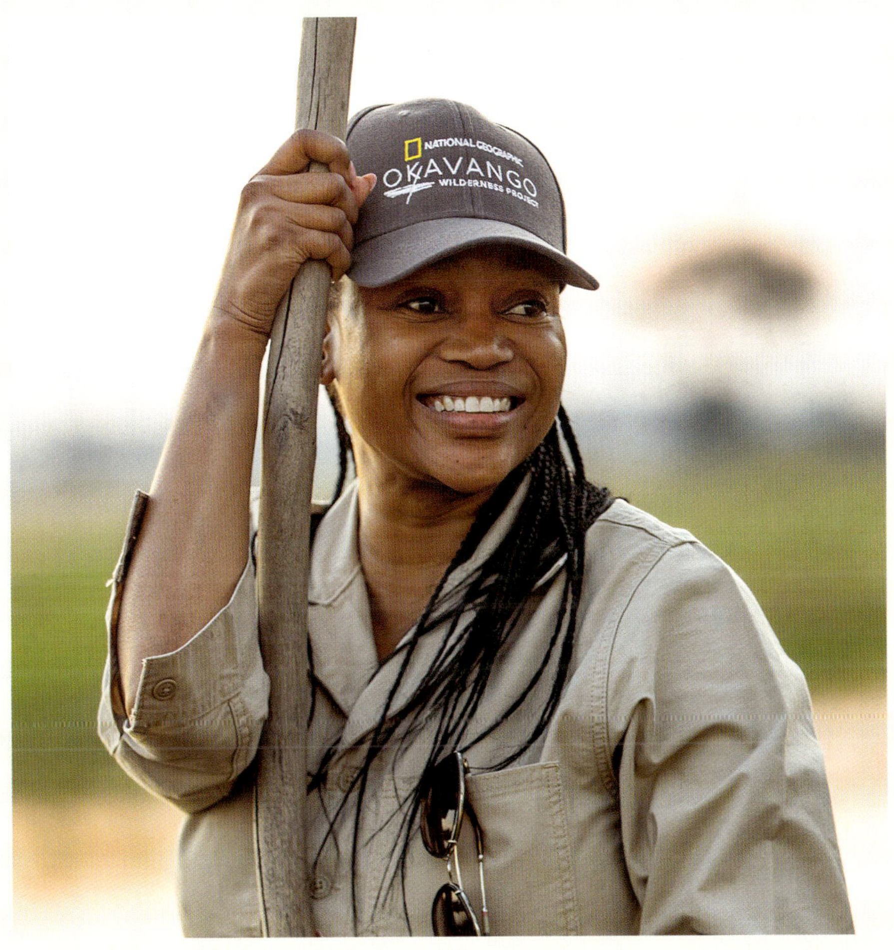

Bogolo Joy Kenewondo has been a guiding light in the Nkashi Trust and the National Geographic Okavango Wilderness Project, serving as a trustee on the board.

BORN A CHILD of the Okavango Delta in Motopi village along the Boteti River, Bogolo Joy Kenewondo is most at peace when she glides in a mokoro across floodplains covered with water lily flowers. In April 2018, Bogolo was sworn in as Botswana's minister of trade and industry, becoming at 30 years old the youngest minister in African history. In 2022, she was honored in *Time* magazine's list of the world's top 100 emerging leaders.

From an unassuming village in the Kalahari Desert to the United Nations, Bogolo is a guardian of the Okavango River. She is a tireless champion of gender equality, climate action, and sustainable development across Africa, and as a trustee of the Nkashi Trust, she helps us focus everything we do on direct benefits to local communities, understanding them to be the primary mechanisms of protection in the Okavango Delta. She knows that the future of conservation is local.

African women, such as Queen Nzinga of Angola and Yaa Asantewaa of the Asante Empire, have long played powerful roles in African diplomacy and resistance. Ellen Johnson Sirleaf, past president of Liberia, and Ngozi Okonjo-Iweala, director general of the World Trade Organization, herald the ascendance of African women into high-level leadership positions in the modern age.

Bogolo represents an entire generation of young African women emerging to take control of Africa's development trajectory. In 2019, Adjany Costa, our country director in Angola, became one of the UN Environment Programme's Young Champions of the Earth and was awarded the Angolan Order of Civil Merit. In April 2020, she was appointed as the minister of culture, tourism, and environment, making her the youngest minister in Angolan history.

African women are rising, and the world is all the better for them.

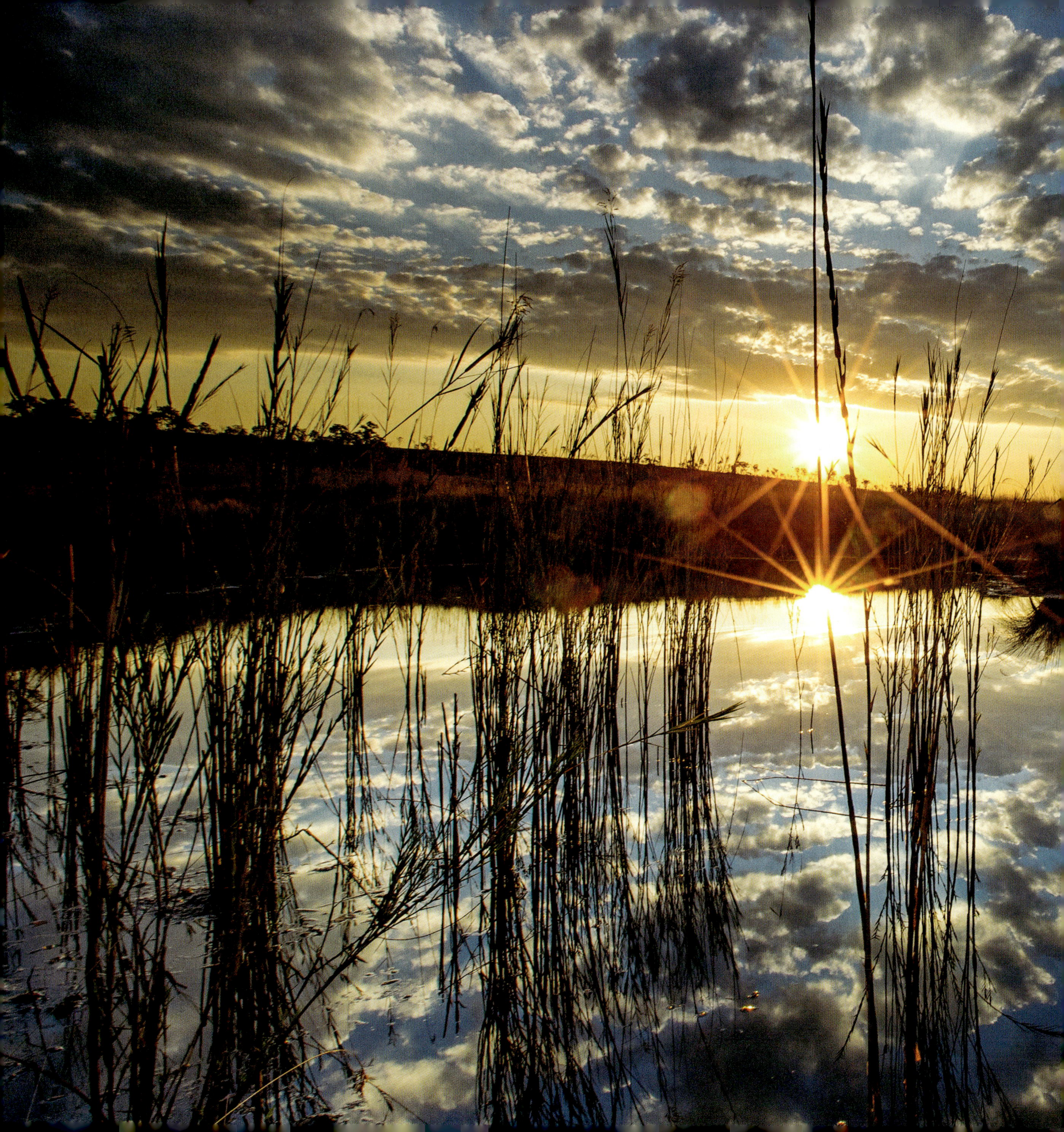

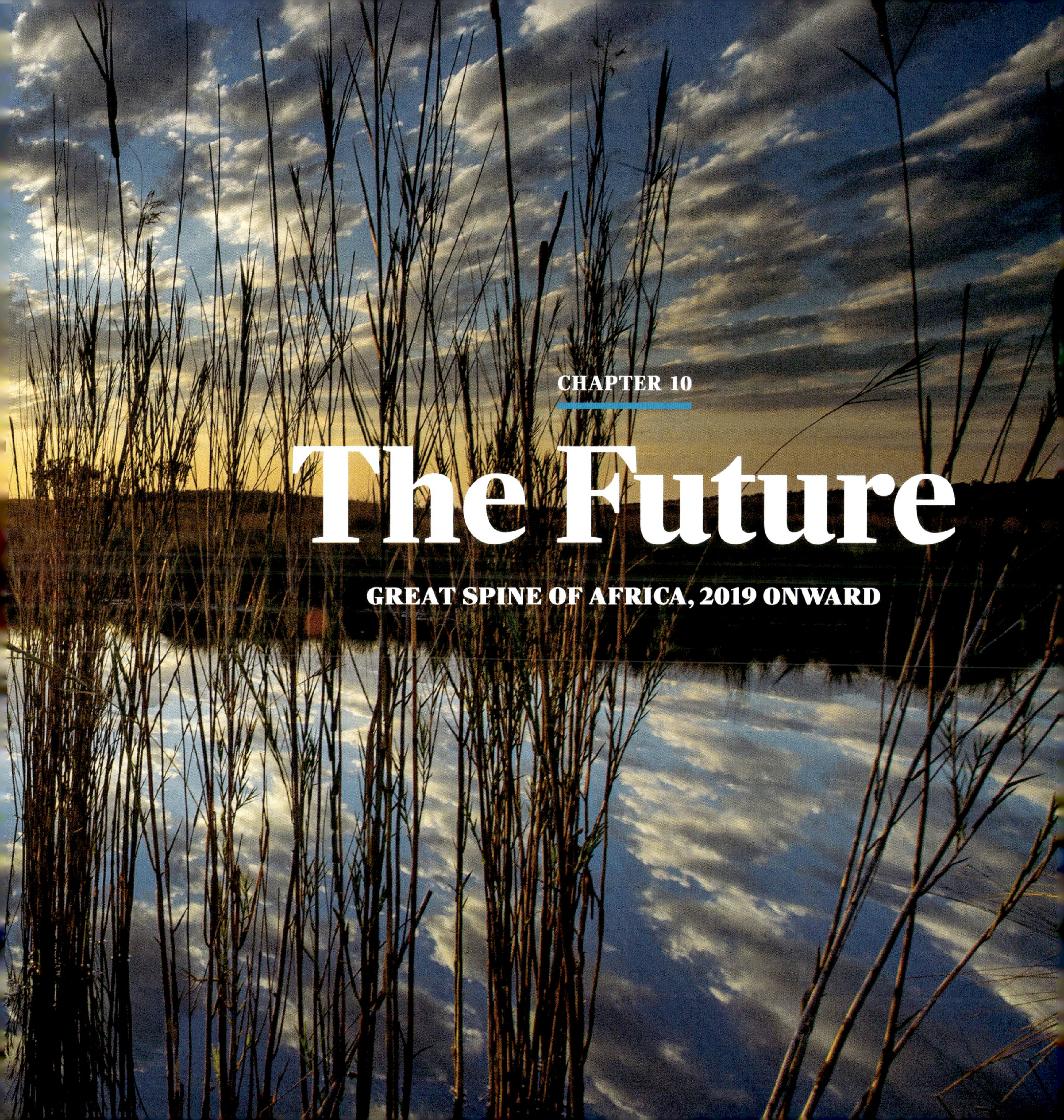

CHAPTER 10

The Future

GREAT SPINE OF AFRICA, 2019 ONWARD

EVERY QUEST into the wild is a warning, whether from the past or about the future. When most of us enter the wild, we enter a place that feels unknown to us—a place where magic and mysticism still exist, a land where daily life centers on natural cycles. While everyone else left, Indigenous people decided to stay because they chose to remain connected, not only to nature but to their ancient cultures and traditions. They are uniquely connected to the past; they remember. We are the explorers that kept on leaving until, after tens of thousands of years, we were almost everywhere.

But a few thousand years ago, the curiosity that drove us to explore turned into conquest, and conquest led to trauma, suffering, and exploitation. Now, in the 21st century, exploration means something different. We all carry aerial views of Earth in our pockets, can fly anywhere on Earth in a day, and have all human knowledge at our fingertips. We now recognize that global resources are finite and languages are going extinct at a faster rate than species. Today, modern explorers are again driven by curiosity, not conquest.

The 21st-century explorers are returning to the Indigenous communities who knew enough to stay, people who are the custodians of traditional knowledge systems that protected us against extinction for millennia. These explorers understand the necessity of humility, respect, and reciprocity when interacting with Indigenous ways of life.

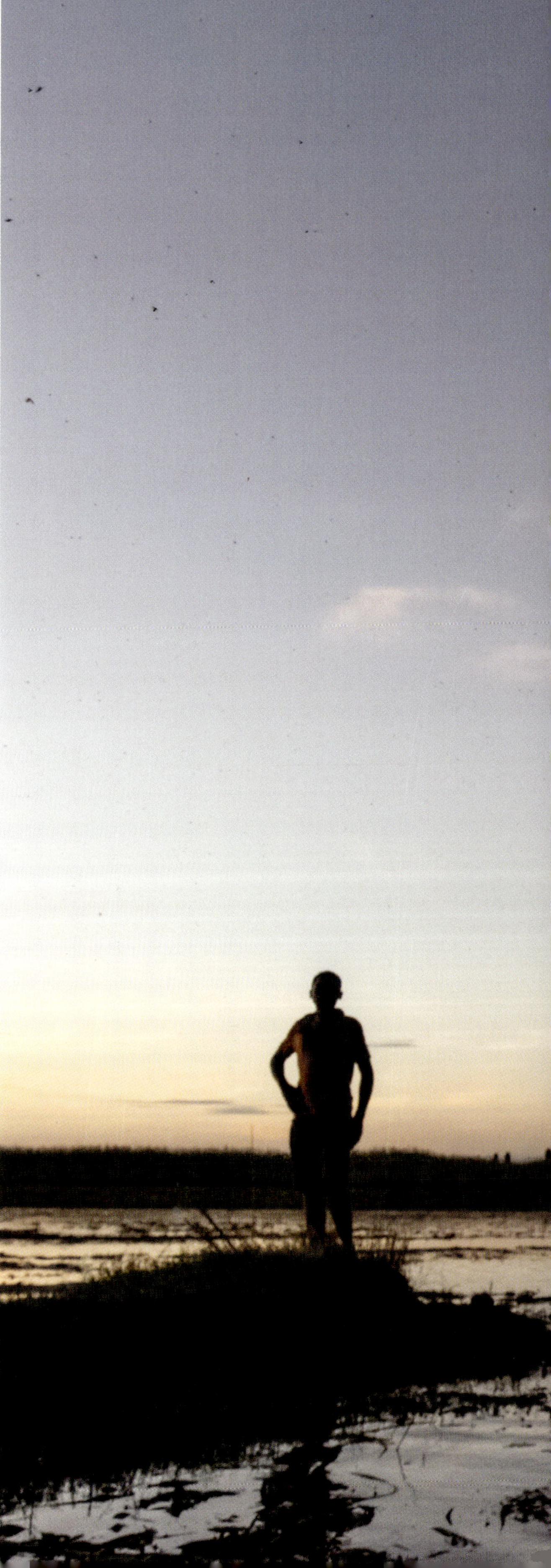

Our work continues on an oxbow lake along the Cubango River. By 2030, we aim to complete 200 river expeditions, establishing early 21st-century scientific baselines for all major rivers in the Okavango, Zambezi, Congo, Nile, Chad, and Niger Basins, spanning the Great Spine of Africa. **PAGES 194–195:** *Life on expedition provides opportunities for reflection on our human experience in the wilderness. We coexist with nature like waves connected to the ocean. We can't exist without it.*

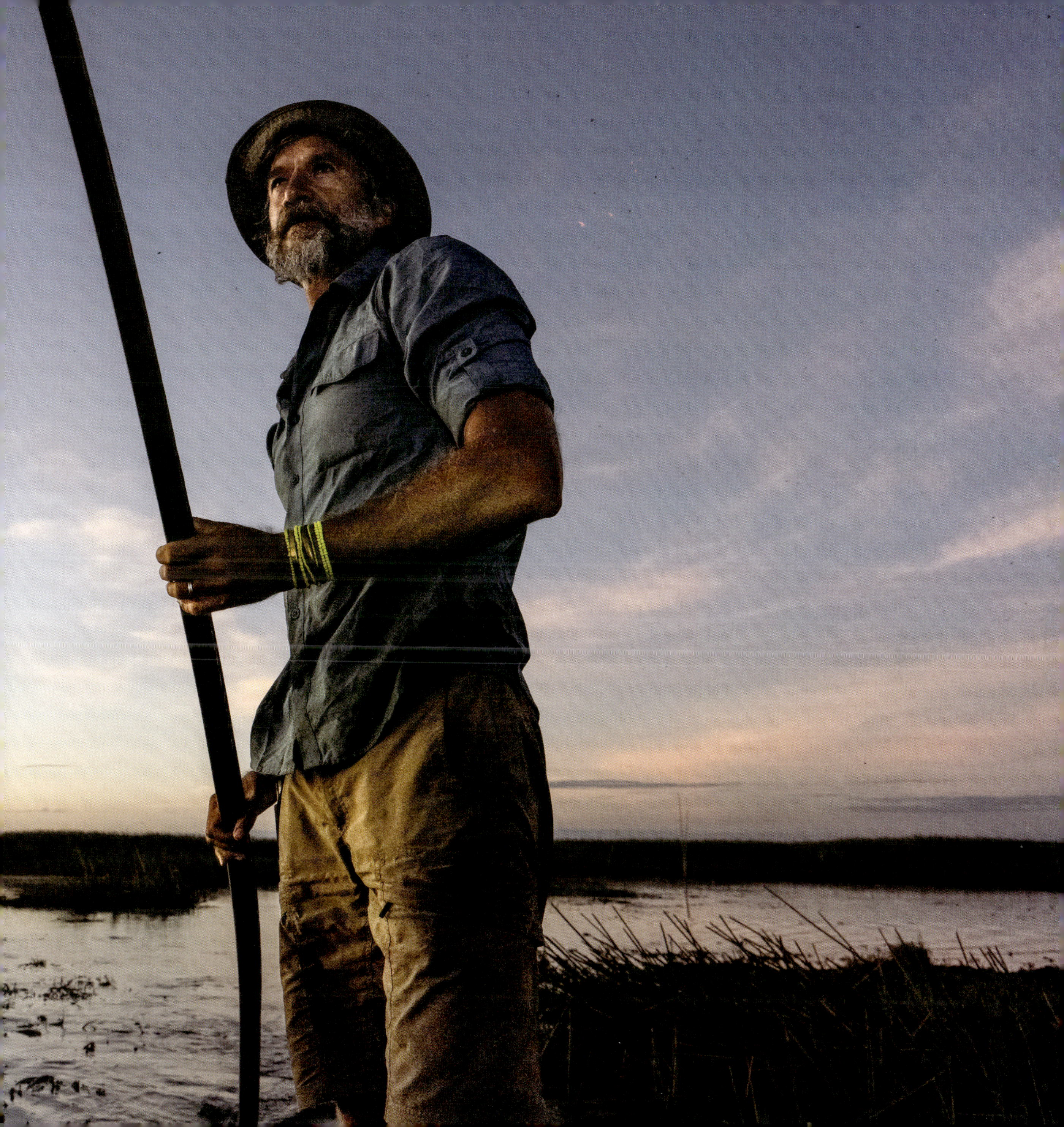

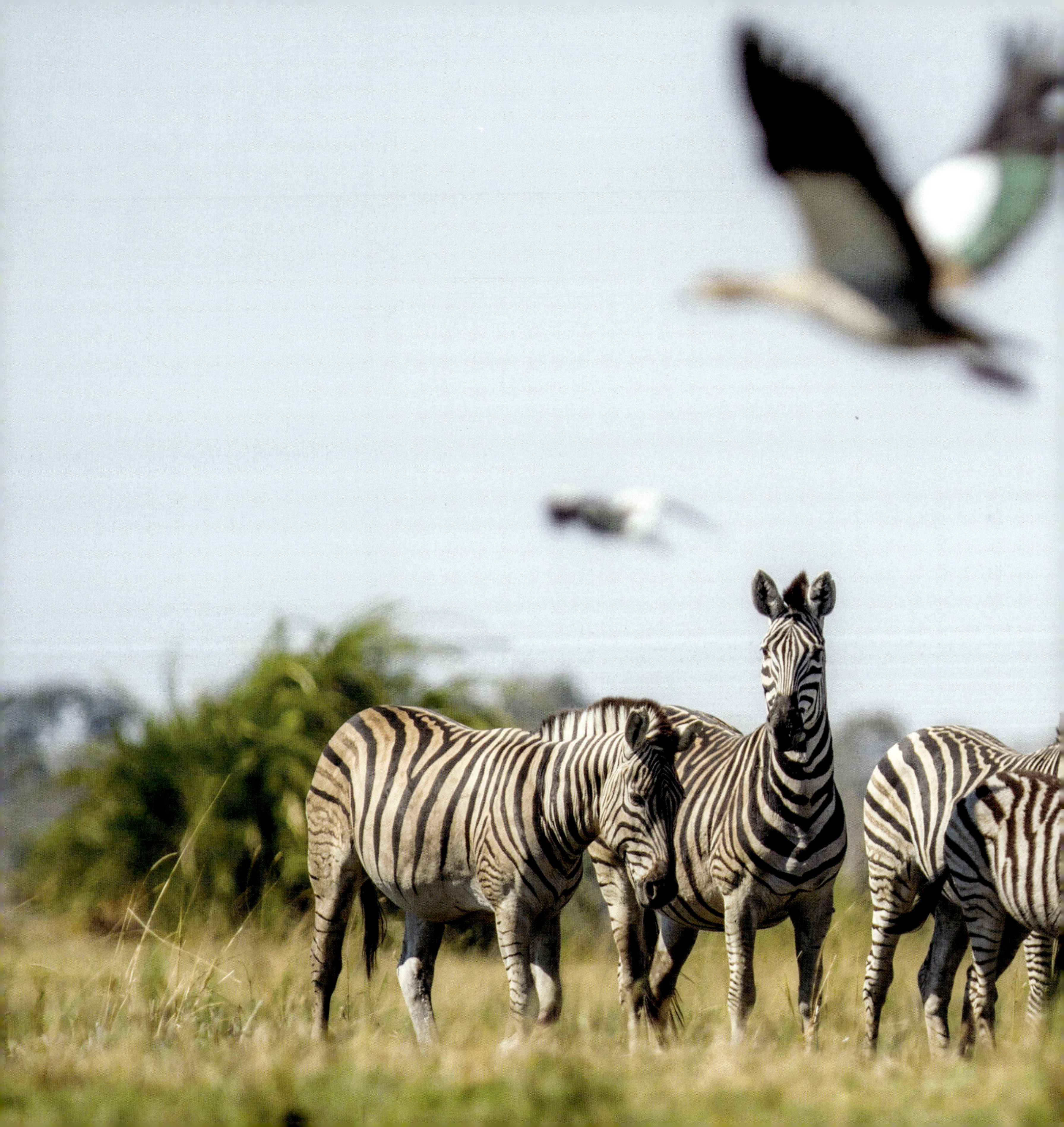

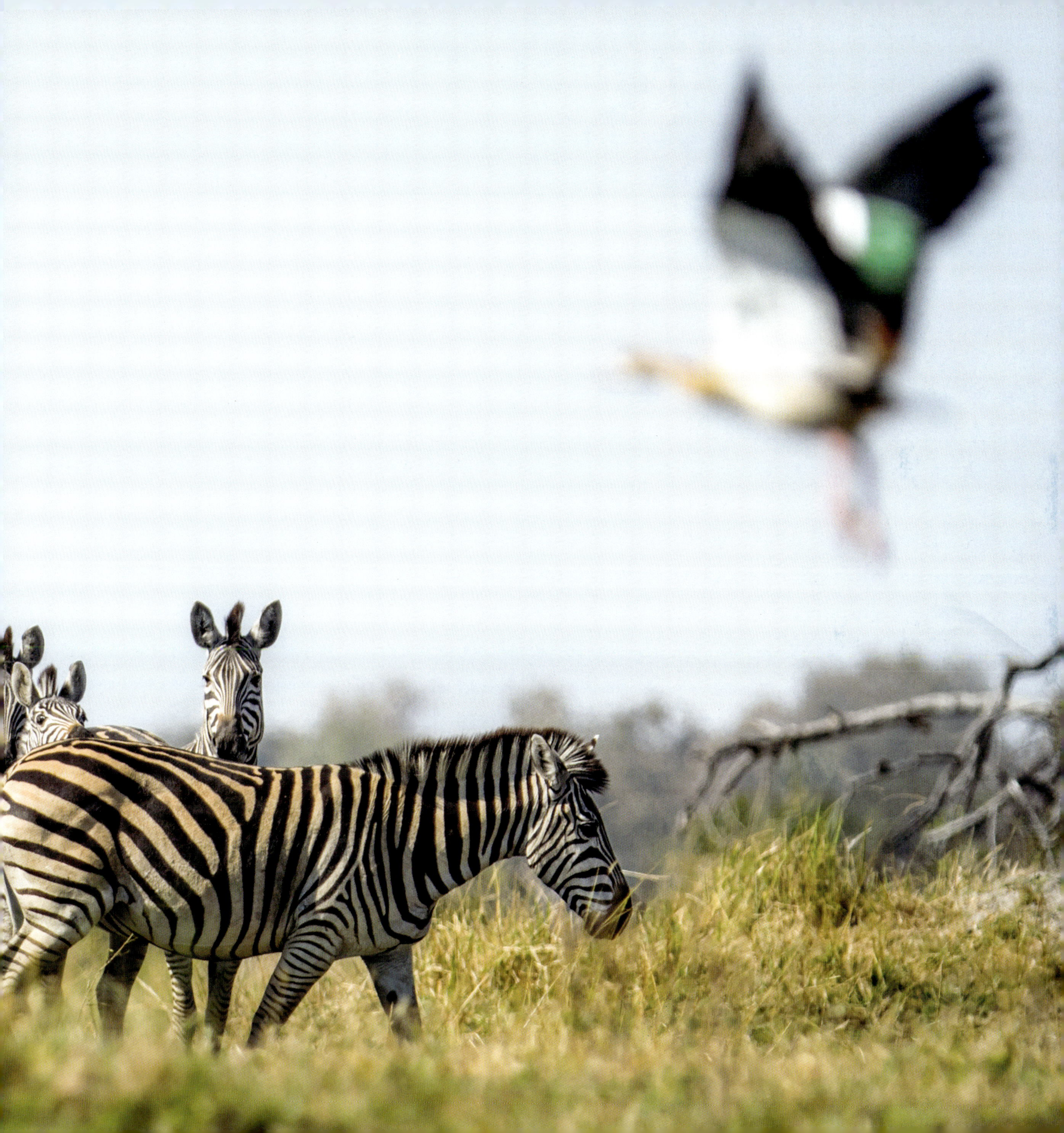

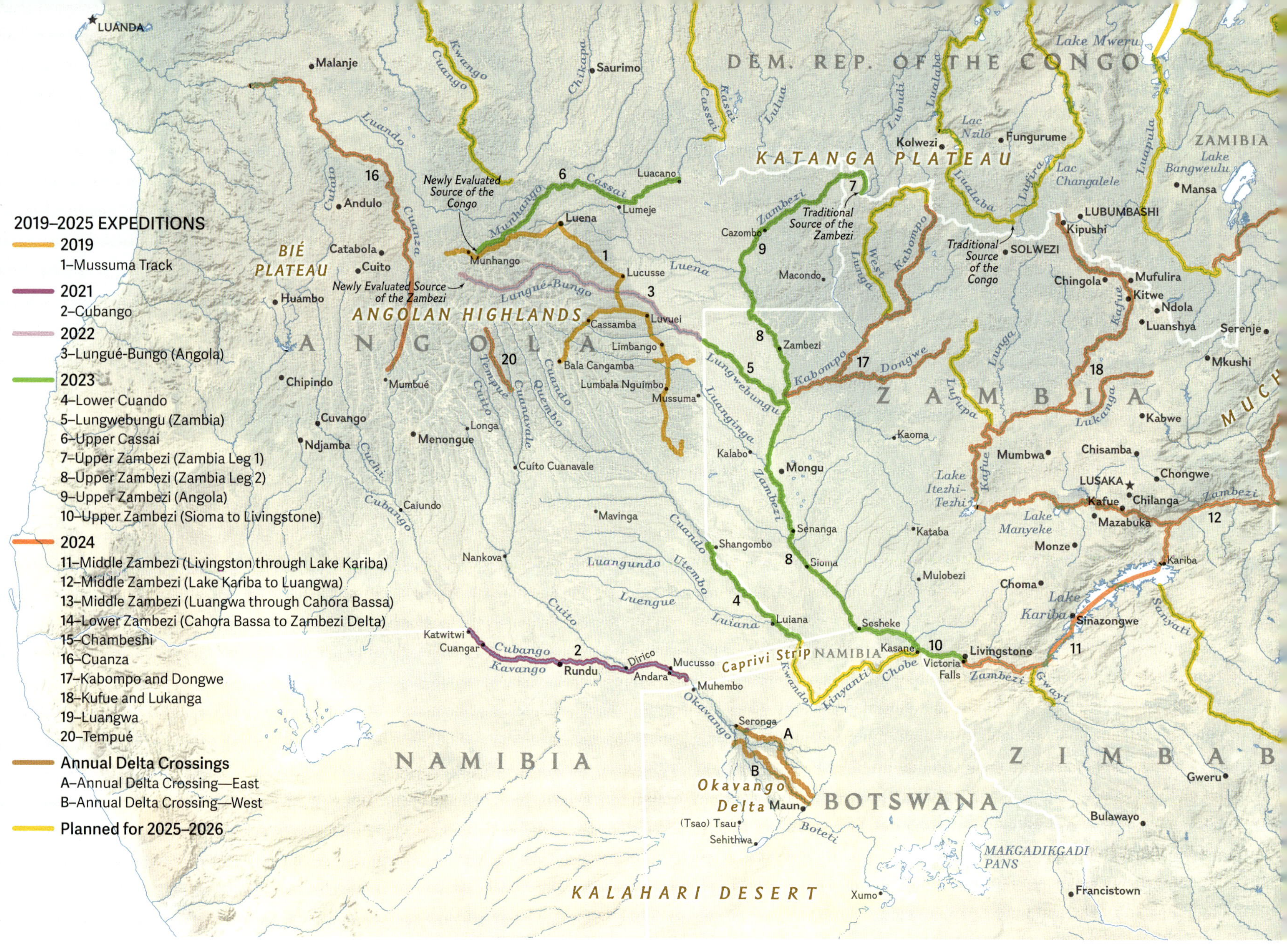

Modern explorers aren't prospectors, traders, pioneers, or conquerors. We have come full circle. Early 21st-century exploration is about returning to the source we left long ago. It's about protecting, not exploiting; documenting, not discovering; and reconnecting with what it means to be human—because, somewhere along the way, we forgot.

IN 2015, through doubt, fire, land mines, impassable roads, and the unknown, we entered the *terra do fim do mundo*. Long called the "land at the end of the earth" by outsiders, it is aptly now known as the Source of Life among the local Luchazi people. We were overwhelmed by its sense of place. It was otherworldly, magical, powerful, and wild beyond imagination.

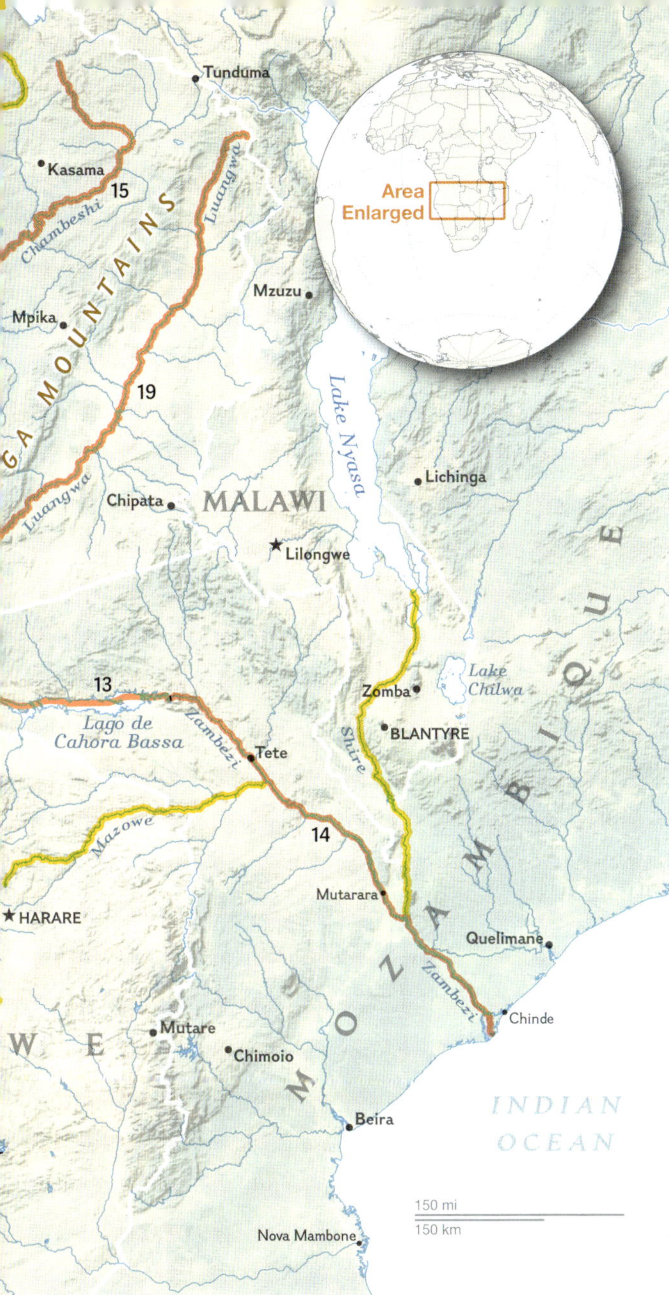

Now, years into our expeditions, it feels familiar. This now familiar place was likely the last undocumented, unphotographed landscape in Africa.

In 1488, Bartolomeu Dias, a Portuguese mariner, sailed around Africa's southern tip, proving the Atlantic and Indian Oceans flowed into each other. In 1805, more than three hundred years later, John Cary's famous map of Africa (page 35) simply identified the sub-Saharan African interior as "Unknown Parts." In 2018, an exhaustive search of the Library of Congress in Washington, D.C., found more than 40 detailed maps of Angola, including the eastern Angolan Highlands, dating back to the 17th century. These intricate, hand-drawn maps were pieces of art portraying imaginary places identified only in stories. The "land at the end of the earth" had rivers that didn't exist flowing in all directions. The maps were akin to the stories brought back by explorers, traders, and missionaries—tales of cryptids like the "Angolan water lion" and a flying reptile that snatched children.

Rivers we had been exploring in the eastern Angolan Highlands—the Cuito, Cubango, Cuanavale, Quembo, Cuando, Lungué-Bungo, Munhango, and Cassai—didn't start to appear on maps until the mid-19th century, but they were often confused with one another and poorly charted. Maps from the late 1950s charted these river systems correctly for the first time, but the source lakes weren't noted at all. The people who made these maps likely did not know they existed.

The war of independence broke out in Angola in 1961, and the civil war that followed continued until 2002. No new maps were made, at least not for the general public. The secrets of Lisima—the Source of Life, and the source of these essential rivers—remained hidden to outsiders until we arrived looking for them with satellite imagery in our pockets.

To the Luchazi, the source lakes and waterfalls are sacred, protected by spirits and

By 2025, we had expanded our operations to explore all major rivers in the Okavango, Kwando, Cuanza, Zambezi, and Congo Basins to highlight the importance of the Angolan Highlands as a source of fresh water. **PAGES 198–199:** *Africa has always been a sanctuary for megafauna during periods of global cataclysm, and the continent still harbors unrivaled wildlife populations. Dazzling zebras and flocks of Egyptian geese are a common sight in the Okavango Delta.*

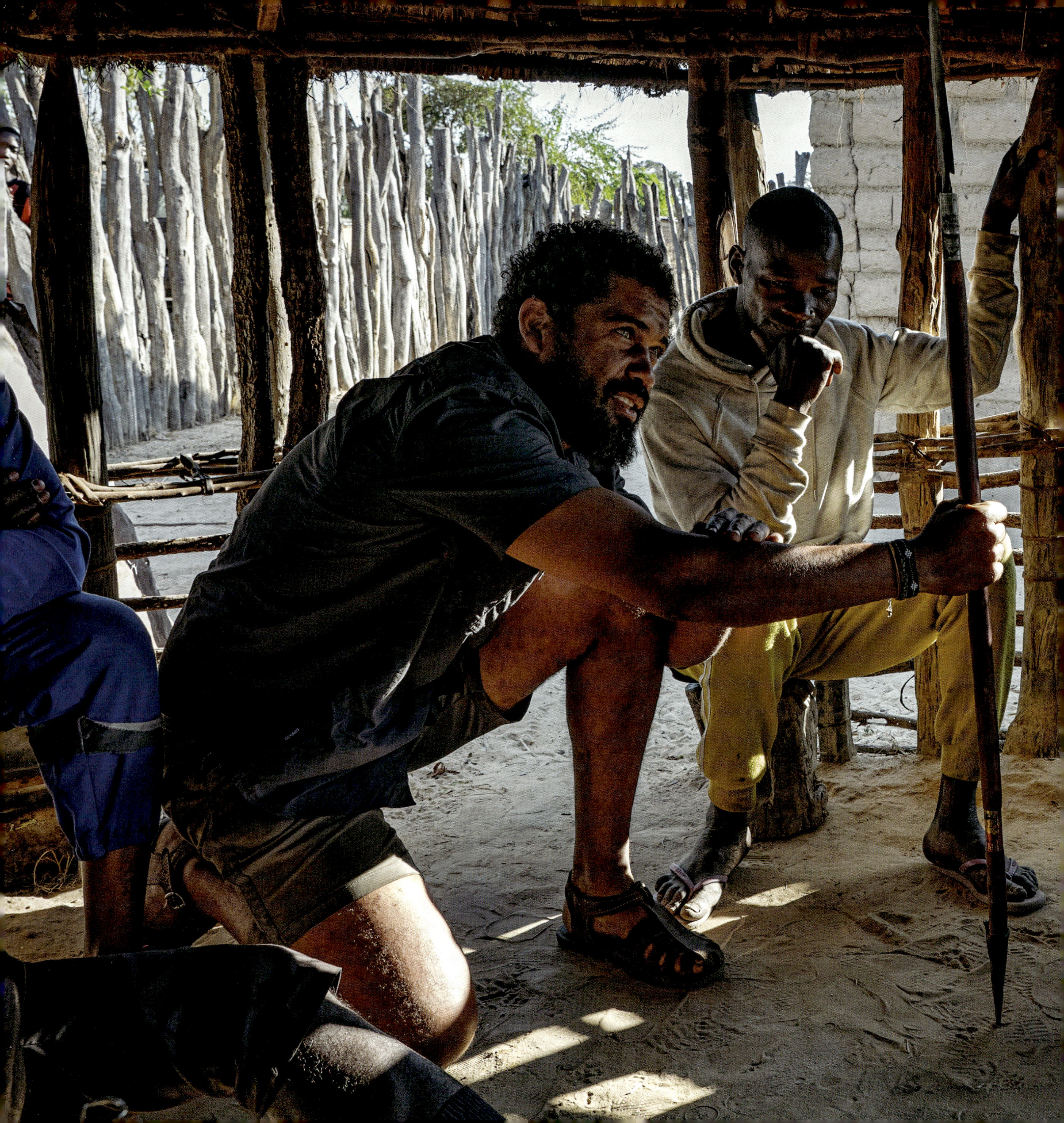

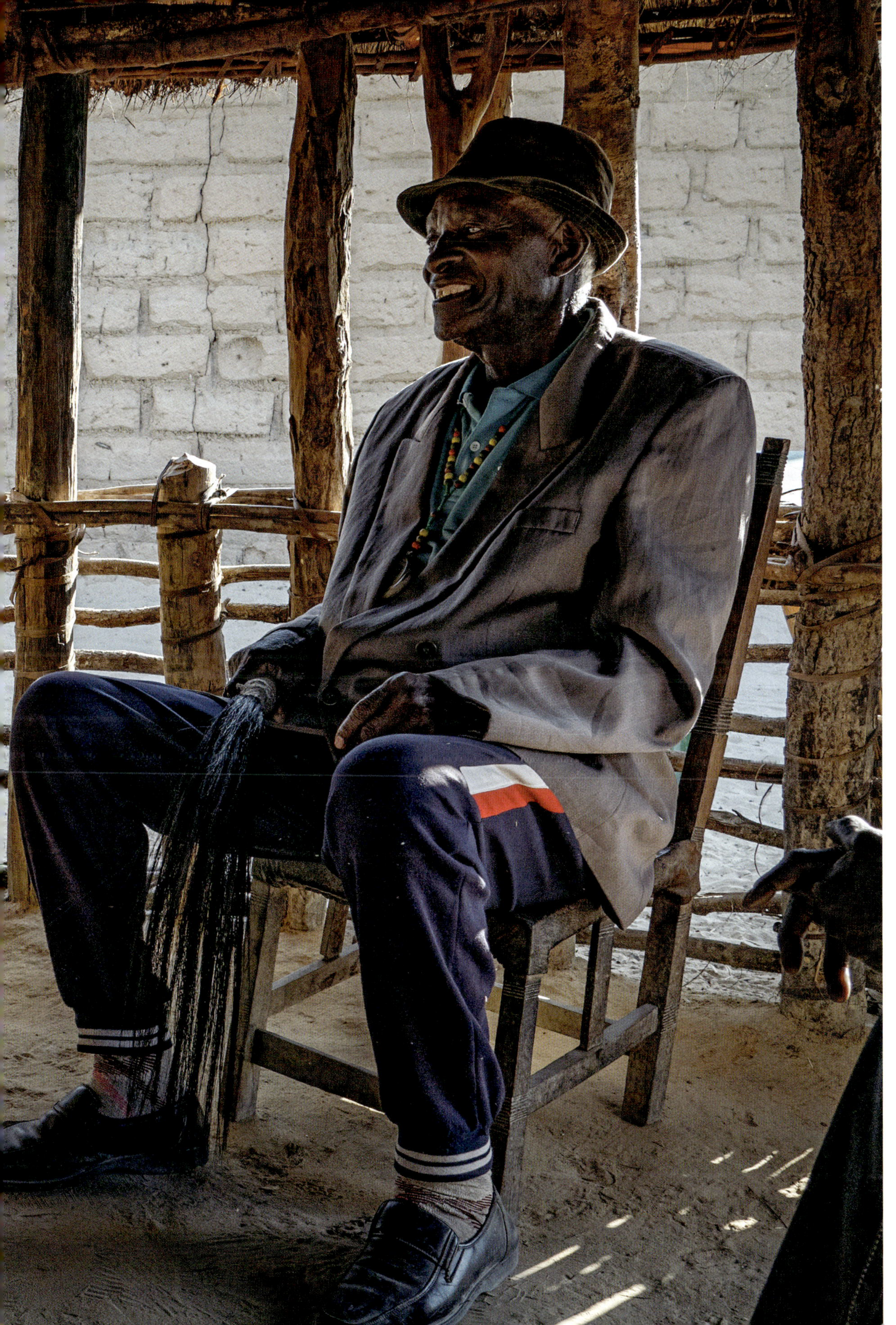

If Africa is to preserve its wildlife populations as it enters a renaissance between now and 2050, we need to focus on using traditional knowledge, culture, and leadership structures as the primary mechanisms for protection. Here, Kerllen Costa visits with traditional leaders in Tempue who offer him the opportunity to handle a ceremonial spear.

The Okavango Basin was just the beginning. We are now going on to explore and survey the great inland basins of Africa, the Great Spine of Africa, the continental divide, and the keystone water towers of Africa.

ancestors. They believe that if you put a village next to a lake, the lake will eat the village. At Lake Sápua, the largest source lake we have documented so far, strange currents cause the winds to make a lapping sound at night, and it's said that you can hear drums and the sounds of a village at the bottom of the lake. For centuries, Luchazi leaders kept the source lakes, together with these water and forest spirits, from outsiders, hidden down secret footpaths traversing vast impenetrable forests.

I OFTEN SIT NEXT TO the magical Cuanavale source lake in the heart of the Lisima Landscape to pause and reflect. I once asked myself out loud: "Do we really need to protect this place?" Mr. Water was sitting unseen nearby and said, "Yes." When I asked him why, he said that he had never been to such a place, where people were still free. He thought that was worth protecting.

Yet only local people know how best to protect places like Lisima. The best thing we can do is learn how to support them. Life began for modern human beings in Africa, and it will be decided here. The wild heart of Africa, the soul of the world, is resolute and alive, even in these bewildering modern times. The synergy of Indigenous wisdom and scientific inquiry now defines early 21st-century exploration.

Getting to know the Lisima Landscape and the Luchazi people over the past decade, and knowing the Wayeyi people of the Okavango Delta, I have come to believe that this isn't the "land at the end of the earth," as mused by the first Portuguese explorers—it is one of the paradises that will remain a sanctuary, primordial and unchecked at the end of time. The meeting place of heaven and Earth. A place where trees talk to stars and the sunset challenges the universe. A tormented Eden, green and abundant, yet depopulated and terrified by decades of war. This is one of the last wild places on Earth.

Through century-long droughts that ended kingdoms downstream, the Luchazi survived in remote inaccessibility, thanks to Indigenous wisdom deeply intertwined with cultural resilience rooted in language and tradition. For this last brief decade, we have focused on understanding and respecting this intricate and delicate web of knowledge, language, and culture that has sustained the Lisima Landscape: vast keystone watersheds and sources for the Okavango, Zambezi, Congo, Cuando, and Cuanza Rivers. Now, we have explored all the major rivers and are ready to work with the Luchazi to better protect this wilderness. Trust is earned, not given, and we are making a multigenerational commitment to this extraordinary landscape.

AFRICA'S GREAT RIVER BASINS are a mosaic of interconnected watersheds, binding plants, wildlife, and people to ancient landscapes set between rivers, rift valleys, mountains, deserts, and coastal plains—the Great Spine of Africa. Rivers unite and divide, connect and isolate. They bring new life and can wash it away. Africa has the most free-flowing, unstudied rivers of any continent. The major river basins cover over 15 million square kilometers (5.8 million sq mi), more than half of the continent. At least 500 million people live in these basins, and more than two-thirds of

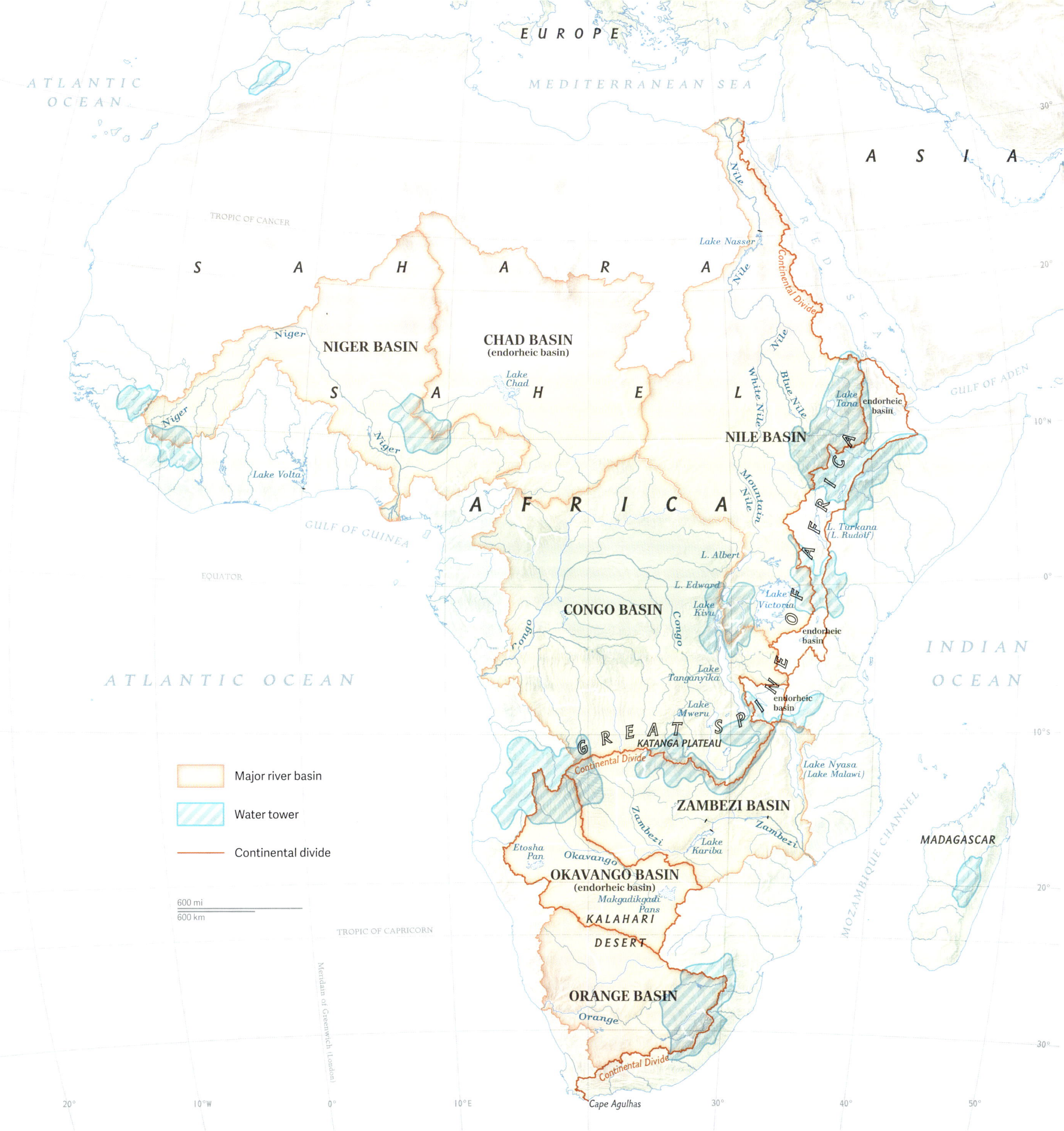
EUROPE
ATLANTIC OCEAN
MEDITERRANEAN SEA
ASIA
TROPIC OF CANCER
SAHARA
Lake Nasser
Nile
RED SEA
Continental Divide
NIGER BASIN
CHAD BASIN
(endorheic basin)
Niger
Lake Chad
SAHEL
White Nile
Blue Nile
Lake Tana
endorheic basin
GULF OF ADEN
NILE BASIN
Lake Volta
AFRICA
Mountain Nile
GULF OF GUINEA
L. Turkana
(L. Rudolf)
L. Albert
EQUATOR
L. Edward
Lake Kivu
Lake Victoria
CONGO BASIN
Congo
GREAT SPINE OF AFRICA
endorheic basin
ATLANTIC OCEAN
INDIAN OCEAN
Lake Tanganyika
Lake Mweru
endorheic basin
KATANGA PLATEAU
Continental Divide
Lake Nyasa
(Lake Malawi)
Major river basin
Water tower
Continental divide
ZAMBEZI BASIN
Zambezi
Lake Kariba
MOZAMBIQUE CHANNEL
MADAGASCAR
Etosha Pan
Okavango
OKAVANGO BASIN
(endorheic basin)
Makgadikgadi Pans
KALAHARI DESERT
600 mi
600 km
TROPIC OF CAPRICORN
Meridian of Greenwich (London)
ORANGE BASIN
Orange
Continental Divide
Cape Agulhas
30°
20°
10° N
0°
10° S
20°
30°
20°
10° W
0°
10° E
30°
40°
50°

The decisions made by young Africans today will determine the future of life on Earth. By 2050, 40 percent of all births will be here.

all Africans depend on them. Rivers, from their sources to their ends, are the lifeline of a continent. They are the water and nutrient pumps that sustain wildlife and humans alike.

Grossly underestimated on world maps, Africa covers more than 30 million square kilometers (11.6 million sq mi)—more than three times the land area of either the United States or China, the world's two largest economies. The African continent is the most resource-abundant continent, with 97 percent of the world's chromium, 90 percent of the cobalt, 85 percent of the platinum, 70 percent of the cocoa, and 60 percent of the coffee. With a population of 1.5 billion, Africa has as many people as India or China, Earth's two most populous countries. But India is 10 times more densely populated, China four times more. Africa has room to grow.

We are entering a new age of African exploration by Africans, driven by pride and ownership, not conquest and exploitation. Africans traverse lands from plains to mountaintops and rivers from their sources using traditional knowledge and modern technologies to monitor, manage, and protect, not exploit and destroy. Africa is Earth's ark, and an African renaissance is coming. By 2050, global stability will be based on food and water security. The African continent has 60 percent of the world's arable land, and it includes the largest carbon sink on the planet: the Congo Basin. Africa is the only continent with intact megafauna populations: elephants, rhinos, hippos, lions, and giraffes. Life retreated into this vast forested refugium during periods of global cataclysm.

The decisions made by young Africans today will determine the future of our iteration of life on Earth. By 2050, there will be 2.5 billion Africans, and 40 percent of all births will be here. Today, half of all Africans are under 17 years old, while half of all Europeans are over 44. The baby boomers in the developed world grew up with landline telephones, gas-guzzling cars, transistor radios, and black-and-white TVs. Africa's boom will be powered by the sun, with smartphones, electric cars, virtual reality, blockchain, and artificial intelligence. Africa is on the rise.

We especially need to pay attention to the women of Africa. They are a force of nature and the life-givers to communities across the continent. When globalization threatens Africa's cultural heritage and traditional knowledge, female leadership preserves local wisdom. When climate change threatens wildlife, crops, and communities, female workers—farmers, gatherers, doctors, healers—bring innovation. When war ignites, female artists—dancers, singers, creators—lead the cry for peace. Africa's women inspire change. They strive for improved health care and education systems, and they advocate for human rights on behalf of the vulnerable.

Yet even as we look ahead to the potential of our future, huge questions remain: Do we have enough water? Are we resilient to climate change? How do we protect natural heritage, preserving language, culture, and wisdom? How do we unite 54 countries in equitable resource use, pride, ownership, peace, and sustainable development?

In our humble way, this is the work of the Okavango Wilderness Project, funded by the

IN MEMORIAM

Leilamang "Snaps" Kgetho

Humility Born of Tradition

SON OF KGETO KGETHO and brother of Gobonamang Kgetho, Leilamang "Snaps" Kgetho had a gentle smile that belied the enormity of his presence. Soft-spoken yet with a booming voice, Leilamang echoed the ancient wisdom of his people. The youngest Wayeyi poler to explore all major rivers and channels in the Okavango Basin, he commanded respect through his actions with a humility born of deep tradition. He passed away suddenly in January 2020.

There are giants in this world, quiet contributors who carry us on their shoulders without fatigue or complaint. Snaps was one of these people, never ostentatious but always ready with swift, silent, life-saving action. In moments of chaos, he showed no fear. He taught me that leaders' strength doesn't overpower, and wisdom doesn't boast. I am a better person for having known him.

Leilamang's life is a testament to the awesome power of understatement in a world that confuses noise with importance. His hands were calloused by labor, and his soft words represented a voice of reason in a cacophony of educated guesses. In crisis, he was always a source of direct, unflinching action. He loved his family, knew his ancestors, and lived for his friends.

This hero was a conflicted soul. Inside him, the tides of tradition and the push for modern relevance waged a silent war, but his strength was born within this conflict. I miss him every day. Our daughter is named for him: Leila River Boyes. May Leila be Leilamang for us all.

Leilamang "Snaps" Kgetho was one of the strongest polers on our expedition team. He passed away suddenly and will never be forgotten by those who knew him.

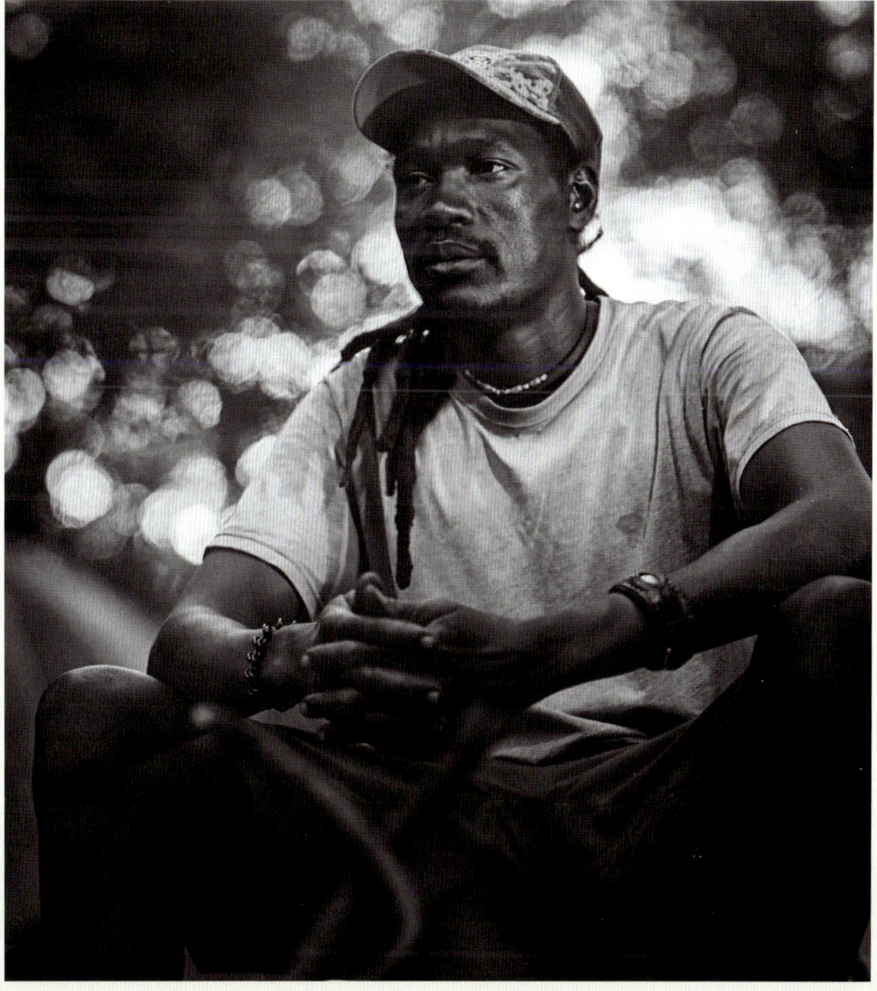

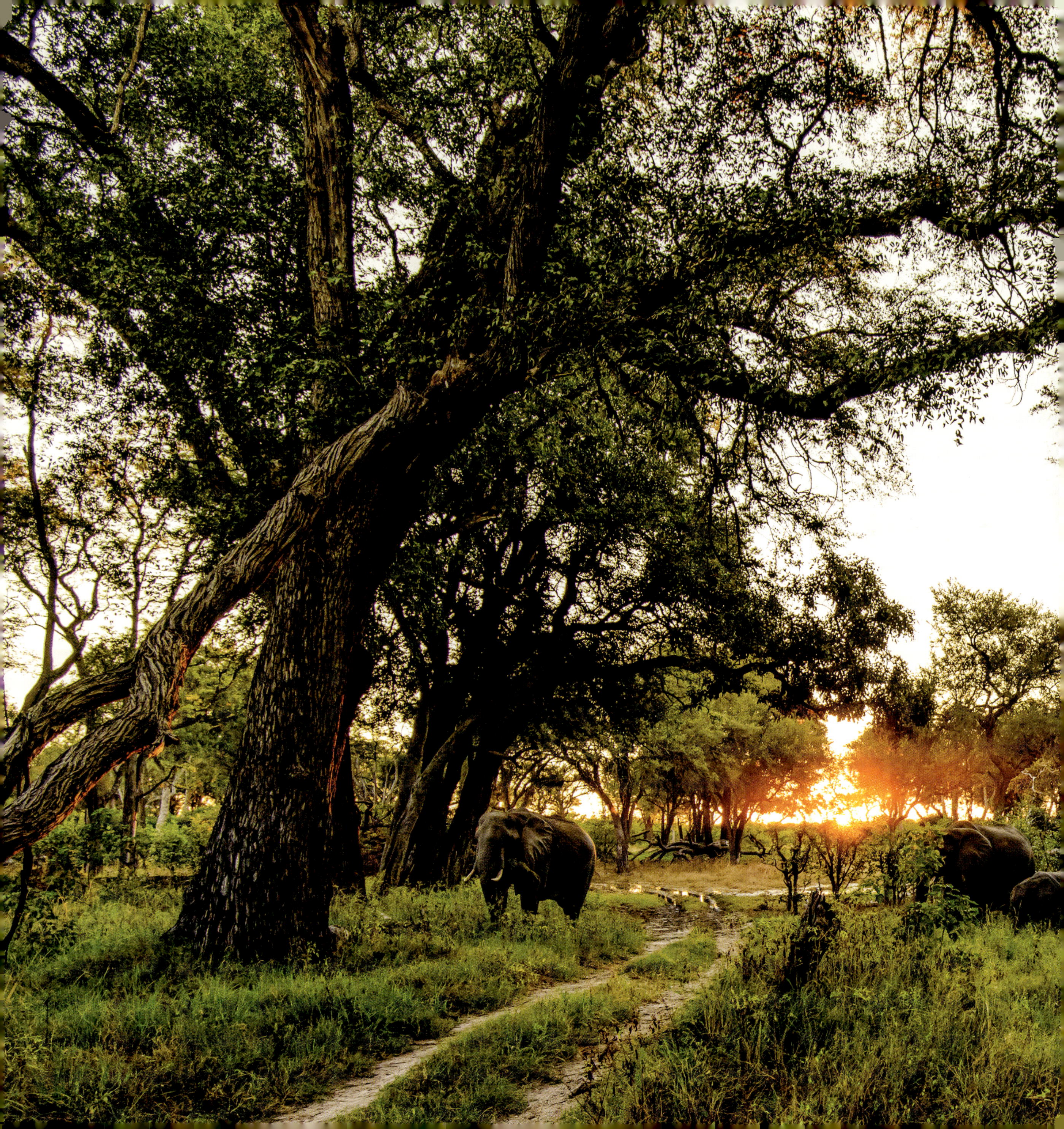

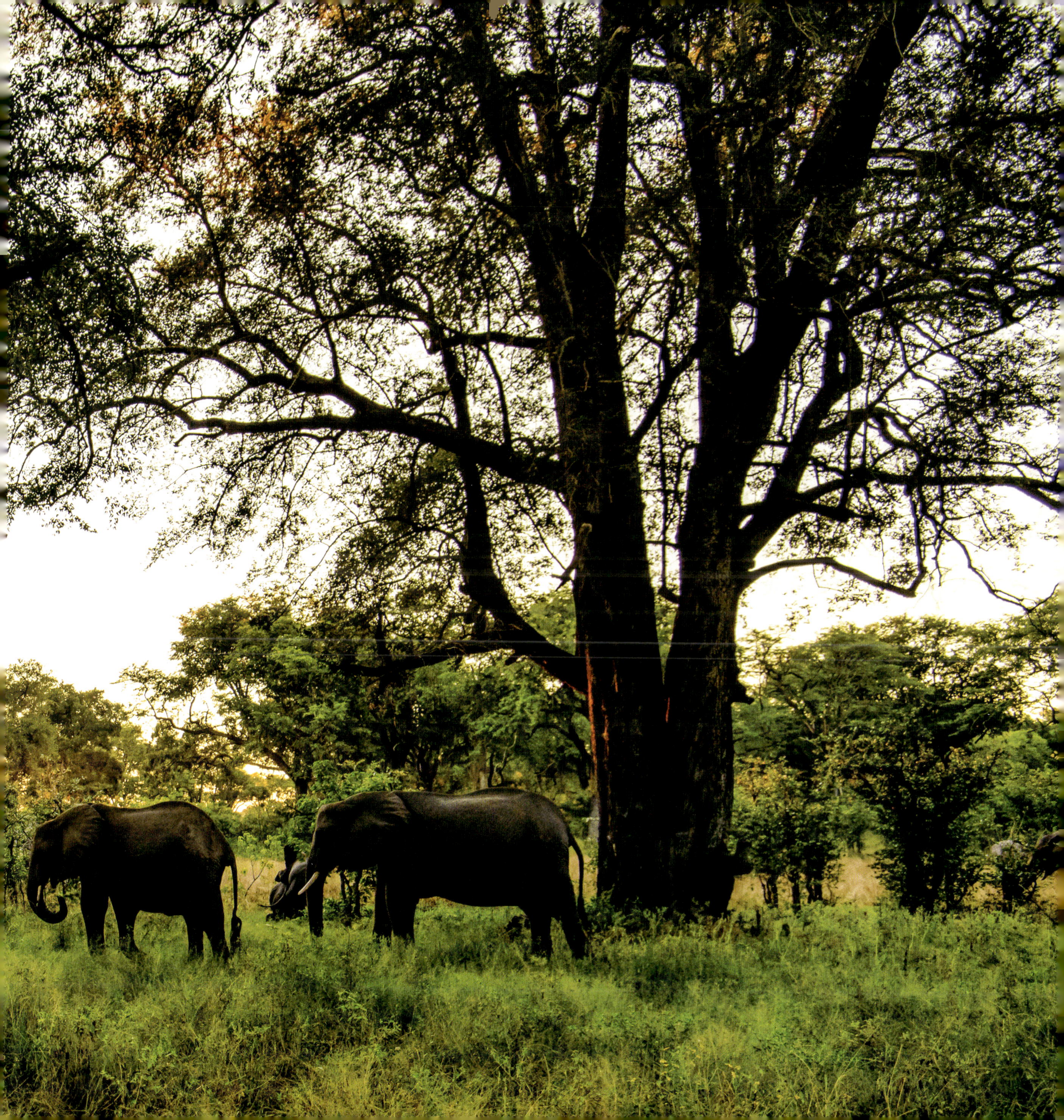

National Geographic Society's Okavango Legacy Fund. The Okavango Wilderness Project has made a forever commitment to protect the forests of Lisima lya Mwono, the Source of Life. We need remote sanctuaries for biodiversity where life can hide during cataclysm, natural or human-made. The Luchazi are the guardians of these rivers and forests, and empowering them to protect these vast woodlands, peatlands, and source rivers is the only path forward.

Since 2022, we have expanded our exploratory work to include all of Africa's great inland river basins—the Okavango, Zambezi, Congo, Nile, Chad, and Niger—spanning the Great Spine of Africa. By 2030, we aim to have completed 200 river expeditions, scientifically surveying all of Africa's most important river systems and working with local people to develop them sustainably. This is the largest nonmilitary expeditionary mobilization in modern African history. By 2035, we aim to establish better protections for 1.2 million square kilometers (463,000 sq mi) of keystone African water towers (such as the Lisima Landscape), watersheds, and plateaus to sustain Africa's swamps, deltas, and rivers for future generations. The Okavango Basin, the world's largest undeveloped river basin, and the Angolan Highlands water tower, including the Lisima Landscape, represent an opportunity for us all. In the crystal clear water of a free-flowing river, we are reborn, guided to a better future connected to our ancestral roots, curious again, wanting to know more about the natural world, wanting to protect it for those who will come after us.

The most profound moments of my life have occurred on expedition. After months paddling down the Cuando River, camping under the stars, and living on rice and beans, you forget the PINs for your bank accounts and passwords for social media. It's a complete systems reboot. **PAGES 208–209:** *What would the Okavango Delta be like without elephants? As human and cattle populations increase and human-wildlife conflict escalates on the edge of the delta, is there room for elephants?* **PAGES 212–213:** *Young Africans such as these boys, on their way to fish along southeastern Angola's Cubango River, are the hope for the future.*

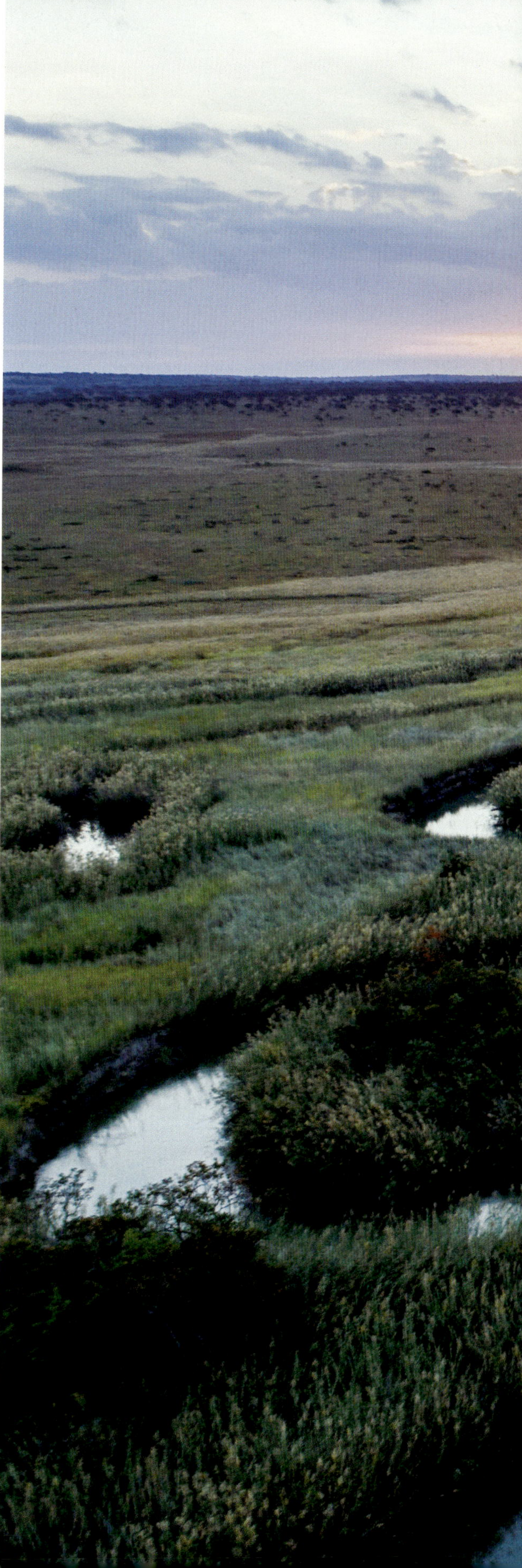

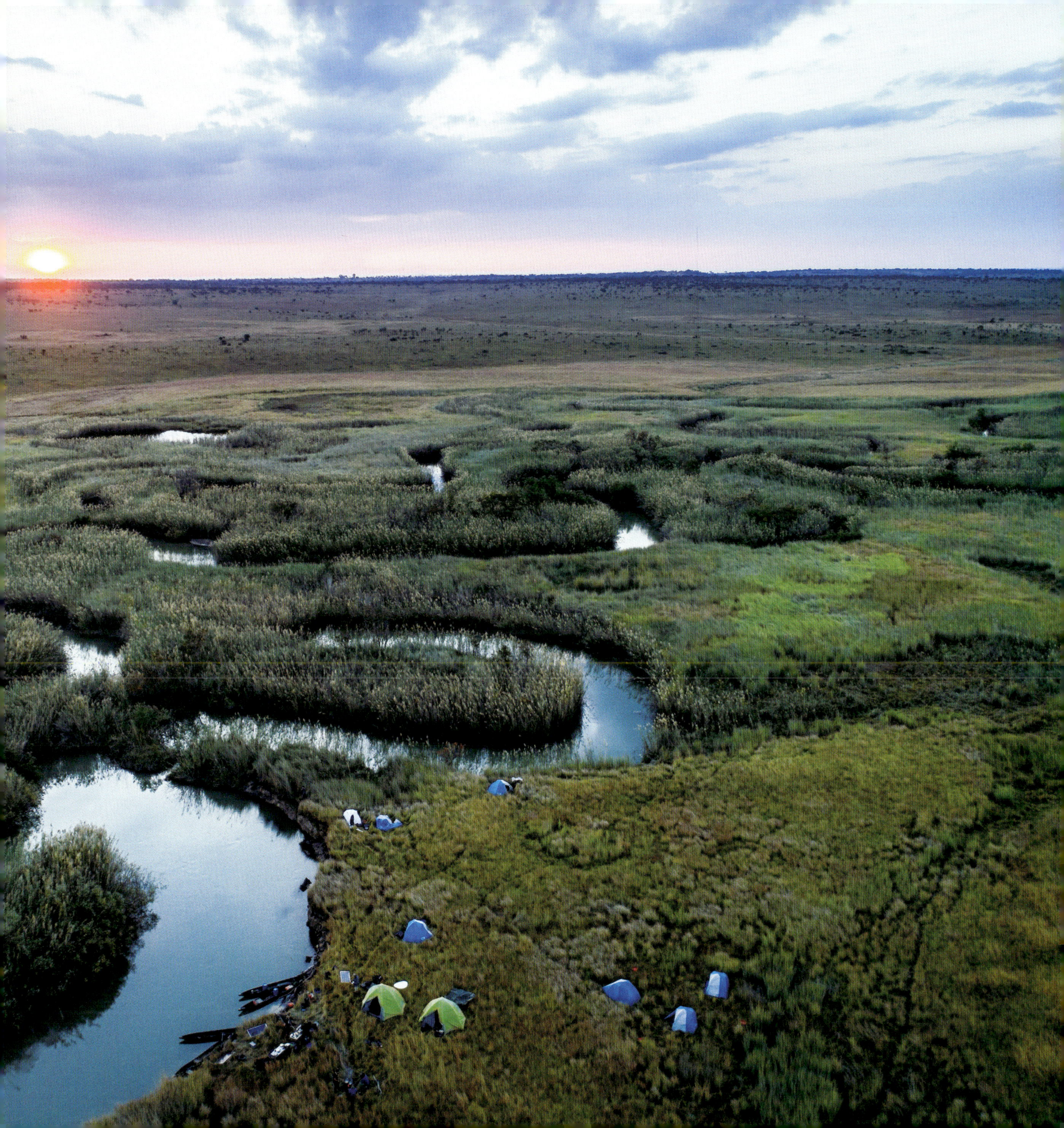

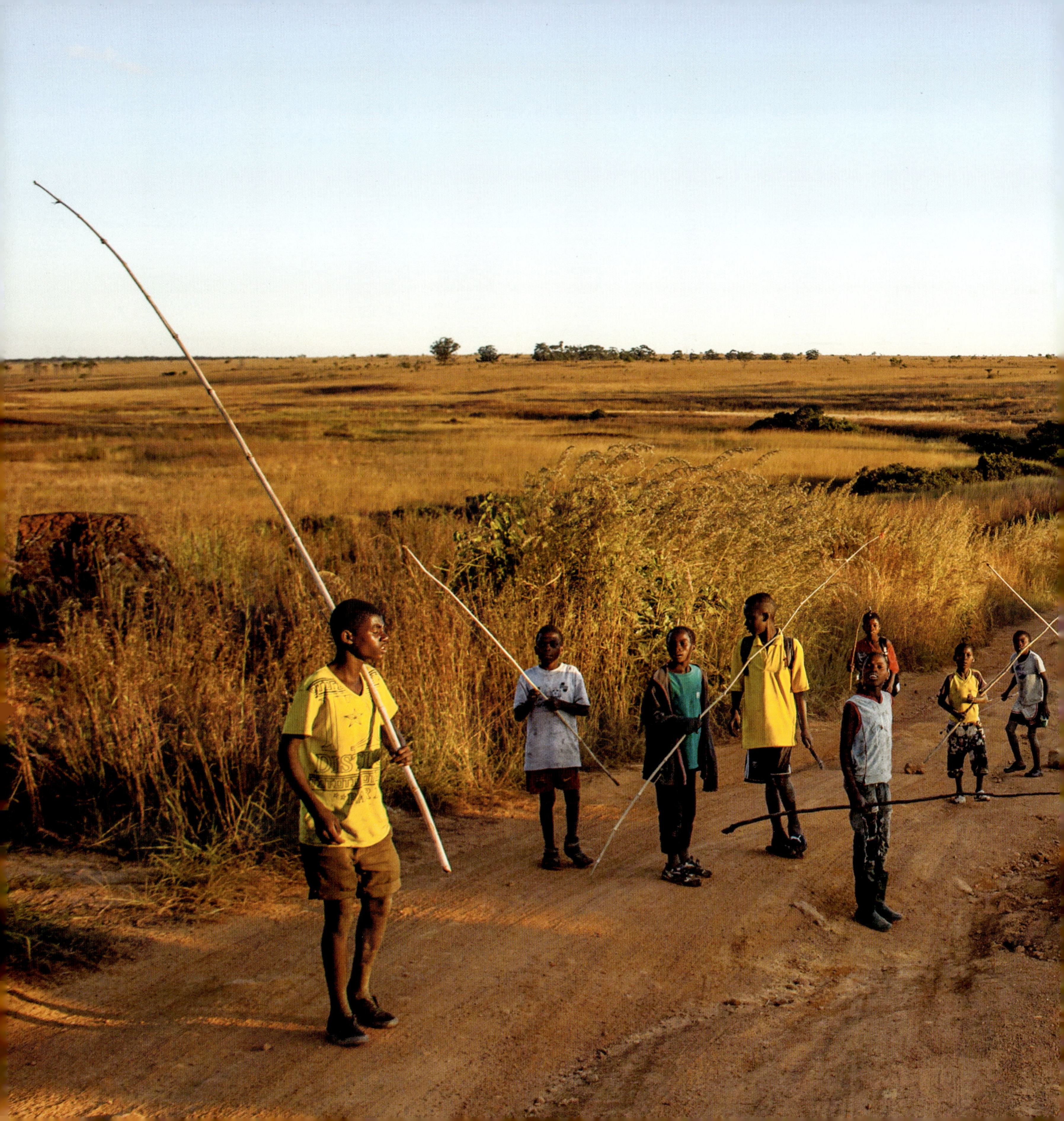

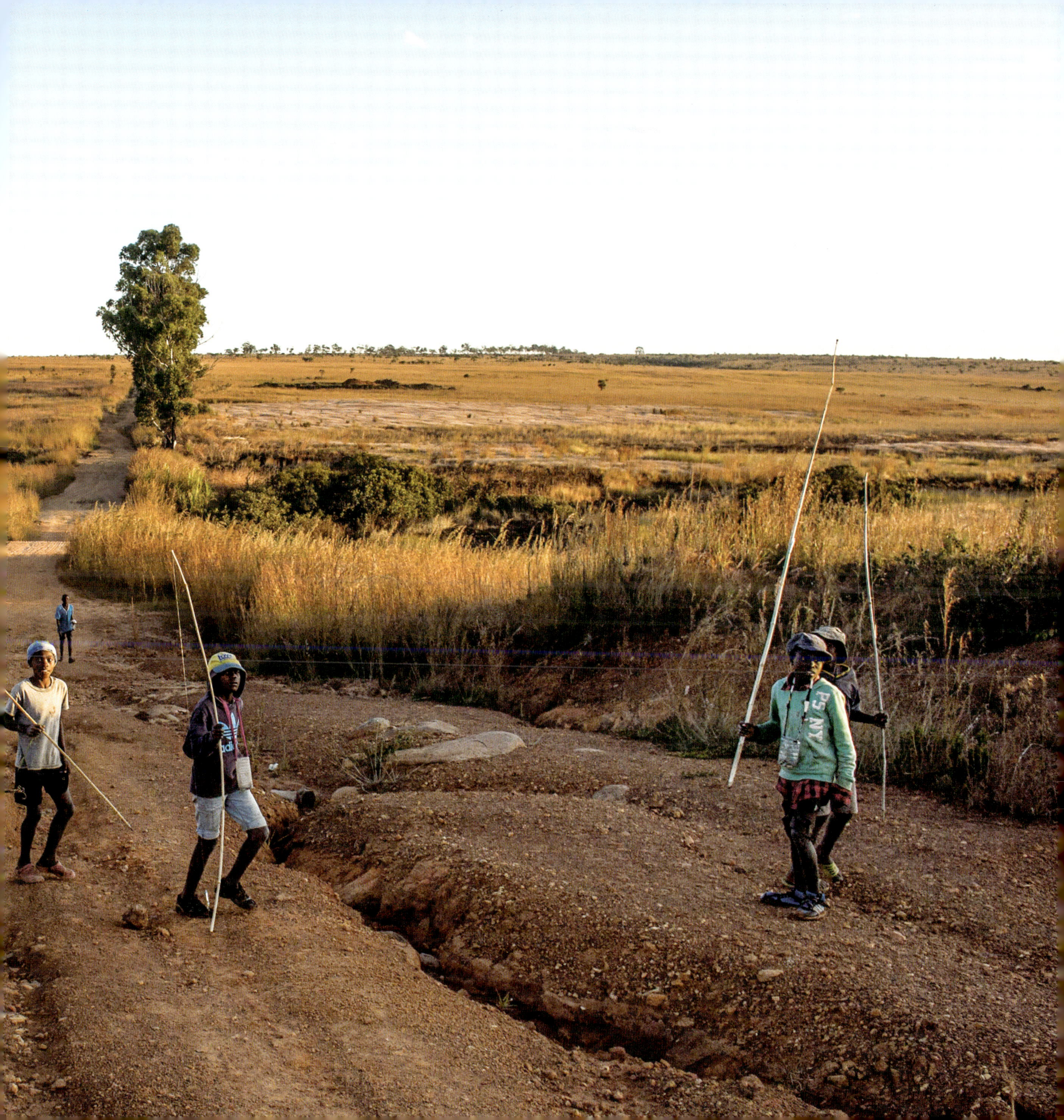
PS NY

Acknowledgments

This book would have been impossible without Susan Tyler Hitchcock, who, over a decade ago, started working with me on the idea for a book on our work. I would also like to thank Lisa Thomas, Tyler Daswick, Katie Dance, Jerry Sealy, Michael O'Connor, Becca Saltzman, Heather McElwain, and Jenny Miyasaki for the wonderful spirit of community and purpose that you brought to putting together this beautiful book. With my scientific background, it was a group effort teaching me to show, not tell.

Behind every resupply, medical evacuation, government meeting, and expeditionary success was John Hilton, our team leader, fixer, explorer, pilot, chief executive officer, and lifesaver. Fueled by passion and an uncanny ability to navigate African bureaucracy, John has been a driving force in the National Geographic Okavango Wilderness Project.

In 2014, John was lost in Menongue, the capital of Angola's Cuando Cubango, and saw "Okavango" written on an unassuming building. Seeking to know more, John found himself in conversation with Gime Sebastião—the beginning of a decades-long partnership with this well-respected government official, who helped us secure all the necessary permissions to explore this remote part of Angola. That year, we also met Gerhard Zank, Tim Gargan, and Christy Phillimore from the HALO Trust, who very quickly activated their teams to help us find the source of the Cuito River. We would have never found safe passage through active minefields without them.

Our work in Angola would have been impossible without Ângela Bragança, Sergio Freitas Costa, Afonso Lussati, António Chavonga, Elves Zambela, Maans Booysen, Stefan van Wyk, Rui Lisboa, Abias Huongo, Marisa Rodrigues, and Vladimir Russo. In Botswana, it would have been impossible without Keith Vincent, Grant Woodrow, and Kim Nixon. Kai Collins played a key role in helping us build partnerships across the Okavango Basin. Nyambe Nyambe, the executive director of the Kavango-Zambezi Transfrontier Conservation Area, has provided crucial support. David Garrett and Lisl Bennett are driving forces in the expansion of our programs across the region. I would like to offer special thanks to the De Beers Group for launching its Okavango Eternal partnership with us.

My thanks and appreciation go out to the following people who made invaluable contributions to making this journey possible: Gary Knell, John Fahey, Dawn Arnall, Barbara and Herb Fritch, Saint Claire Siefert, Richard Sneider, Laura Ng, Ann Luskey, Julie Folger, Kristin Rechberger, Olga Albuquerque, Fernando Matias, Brooke Runnette, Mike Ulica, Jill Tiefenthaler, Kara Mullins, Crystal Brown,

The 2018 expedition team of the National Geographic Okavango Wilderness Project

Kaitlin Yarnall, Shoham Arad, Geoff Daniels, Shannon Bartlett, Tracy Wolstencroft, Jonathan Bailey, Ian Miller, Amy Pokempner, Michael Beckner, Enric Sala, Neil Gelinas, Mia Maestro, Sandesh Kadur, Corey Jaskolski, Zach Vincent, Karolina Ensor, Steve Spence, Kostadin Luchansky, James Kydd, Jer Thorp, Cory Richards, Pete Muller, Paul Skelton, Bill Branch (in memoriam), Rob Little, David Goyder, Fenton "Woody" Cotterill, Steve Beissinger, Werner Conradie, Helen James, Moreangels Mbizah, and Andrew Parker.

I would also like to thank the National Geographic Society Board of Trustees, who visited the Okavango Delta and committed to establishing an endowment to support our forever commitment to the communities of the Okavango Basin.

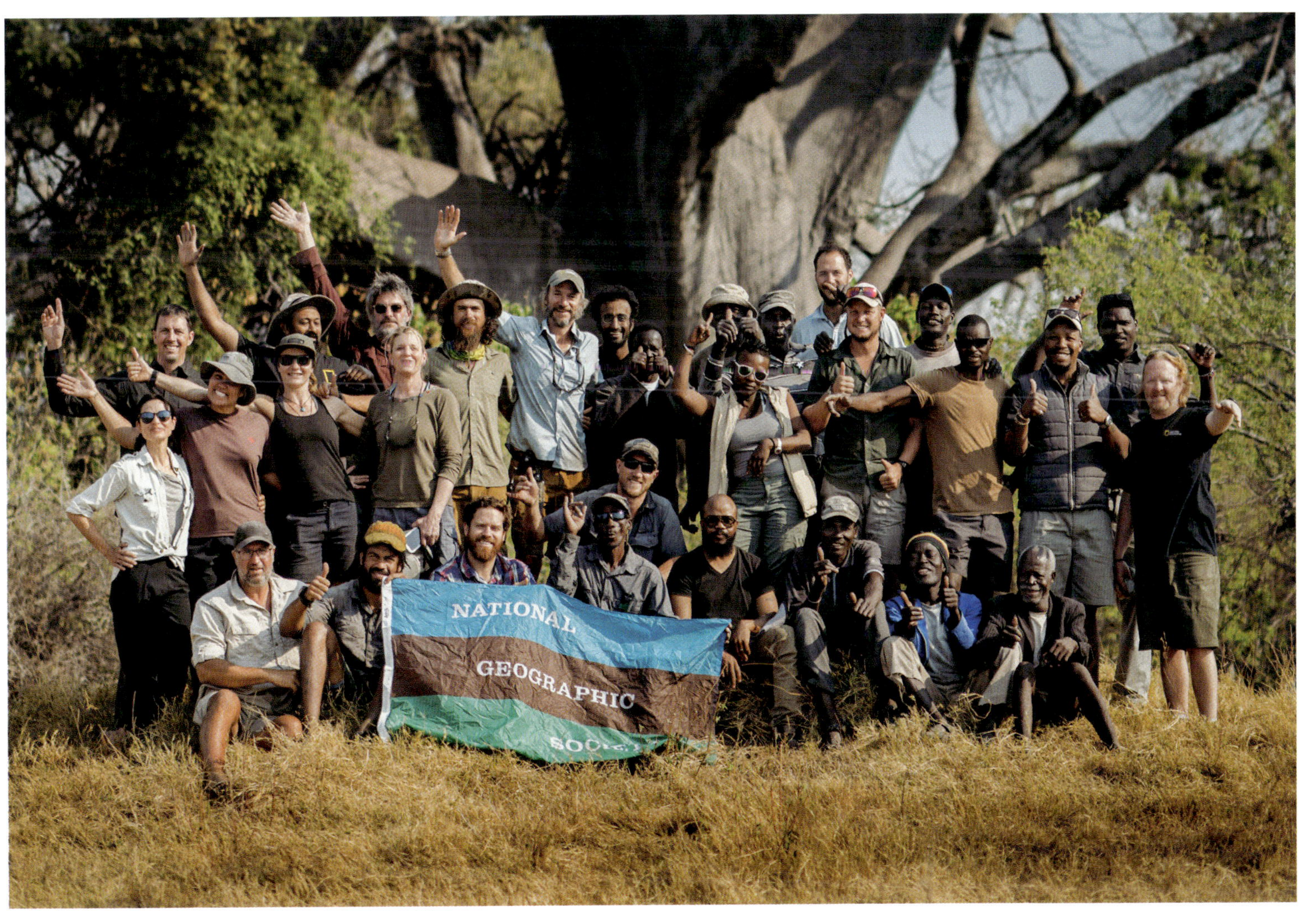

Illustrations Credits

All photographs made by people contracted by National Geographic (NG), National Geographic Okavango Wilderness Project (NGOWP), or The Wilderness Project (TWP) Foundation, except for pages 41, 51, 56-7, and 208-9.

Cover, Aaron Huey/NG; 2-3, Mark Stone/NG; 4, Cory Richards/NG; 6, Mark Stone/NG; 8, Kostadin Luchansky/NG; 10-13, Cory Richards/NG; 14-5, Kostadin Luchansky/NGOWP; 16-7, Mauro Sérgio/NG; 18-9, Chris Boyes/NG; 20-1, Kostadin Luchansky/NG; 22-3, Cory Richards/NG; 24, Kostadin Luchansky/NG; 25, Kostadin Luchansky/NGOWP; 26-7, Rainer von Brandis/NGOWP; 27, Koketso Mookodi/NGOWP; 28, Kostadin Luchansky/NG; 30-1, Cory Richards/NG; 32-3, Karabo Moilwa/NGOWP; 34, "A New Map of Africa" by John Cary, 1805/Yale University Library; 36-7, Kostadin Luchansky/NGOWP; 41, Ami Vitale; 42-3, Cory Richards/NG; 44-5, Kostadin Luchansky/NG; 46, Pete Muller/NG; 47, Kostadin Luchansky/NGOWP; 48-9, Kostadin Luchansky/NG; 51, Sergio Pitamitz/NG; 53, Rufusiah Molefe/NGOWP; 54-5, Cory Richards/NG; 56-7, Chris Schmid/NG; 58, Kostadin Luchansky/NGOWP; 60-1, Pete Muller/NG; 62, Kostadin Luchansky/NG; 63, Pippa Ehrlich/NGOWP; 64-7, Kostadin Luchansky/NGOWP; 69, Pippa Ehrlich/NGOWP; 71, Cory Richards/NG; 73, Mauro Sérgio/NGOWP; 74, Cory Richards/NG; 77, Karabo Moilwa/NGOWP; 78-82, James Kydd/NG; 83, Kostadin Luchansky/NG; 85-8, Cory Richards/NG; 89, Kostadin Luchansky/NGOWP; 90-1, Kostadin Luchansky/NG; 92-5, Cory Richards/NG; 97, Jesse Manuel/NGOWP; 98-9, Kostadin Luchansky/NGOWP; 100-1, Karabo Moilwa/NGOWP; 102-3, Kostadin Luchansky/NGOWP; 104, Mike Beckner/NGOWP; 105, Chris Boyes/NGOWP; 106-7, Chris Boyes/NG; 108, Kostadin Luchansky/NG; 109, Karabo Moilwa/NGOWP; 110-1, Pete Muller/NG; 112, Kostadin Luchansky/NGOWP; 113, Cory Richards/NG; 114, Pete Muller/NG; 115, Chris Boyes/NGOWP; 117, Mauro Sérgio/NGOWP; 118-9, Jen Guyton/TWP; 120, Kostadin Luchansky/NGOWP; 121, Mauro Sérgio/NGOWP; 122-3, Pete Muller/NG; 125, Rainer von Brandis/NGOWP; 126-9, Cory Richards/NG; 130, Carinè Müller/Dan Snyders/NGOWP; 131, Carinè Müller/Dan Snyders/NG; 132-7, Kostadin Luchansky/NGOWP; 138-9, Cory Richards/NG; 141, Jen Guyton/NGOWP; 142-3, James Kydd/NGOWP; 144-5, Sandesh Kadur/NGOWP; 146, Kostadin Luchansky/NGOWP; 148-9, Eric Averdung/NGOWP; 151, Johann Vorster/NGOWP; 152, Chris Boyes/NGOWP;

154-5, Cory Richards/NG; 156-7, Rainer von Brandis/NG; 158-9, Jasper Doest/NG; 161, Lopez Tucunare/NGOWP; 162-3, Pete Muller/NG; 164-5, Cory Richards/NG; 167-71, Pete Muller/NG; 172, Chad Keates/NGOWP; 173, Rainer von Brandis/NGOWP; 174-5, James Kydd/NGOWP; 175, Jesse Sartes/NGOWP; 176, Cory Richards/NG; 177, Mauro Sérgio/NGOWP; 178 (UP LE), Carinè Müller/Dan Snyders/NGOWP; 178 (UP RT), Madeleine Foote/NGOWP;178 (LO LE), Pippa Ehrlich/NGOWP; 178 (LO RT), Rainer von Brandis/NGOWP; 180, Pete Muller/NG; 182-3, Kostadin Luchansky/NGOWP; 184-9, Cory Richards/NG; 190-1, Carinè Müller/Dan Snyders/NGOWP; 192, Pete Muller/NG; 193, Thalefang Charles/NGOWP; 194-5, Cory Richards/NG; 196-7, James Kydd/NGOWP; 198-9, Kyle Neil Gordon/NGOWP; 202-3, Jen Guyton/TWP; 207, Kostadin Luchansky/NGOWP; 208-9, Chris Schmid/NG; 210-1, Kostadin Luchansky/NGOWP; 212-3, Pete Muller/NG; 215, Kostadin Luchansky/NGOWP; 223, Jen Guyton/TWP; back cover: (UP LE), Karabo Moilwa/NGOWP; (UP CT), Kostadin Luchansky/NG; (UP RT), Cory Richards/NG; (LO LE), Kostadin Luchansky/NG; (LO CT), Kyle Neil Gordon/NGOWP; (LO RT), Jesse Sartes/NGOWP.

Index

Boldface indicates illustrations.

About the Author

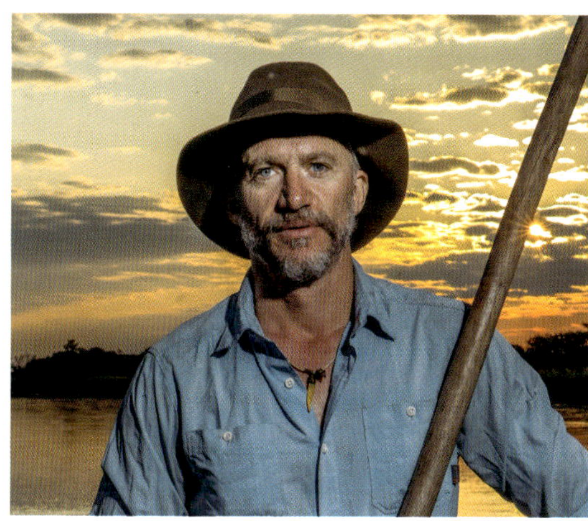

Steve Boyes is a passionate conservationist, scientist, and modern-day explorer dedicated to protecting one of Africa's most vital ecosystems—the Okavango Delta. A native of South Africa with a Ph.D. in zoology, Boyes has been a National Geographic Explorer since 2010. With decades of fieldwork experience, he is a founder and chairman of the Wild Bird Trust in South Africa, the Nkashi Trust in Botswana, and the Wilderness Project Foundation in the United States. Boyes is also a founding board member of Fundação Lisima in Angola. In 2015, Boyes launched what has become the National Geographic Okavango Wilderness Project to support the establishment of community-driven systems of protection for the Okavango Basin in Angola, Namibia, and Botswana based on detailed, repeatable ecological surveys and long-term environmental monitoring systems. Boyes and his research team focus on establishing detailed hydrological and ecological baselines as part of their mission to help local governments establish long-term resilience to the impacts of climate change. In 2023, they described the Angolan Highlands water tower, one of Africa's largest sources of fresh water, for the first time in history, and they have discovered more than 150 species new to science in these highlands so far. Boyes is a fellow of the Royal Geographical Society and the recipient of many accolades. In 2019, Boyes and his team were awarded the prestigious Rolex National Geographic Explorer of the Year award.

Since 1888, the National Geographic Society has funded more than 15,000 research, conservation, education, technology, and storytelling projects around the world. National Geographic Partners distributes a portion of the funds it receives from your purchase to National Geographic Society to support their mission to illuminate and protect the wonder of our world.

National Geographic Partners, LLC
1145 17th Street NW
Washington, DC 20036-4688 USA

Get closer to National Geographic Explorers and photographers, and connect with our global community. Join us today at nationalgeographic.org/joinus

For rights or permissions inquiries, please contact National Geographic Books Subsidiary Rights: bookrights@natgeo.com

Financially supported by the National Geographic Society.

ISBN: 978-1-4262-2407-2

The authorized representative in the EU for product safety and compliance is Disney Trading B.V., Asterweg 15S, 1031 HL, Amsterdam, The Netherlands
email: DCP.DL-EU.bookscontact@disney.com

Printed in Malaysia

25/IVM/1

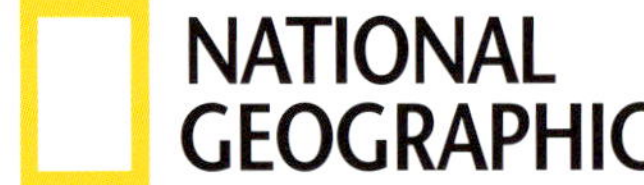

The National Geographic Society is a global nonprofit organization that uses the power of science, exploration, education, and storytelling to illuminate and protect the wonder of our world.

Since 1888, the Society has pushed the boundaries of exploration to better understand our world. We have awarded more than 15,000 grants to National Geographic Explorers—scientists, conservationists, innovators, educators, and storytellers—for work across all seven continents. Today, Explorers are advancing knowledge and leading conservation programs with outsize impact to protect nature, wildlife, historical places, and communities. They're documenting the wonder of our world—including its beauty, its mystery, and the threats it faces—and inspiring people to care and act on behalf of our planet and its people.

South African conservation biologist and National Geographic Explorer Steve Boyes is the founder and chairman of the Wild Bird Trust and leader of the National Geographic Okavango Wilderness Project (NGOWP). Collaborating with the people who live in the Okavango Basin, who have sustained this landscape through their traditions and knowledge, is a central part of his team's work to help safeguard this landscape. This work is vital to community-led conservation management across the project area in southeastern Angola and to ensure the future sustainability of the ecosystem's health for Namibia and Botswana.

The NGOWP is committed to securing permanent, sustainable protection for the greater Okavango River Basin: the source waters that originate in the highlands of Angola then flow through Namibia and into the Okavango Delta in Botswana. Since 2015, NGOWP has been working with local communities, NGOs, and the governments of Angola, Namibia, and Botswana to realize this vision. In 2023, for the first time within academic science, the NGOWP team defined the boundaries of the Angolan Highlands water tower. This is a critical step toward preserving an area that provides water to millions of people and species downstream, holds profound cultural and spiritual significance, and represents the source of 95 percent of the water that fills the Okavango Delta.

By combining the power of academic science and traditional knowledge, community education, and powerful storytelling, we hope to help the communities of the Okavango Basin continue to preserve this unique ecosystem for current and future generations.

To learn more about our Explorers and how you can join us, visit natgeo.com/impact.